Formation, Dynamics and Statistics of Patterns

Volume 1

Formation, Dynamics and Statistics of Patterns

Volume 1

Editors
Kyozi Kawasaki
Kyushu University
Masuo Suzuki
University of Tokyo
Akira Onuki
Kyoto University

World Scientific
Singapore • New Jersey • London • Hong Kong

Published by

World Scientific Publishing Co. Pte. Ltd.
5 Toh Tuck Link, Singapore 596224
USA office: 27 Warren Street, Suite 401-402, Hackensack, NJ 07601
UK office: 57 Shelton Street, Covent Garden, London WC2H 9HE

British Library Cataloguing-in-Publication Data
A catalogue record for this book is available from the British Library.

FORMATION, DYNAMICS AND STATISTICS OF PATTERNS
Volume 1

ISBN-13 978-981-02-0170-8
ISBN-10 981-02-0170-2

PREFACE

During the decade that will soon end, we have witnessed increasing importance of the role played by patterns in physical sciences, especially in condensed matter physics. There are many reasons for this trend. For instance physicists are facing increasingly tough problems to solve, which are mostly nonlinear and beyond the reach of analytical methods. In many cases, however, clues to or even deep insight into the problems are obtainable by thinking in terms of patterns. Availability of more advanced computers permits easier access to visualization. Whatever the reasons are, we can no longer talk about recent important advances in physical sciences without referring to patterns. With this circumstance in mind, we, together with the help of our colleagues, have initiated in the year 1986 the interdiscriplinary research project "Formation, Dynamics and Statistics of Patterns" under the sponsorship of Research Institute for Fundamental Physics (RIFP) at Kyoto University. So far about 100 scientists participated and 4 scientific meetings were held at RIFP. The present volume is the first of the series and contains reviews of the results obtained with introductions which are written by selected participants of the project.

A wide range of topics is covered in this volume, reflecting multifaceted nature of the pattern research. Examples treated are: use of coupled map in pattern research, patterns in reaction-diffusion systems, random interface systems, diffusion-limited aggregation, patterns in an evaporating liquid drop, crystal growth, and elasticity effects on phase transitions.

The editors are grateful to World Scientific, Singapore, for making publication of this volume possible and apologize for delay in submitting the complete manuscript.

November 10, 1989

Kyozi Kawasaki (*Fukuoka*)
Akira Onuki (*Kyoto*)
Masuo Suzuki (*Tokyo*)

CONTENTS

Simulating Physics with Coupled Map Lattices

Pattern Dynamics, Information Flow, and Thermodynamics of Spatiotemporal Chaos

Kunihiko Kaneko
Center for Nonlinear Studies,
Los Alamos National Laboratory,
Los Alamos, NM 87545
and
Institute of Physics,
College of Arts and Sciences,
University of Tokyo,
Komaba, Meguro-ku, Tokyo 153, Japan

Contents

1 Introduction — Why Coupled Map Lattices?

Modelling and characterization of complex phenomena in spacetime is important in the study of turbulence in a general sense, not only in fluid dynamics but also in solid-state physics, optics, chemical reaction with diffusion, pattern formation problems, and possibly in biology. This kind of phenomena is called as "spatiotemporal chaos" [1], in an attempt to understand it on the basis of knowledge on dynamical systems theory, especially, chaos.

Several years ago, the author proposed a simple model with essential features of spatiotemporal chaos. In the present review, we first discuss the qualitative features of the model with some visualization, and then try to explore the universal scenario for pattern dynamics. We introduce some quantifiers to characterize the pattern dynamics, such as spatial-temporal power spectra, domain distribution as order parameters, Lyapunov spectra and mutual information flow and so on.

We briefly discuss some trials towards the construction of thermodynamics with the use of Perron-Frobenius operator and self-consistent approximation. Also we discuss the mean-field version of the present model briefly.

Some applications to physical, chemical and biological systems are also discussed, such as Benard convection, convection in liquid crystals, Taylor flow, open flow in fluid systems, chemical reaction with diffusion, some solid-state systems such as Josephson junction array, charge density wave, spin wave turbulence, spinodal decomposition, and possibly some biological networks [2].

The model we use here is a coupled map lattice (CML) [1,2,3,5]. It has been recently investigated extensively in various contexts [1-43].

A CML is a dynamical system with a discrete time, discrete space, and continuous state [1,2,3,5,6] see also, [9,10,12]). A modelling of physical phenomena by CML is based on the following steps:

Our strategy of studying dynamical phenomena in spatially extended systems by CML is based on the following steps:

(A) Take a (set of) field variable(s) on a lattice.

(B) Decompose the phenomena into independent units (e.g., convection, reaction, diffusion, and so on)

(C) Replace each unit by the possible simplest parallel dynamics on a lattice: the dynamics consists of nonlinear transformation of the field variable on each lattice point and/or a coupling term among suitably chosen neighbors.

(D) Carry out each unit dynamics ("procedure") successively.

Let us take the simplest example, that is the local nonlinear process and diffusion, as can be seen in the reaction-diffusion system. As the independent units, we take local nonlinear process $x(i) \to x'(i) = f(x(i))$ and the discretized diffusion process $x'(i) \to (1-\epsilon)x'(i) + (\epsilon/2)[x'(i+1) + x'(i-1)]$. Combining these two procedures, our dynamics can be written as

$$x_{n+1}(i) = (1-\epsilon)f(x_n(i)) + \epsilon/2[f(x_n(i+1)) + f(x_n(i-1))] \qquad (1)$$

where n is a discrete time step and i is a lattice point ($i = 1, 2, \cdots, N$=system size)

[1]This term is often used incorrectly, that is for the low-dimensional chaos with some spatial structure. Here we use the term for the high-dimensional chaos whose dimension (effective degrees of freedom) diverges with a system size.

[2]This area of the field is rapidly growing recently and it is impossible to cover in a short review.

with a periodic boundary condition. Extensions to a higher dimensional lattice are straightforward which will be discussed in §3.

Here the mapping function $f(x)$ is chosen to be the logistic map

$$f(x) = 1 - ax^2, \tag{2}$$

since the map has already been investigated in detail. Features in the logistic map lattice, however, can be seen in a wide class of mappings and in other types of couplings, and are thought to be rather universal in a system with local chaos and diffusion. If we adopt a different class of mappings like circle map $f(x) = x + (K/2\pi)sin(2\pi x) + \Omega$, or a bistable map $f(x) = tanh(\beta x)$, different behaviors can be seen (soliton turbulence and pattern formation respectively).

In the model (1), the independent procedures in (B) are local transformation (eq.(2)) and the diffusion process, which are separated parallel procedures. The model consists of the sequential repetition of these two procedures. This argument leads to the following equivalent form with the above model:

$$y_{n+1}(i) = f((1-\epsilon)y_n(i) + \epsilon/2(y_n(i+1) + y_n(i-1))). \tag{3}$$

If we adopt different procedures, we can construct models for different types of dynamical behavior of spatially extended systems. Examples with a different type of coupling ("convective coupling") will be studied in §8. Other choices of nonlinear process for the construction of CML corresponding to the spinodal decomposition, excitable media, Josephson junction and so on are also discussed.

Let us compare the merits and demerits of our model with other models for spatially extended systems, in particular, partial differential equation (PDE) and cellular automata (CA)[44].

PDE vs. CML

A merit of PDE over CML is that it is easily related with physical phenomena[3], starting from macroscopic equations such as the Navier-Stokes. Reduced equations can be obtained with the use of some perturbative techniques (e.g., reductive perturbation method). Analytic methods developed since Fourier are available for PDE.

Merits of CML over PDE are that (i) straightforward application of dynamical systems theory, in particular, chaos (ii) straightforward connection with the studies in statistical mechanics on a lattice system and (iii) numerical efficiency which enables us a global search in the parameter space.

CA vs. CML

Since there is no inherent continuos parameter in CA, it is rather difficult to see a change of a state with a change of parameter. Applications of dynamical systems theory are difficult in CA. Another drawback in CA is that it cannot create information in the sense of Shannon [22]. In a finite lattice, information in CA is finite, of course, and there is no "creation of information" in the bit space. Another technical drawback in CA is that an aperiodic state exists only in a lattice of an infinite size. One has to discuss an asymptotic behavior as a system size gets large. However, it is not always

[3]This, however may be just a prejudice produced through long-term history.

easy to distinguish if the relevant behavior in an infinite system comes from a transient or a periodic attractor in a finite lattice [4].

In CML, notions developed in dynamical systems theory can be extended, while the straightforward application of notions in information science [45] is possible in CA.

As for the efficiency: In recent days, the use of "lattice gas automata" to fluid dynamics is becoming popular [46]. Numerical efficiency is stressed in this kind of approach in the term of "bit democracy", which means that each bit is used almost equally in the course of computation. It is thought that this method is very efficient if a special purpose machine is constructed. This is not necessarily true.

Indeed, there are two merits of CML over CA.

First, a field variable in CML is a quantity after a coarse graining. In CA, each cell corresponds to a microscopic variable. Thus we need a very large lattice size to have a realistic simulation. On the other hand, we can get a suitably statistical behavior within a small lattice system in CML. Statistical physics is science of coarse-graining, and it is obvious that a suitably coarse-grained model is in higher level description of the nature.

The second point is more technical. In the present computer with floating processors, hierarchical use of bits is quite fast, where bits are not used in an equal weight. On the other hand, in the architecture of digital computers, discreteness of time and space are not overcome. This leads to an efficiency of models with continuous state, discrete time and space [5].

2 Pattern Dynamics in a 1-dimensional Lattice

2.1 Global phase diagram

First we show a global phase diagram in our coupled logistic lattice, which is obtained by the simulation of (1) with random initial conditions. The meaning of each phase in the figure is explained in the course of the paper. Some examples of the space amplitude plots are shown in Fig.2, where patterns $x_n(i)$ from many time steps are overlayed. Also, examples of space-time diagrams are given in Fig. 3.

Let us briefly survey the characteristic property of each phase, with the use of spatial power spectra, and Lyapunov spectra.

Detailed account of a feature of each phase will be shown in the following sections.

Spatial power spectra

Spatial power spectra (⇒ Appendix) for each phase are given in Fig. 4. We note the following changes:

i) In the frozen random phase, the spectra decay slowly (roughly by $exp(-k)$). This comes from a random configuration of domains (Fig.4a).

ii)Few peaks grow at the pattern selection regime. The peak(s) are located at the wavenumber of selected patterns (Fig.4b).

---

[4] We may distinguish the two by growth rates of transients and period of attractors in a finite lattice. It depends on rules which of the period or transients grows faster with the system size [50].

[5] The purposes of the present paper are two-fold. One is to see the phenomenology of spatiotemporal chaos studied in CML, and the other is the application of CML to physical phenomena. If the reader is interested only in the latter aspect, (s)he can skip the details of §2 -7 and jump to §8.

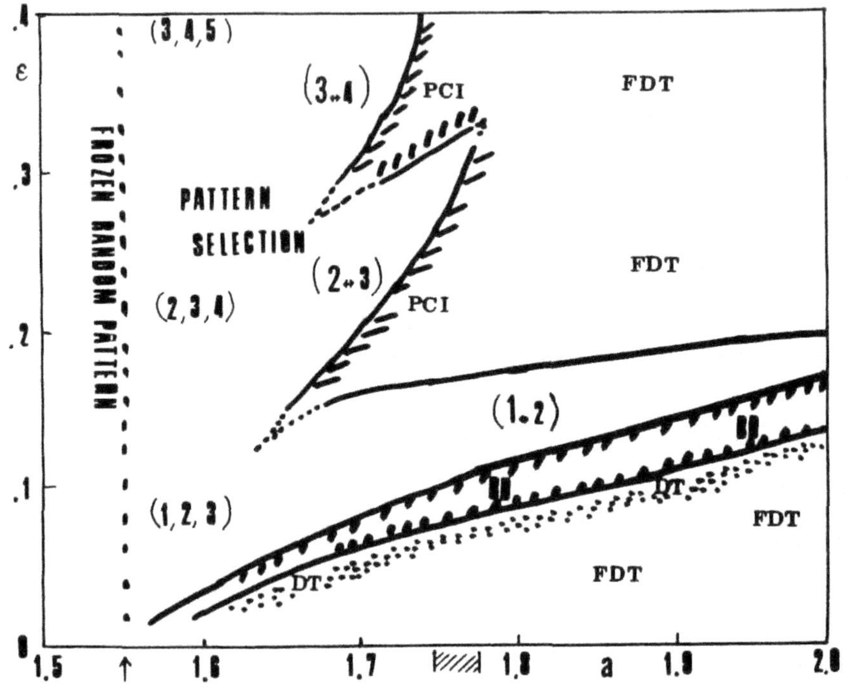

Fig.1: Phase diagram of the coupled logistic lattice (1): The phases are determined by the spatiotemporal patterns and the distribution function of pattern $Q(k)$ to be defined later. Simulation has been carried out for (1) with random initial configurations and N=100. The parameters are changed from a=1.5 to 2.0 by 0.01 and $\epsilon = 0.02$ to 0.4 by 0.02. Here BD, DT, PCI, and FDT are the abbreviations of Brownian motion of Defect, Defect Turbulence, Pattern Comepetition Intermittency, and Fully Developed Turbulence, respectively, which are discussed in detail later. The numbers such as 1,2,3 represent the selected domain sizes. See text for details. The arrow at the bottom line shows the band merging point for the single logistic map, while the region with oblique lines corrsepond to the period-3 window in the logistic map. (Reproduced from Ref. [3])

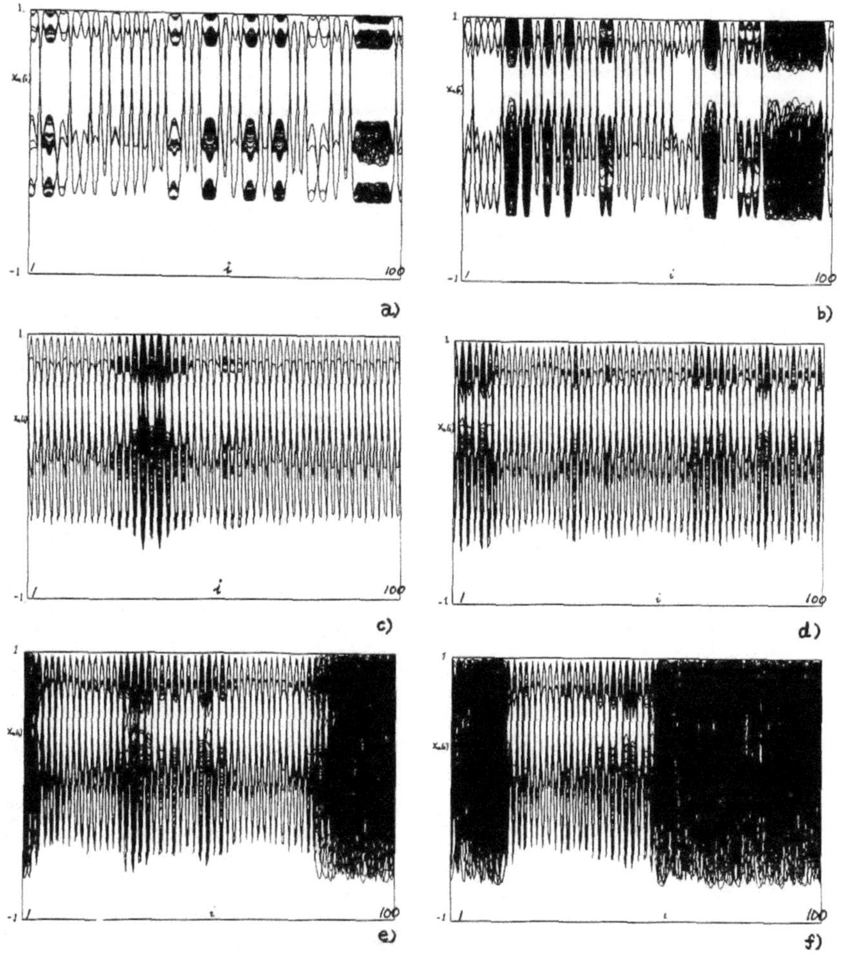

Fig.2: Space-amplitude plot for the coupled logistic lattice (1). Amplitudes $x_n(i)$'s are overlayed for 250 time steps after the 10000 transients. $\epsilon = 0.4$, N=100, and random initial condition.

(a) a=1.42 (b) a=1.51 (c) a=1.65 (d) a=1.735 (e) a=1.737 (f) a=1.75

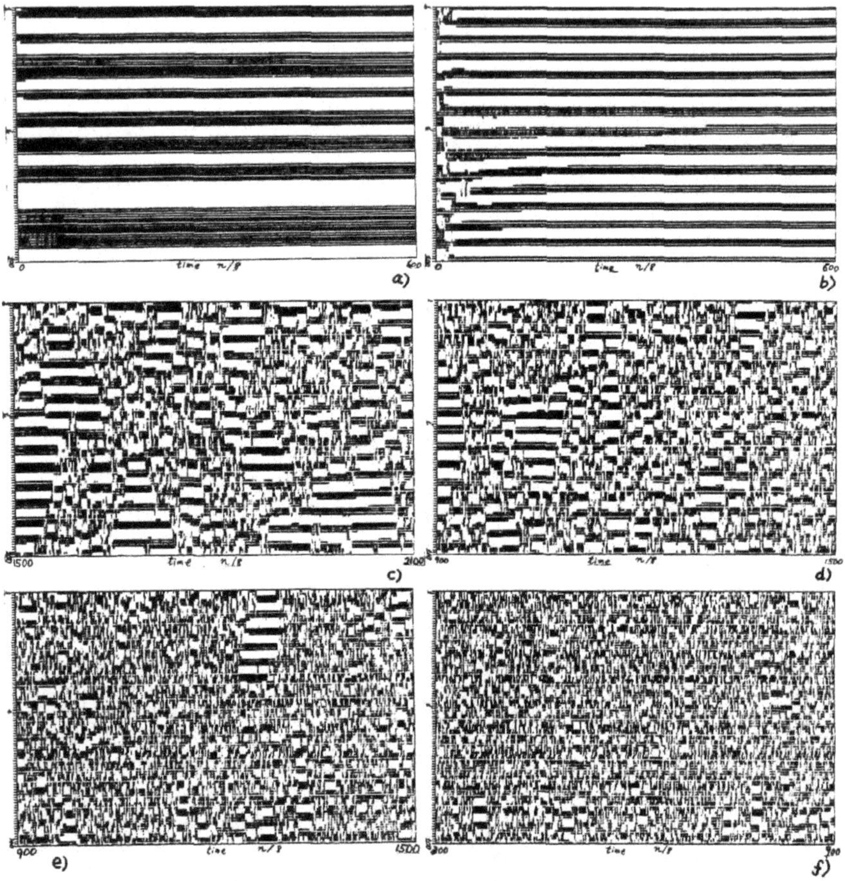

Fig.3: Space-time diagram for the coupled logistic lattice (1), with $\epsilon = 0.4$, N=100 and starting with a random initial condition. Every 8th step is plotted. If $x_n(i)$ is larger than $x*$(unstable fixed point of the logistic map), the corresponding space-time pixel is painted as black (if $x > x_2^* \equiv (\sqrt{1+7a} - 1)/(2a)$ painted darker), while it is left blank otherwise.

(a) a=1.5 (b) a=1.68 (c) a=1.75 (d) a=1.77 (e) a=1.83 (f) a=1.88

iii) Coexistence of the sharp peak(s) and the broad band noise at $k \approx 0$ for the defect turbulence and the pattern competition intermittency. In the defect turbulence, the peak at $k = 1/2$ and the broad band noise around $k = 0$ coexist. The latter part increases as the nonlinearity is increased. In the intermittency case also, the burst brings about the broad band spectra at $k = 0$ for $S(k)$. The whole spectra are composed of the broad-band and the sharp peaks at the wavenumbers corresponding to the selected patterns. As the nonlinearity is increased, the portion of the broad band noise increases (see Fig.4c).

iv)Fully developed spectra: As the nonlinearity a is increased further, sharp peaks disappear completely (see Fig.4d). The spatial spectra decay monotonically as k. They are roughly fitted by the form of $exp(-const. \times k^2)$. This Gaussian form arises from the diffusion nature of our model.

Temporal power spectra

Temporal power spectra (⇒ Appendix) are useful to characterize the temporal behavior of the system.

In the frozen random pattern, the peak at $\omega = 1/2$ is seen as is expected since a single logistic map exhibits the period-2 band motion for this parameter range. In the pattern selection regime, the above peak (and some other peaks in some cases) are observed, which is due to the regular motion by the pattern selection mechanism, to be discussed in §2.3. At some parameters, the motion is quasiperiodic in time and the peaks are located at $n(1/2) + m\omega$ (n, m are integers, and ω is an irrational number). In the fully developed turbulence, the peaks disappear and are replaced by some Lorentzian band noise around it and around $\omega \approx 0$.

Lyapunov analysis

The stepwise structure is seen in the Lyapunov spectra(⇒ Appendix) in the frozen random phase, which reflects upon the degeneracy by the existence of separated domains (see Fig.5a)b). Localized chaotic motion exists only in large domains, which gives positive Lyapunov exponents. The number of the positive exponents is proportional to the number of such domains. Here KS entropy (⇒) increases with a (Fig.6).

The values and the number of positive Lyapunov exponents decrease by the pattern selection, as is seen in Fig.5 c)d). Also, we can see that the KS entropy in Fig.6 shows a sudden decrease at $a \approx 1.56$. These show the suppression of chaos by the pattern selection clearly. In the pattern selection regime, still the stepwise structure is seen since there are degeneracies due to the existence of domains of the same size.

In the intermittency region, there are some positive exponents which are not very close to zero. The stepwise structure disappears here since the burst destroys the separated domain structure. The spectra as a whole take a rather smooth shape.

In the fully developed turbulence (Fig.5 e)f)), spectra are smooth in form. At the onset of spatiotemporal chaos, the spectra shape are close to be linear (Fig.5 e) ($\lambda_i = const. - ri$). As the nonlinearity is increased the shape increases its concavity,indicating

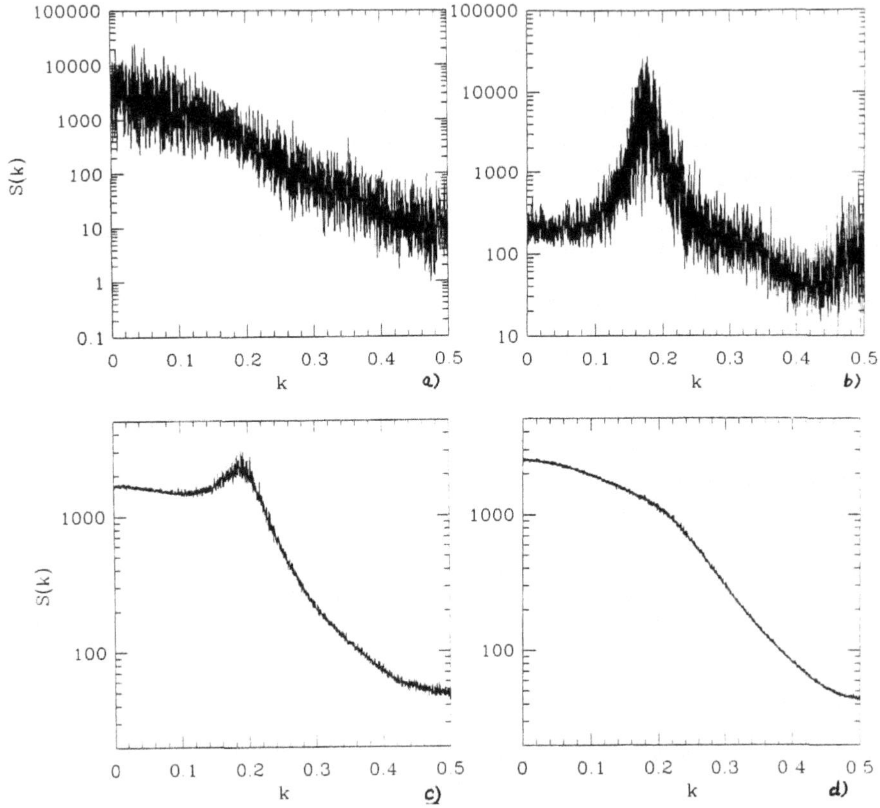

Fig.4: Semi-log plot of spatial power spectra S(k) for the model (1), with $\epsilon = .3$, N=4096, and starting with a random initial condition. Calculated from 10000 time step averages after discarding 10000 transients.

(a) a=1.53 (b) a=1.69 (c) a=1.78 (d) a=1.93

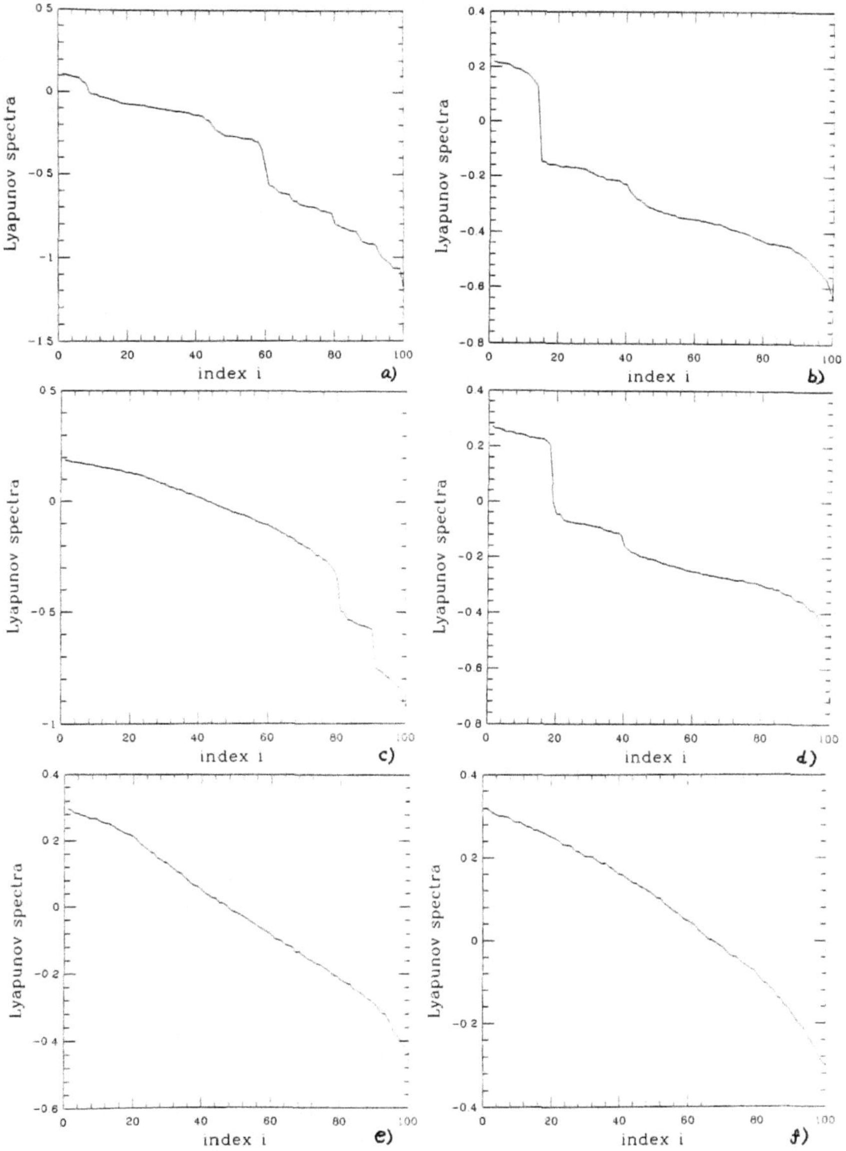

Fig.5: Lyapunov spectra of our model with $\epsilon = .2$, starting with a random initial condition. The calculation was carried out through the products of Jacobi matrices of the time steps 2000 to 5000. N=100.

(a) a=1.45 (b) a=1.55 (c) a=1.65 (d) a=1.7 (e) a=1.75 (f) a=1.85

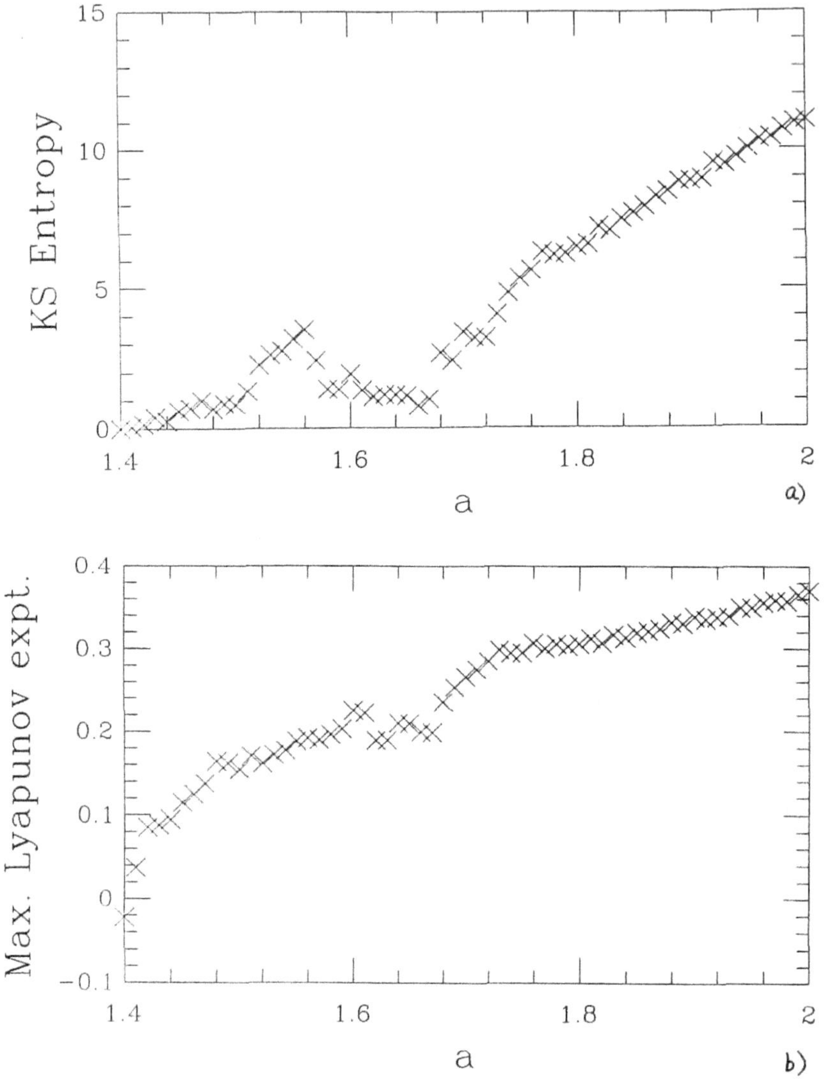

Fig.6: KS entropy (a) and maximal Lyapunov exponent (b) as a function of a, calculated from the Lyapunov spectra in the same way as in Fig.5. $\epsilon = .2$, N=64. To get the KS entropy density divide the values in the figure by $N = 64$.

the approach to independent chaotic motions of elements.

In the following sections we discuss the nature of each phase in a little more detail.

2.2 Frozen random pattern

CML (1) exhibits the period-doubling of kinks with the increase of the nonlinearity a. By the doublings, domains of various sizes are formed. After some numbers of doublings, the system exhibits a chaotic behavior at large domains. The domain boundary does not move in space (Fig. 2a)b), Fig.3a)).

As is seen in the phase diagram (Fig.1), the single logistic map here is in the parameter region for the band splitting with a period-two-band motion. The domain separation is assured by this band splitting.

In a large domain, the motion is quite chaotic, while it is almost period-8 at smaller domains, period-4, for much smaller domains, and period-2 for the smallest ones. We note that the motion of the lattice point differs in space ("spatial bifurcation"), although the dynamics itself does not include any inhomogeneity in space.

Distribution of domain sizes can differ by initial conditions. We can choose initial conditions so that attractors have an arbitrarily large domain. Thus we expect that the number of attractors increases exponentially to the system size.

2.3 Pattern selection and suppression of chaos

As the nonlinearity is increased further, larger domains start to be unstable and split into smaller domains. Initial conditions are no longer preserved (Fig.3b). Through the transient process, domains of few special sizes are selected. As the nonlinearity is increased, the number of possible patterns decreases and at the medium nonlinearity, only $1 \sim 3$ patterns are selected (Fig.2c)d)e)).

After the selection, the pattern of domains is frozen and does not move in space. Selected sizes of domains are such that the dynamics in the domains is less chaotic, that is, the motion with shorter periods. In the frozen random pattern, chaos is suppressed only in domains of small sizes. Here, only such small domains are selected.

A precise definition which distinguishes the frozen random state from the pattern selection is given by the following: In the frozen random phase ($a < a_p \approx 1.54$), there appears an arbitrary large domain if the system size goes infinity, while there exists a critical size ℓ_c, such that a domain of the size larger than ℓ_c cannot exist. Decrease of the variety of domain size is clearly seen in the decrease of static pattern entropy ($\Rightarrow$ Appendix).

As has been discussed in the temporal power spectra, the motion here is quasiperiodic or slightly chaotic, which is clearly seen in the calculation of Lyapunov exponents. In fact the maximal Lyapunov exponent is close to be zero or very small. Suppression of chaos by the pattern selection is seen in the decrease of KS entropy ($\Rightarrow$Appendix) at $a > a_p$ [3].

The diffusion term has a tendency to produce the homogeneity in space, while the chaotic motion makes the system inhomogeneous due to the sensitive dependence on initial conditions. These two tendencies conflict with each other. The conflict leads to the splitting of a larger domain. (One may regard this as the splitting by the "chaos pressure".) On the other hand, there is no conflict if the domains are selected in which chaos is suppressed. In this sense, we may term the present phase as " pattern selection through transient chaos and diffusion". Domains with larger positive exponents is easily collapsed by the boundary effect. This picture leads to the conjecture that *a pattern with smaller Lyapunov exponents is selected.* We have calculated Lyapunov exponents for various domain sizes by taking a small lattice size (2-10). Numerical results show that Lyapunov exponents are smaller (sometimes negative) in the selected domain sizes.

2.4 Selection of zigzag pattern and chaotic diffusion of defects

The simplest example of pattern selection is the selection of a zigzag pattern (domain size =1, i.e., $k = 1/2$). The zigzag pattern is easily characterized by the condition;

$$(x_n(i+1) - x_n(i))(x_n(i) - x_n(i-1)) < 0 \tag{4}$$

or the condition $(x_n(i+1) - x*)(x_n(i) - x*) < 0$ with $x* = (\sqrt{1+4a} - 1)/(2a)$, an unstable fixed point of the logistic map $f(x)$.

As time goes to infinity, a single domain of zigzag pattern covers the whole space. In the transient time regime, we have seen defects as a domain boundary between two zigzag patterns with different phases of oscillations. The defect is localized in space and moves around. The motion of defects is chaotic in time, as is checked by positive Lyapunov exponents (⇒ Appendix).

An example of space amplitude plot for a single defect state is shown in Fig. 7, while an example of space-time diagram is shown in Fig. 8a). Defects pair-annihilate and a domain size of a connected zigzag region increases with time. The system finally settles down to the completely zigzag state if the system size is even ($Q(1) = 1, Q(j) = 0$ for $j > 1$), while only a single defect moves around the space if $N = odd$.

Diffusion coefficient of defects

The locus of the defect looks like the Brownian motion. We note that the diffusion process here takes place *not* in the phase space (as is often discussed in the study of chaos), but in the real space [53]. Here, it is not evident that the motion is really Brownian motion, since our system is deterministic. Let us check the property of the motion of defect in a little more detail.

By the pair-annihilation, the number of defects decreases as $n^{-1/2}$, for a large time step n, if we start from arbitrary chosen initial conditions. This decay is consistent with the Brownian motion picture of defects.

A way to see the motion of a single defect is to take a system with an odd number of size so that a single defect always exists. The position of defect I_n is calculated through the condition (4). The numerical data are well fitted by the following expression

$$< (I_n - I_0)^2 >= 2Dn, \tag{5}$$

from which the diffusion coefficient D is calculated. Here the bracket $< ... >$ represents an ensemble average for a set of initial conditions.

The diffusion of a kink is observed in a cellular automaton problem [53], where the randomness comes from an initial random condition. The important difference here is that the randomness is created by the defect itself.

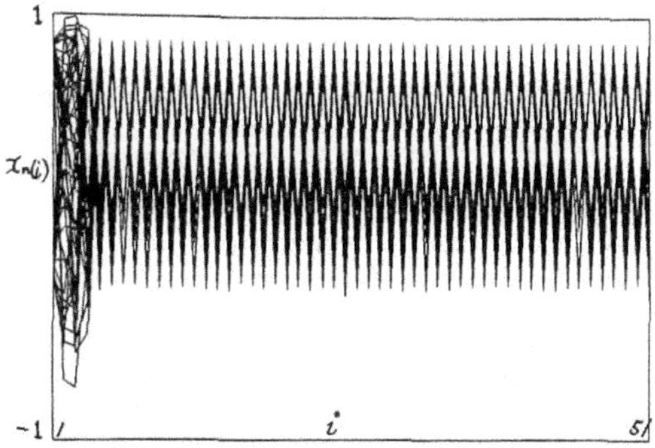

Fig.7: Space-amplitude plot for the coupled logistic lattice (1). Amplitudes $x_n(i)$'s are overlayed for 50 time steps after 1000 transients, starting from a random initial condition. N=51, $a = 1.83, \epsilon = .1$

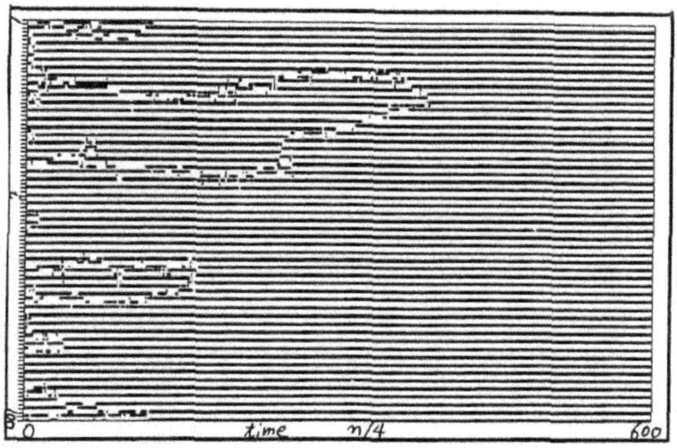

a)

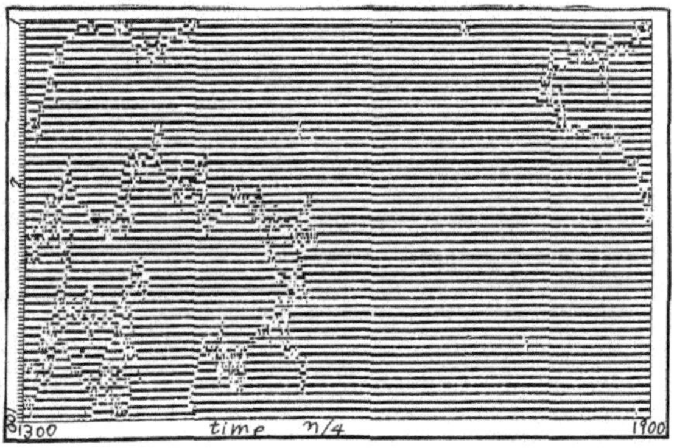

b)

Fig.8: Space-time diagram for the coupled logistic lattice (1), with $\epsilon = 0.1$, N=100 and starting with a random initial condition. Every 16th step is plotted. See Fig.3 for the way of digitization. (a) a=1.79 (b) a=1.885

The diffusion coefficient D is calculated as a function of a. It increases with a ($log(D)$ is roughly proportional to a).

To see if the defect is really chaotic, we calculate Lyapunov spectra for systems with a single defect and without any defects. If there is a defect, Lyapunov spectra have large positive parts, which are not seen in the spectra for a system without a defect. Kolmogorov-Sinai(KS) entropy of a defect is estimated by the sum of these positive parts. A logarithm of KS entropy increases linearly with a. Furthermore, the increase rate of a logarithm of the KS entropy and the diffusion coefficient agree if the zigzag region exhibits the chaotic motion.

KS entropy gives the rate of memory in the phase space. If the diffusion is triggered by the chaotic motion of a defect, the present Brownian motion can roughly be represented by a "coin tossing" per some time steps which are inversely proportional to the KS entropy. If this picture is valid, the KS entropy and the diffusion coefficient are proportional, which are consistent with our data in the chaotic regime.

2.5 Spatiotemporal intermittency transition: Defect turbulence and pattern competition intermittency

The selected pattern collapses at a critical parameter $a = a_c(\epsilon)$. Such a collapse of pattern occurs intermittently in space and time.

(I)Defect turbulence

At the region of $a \gtrapprox a_c$, the spontaneous collapse of zigzag pattern hardly occurs and the defect picture is still valid. At these parameter regions, the dynamics is governed by the pair-creation of defects, their Brownian motion, and collisions of defects, which may cause the pair- annihilation or creation of defects, or complicated transient patterns. The collision mechanism is quite similar to the "soliton turbulence" observed in coupled circle map lattices [11,5] and some cellular automata [50,52]. An example of space-time diagram is shown in Fig. 8b), where a pair-creation is clearly seen.

The phenomenon here can be understood as the crisis in a high-dimensional space (for crisis, see [54,55];see also [56]). The simplest way to see the relation with the crisis is the use of two-coupled logistic map ($N = 2$) [51]. If we plot the 2-dimensional attractor by $(x(1), x(2))$, the plot exhibits sudden broadening from two separated regions to a single connected region, at $a = a'$ which is slightly larger than a_c. As the size of our system N is increased, the onset parameter for the collapse of zigzag pattern decreases from a' to a_c. The reason for the suppression of the onset is that the gate for the crisis is increased as the size, since possible configurations of deviations from the zigzag structure are enriched. In other words, the discrepancy between a_c and a' arises from spatially chaotic modulations of the zigzag pattern.

(II)Selective flicker noise

A characteristic dynamical feature in the defect turbulence lies in its long-time correlation. To study this feature, we use the dynamical form factor (spatiotemporal power spectra (⇒Appendix).

In the parameter region of the defect turbulence, our system exhibits the following flicker noise for the modes with $k \approx 1/2$. In Fig.9, $P(k,\omega)$'s are plotted for k =0, 2/8, 3/8, and 1/2. As k approaches 1/2, low-frequency parts grow and $P(k,\omega) = \omega^{-\alpha}$ is

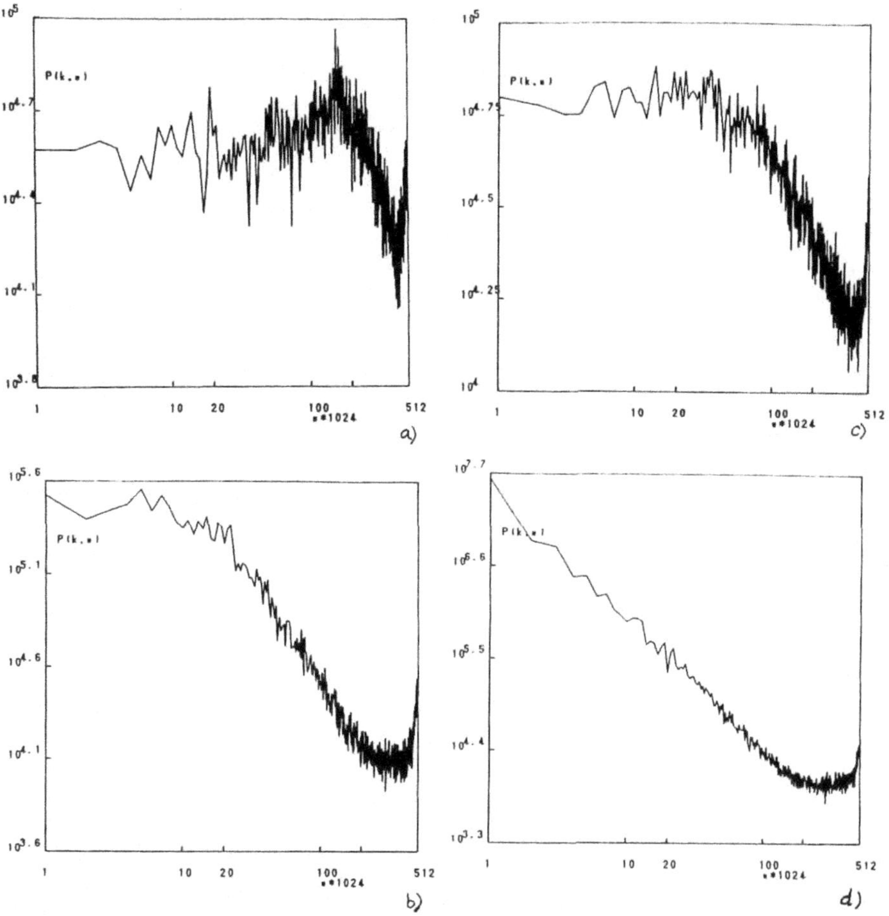

Fig.9: Log-log plot of space-time power spectra $P(k,\omega)$ as a function of ω for a=1.89, $\epsilon = 0.1$: (a)k=0, (b)k=2/8, (c)k=3/8 (d)k=4/8. These power spectra are calculated from the data of $x_{2n}(i)$ for 512×2 time steps, after 10000 steps of transients, and from 50 sequential samplings, starting from a random initial condition (reproduced from [3]).

clearly seen for $k = k_1 (= 1/2)$. Note that the flicker noise is selectively observed only for the modes $k \approx 1/2$.

At the onset of the collapse of zigzag pattern ($a = 1.88$), α is close to 2, which means that the relaxation time diverges. As the nonlinearity parameter a is increased, the exponent α decreases from 2. The $\omega^{-\alpha}$ behavior is observed selectively for $k \approx 1/2$ at $a_c \approx 1.88 < a < 1.92 \approx a'$. Collapse of the zigzag pattern occurs more frequently for larger a, which leads to a faster decay of the correlation function (and small α).

The reason for this flicker-like noise is that the zigzag mode exists with a long range correlation, and it is destroyed only through a long-ranged effect as is seen in the above explanation by the high-dimensional crisis in the region of $a < a'$. Thus the dynamics of the mode of wavenumber 1/2 includes the motion of a very long time scale. On the other hand, the mode with $k \approx 0$ corresponds to the chaotic motion of defects, which has a short-time memory.

A reason for this critical behavior in a wide parameter regime may be attributed to the finiteness of system size. If we follow the common-sense of phase transitions, it is expected that a singular behavior is seen just at a critical point. The above power-law-type behavior in a large parameter space in a finite system, however, may be relevant in experiments, since most "large" nonequilibrium experimental systems (e.g., Benard convection, liquid crystal convection) contain $10^2 \sim 10^3$ effective degrees (e.g., number of rolls) which are much smaller than the Avogadro number.

Also, it is interesting to note that the critical behavior is easily observed in *temporal domains*, not in spatial domains (like correlation length).

As the wavenumber is decreased from $k = 1/2$, the power decreases gradually. As it goes much smaller, the plateau at $\omega = 0$ appears and the spectra approach the Lorentzian form.

The flicker-like noise has been observed in a low-dimensional dynamical system at the onset of chaos through the intermittency [57,58]. The flicker-like noise here should be noted for its selectivity to the wavenumber.

(III)Quantitative analysis of pattern dynamics

Pattern distribution function $Q(k)$'s ($\Rightarrow$ Appendix) are useful as order parameter(s) to distinguish the different phases observed in our model.

In the frozen random pattern, $Q(k) \neq 0$ for various k's. Furthermore, the distribution can depend on a choice of initial conditions. In the pattern selection regime, $Q(1)$ and $Q(2)$ get larger and ratios for larger domains are suppressed ($a > 1.58$). At $a > 1.64$ $Q(k) = 0$ for $k \neq 1, 2$. In these regions, the dependence of the distribution on initial conditions is very small if we restrict ourselves to random initial conditions and take a large N (say, > 100). In particular, the selection of $k = 1$ and 2 at $a > 1.64$ is not affected by the change of initial conditions. In the zigzag pattern selection phase, $Q(1) = 1$ and $Q(k) = 0$ for $k \neq 1$.

At $a > a_c$, the zigzag pattern collapses spontaneously. Thus $Q(k)$ ($k \neq 1$) does not vanish. The value $(1 - Q(1))$ gives a measure of the destructed pattern.

Critical phenomena of the collapse of patterns are investigated through the disorder parameter $1 - Q(1)$. Near $a \approx a_c$, the data are fitted roughly by

$$1 - Q(1) \approx (a - a_c)^{\beta}, \tag{6}$$

with $\beta \approx 1.0$. These critical phenomena are explained as the crisis in a high- dimen-

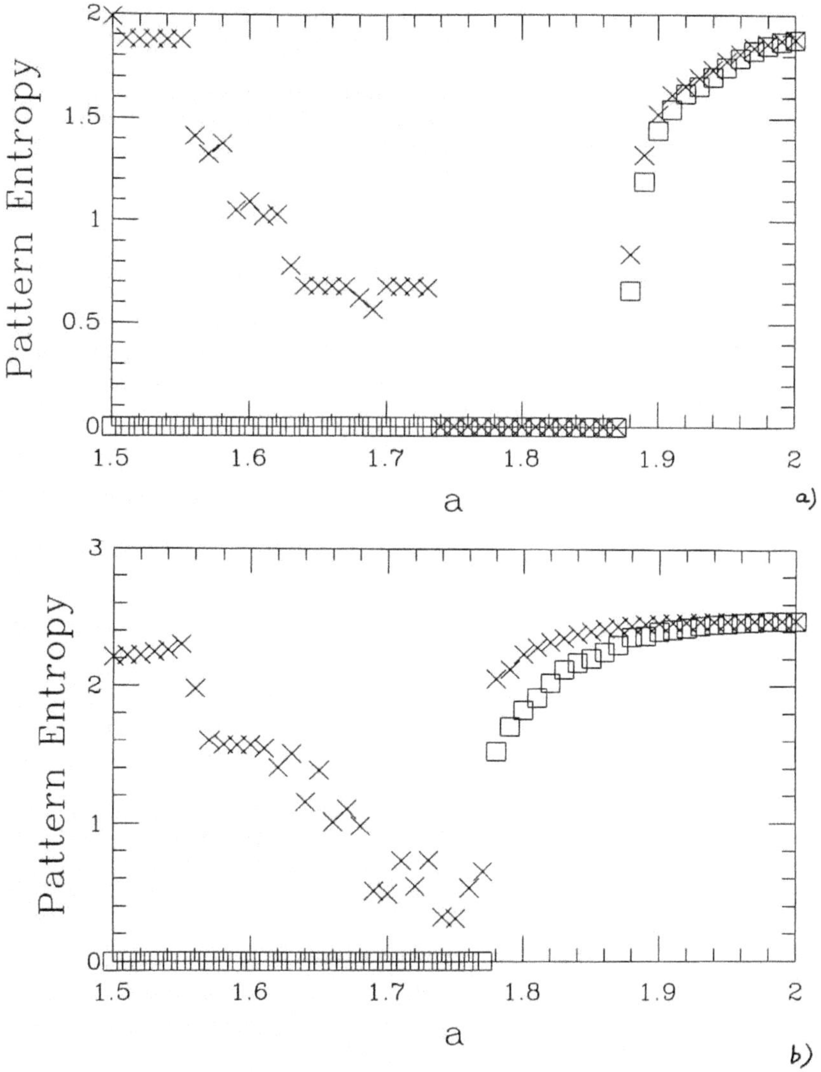

Fig.10 Pattern (static) entropy(×) and pattern dynamical entropy (□) for our model with N=128, calculated from the average for 30000 time steps after 10000 transients. (a) $\epsilon = .1$ (b) $\epsilon = .3$ (essentially reproduced from [3])

sional dynamical system in the terminology of dynamical systems. For the exponents in the crisis in low dimensional dynamical systems, see [55], where the exponents are related with the eigenvalues of saddles.

A single quantity to characterize the pattern complexity is an entropy(⇒Appendix). The change of static pattern entropy is shown in Fig.10. The entropy is large in the frozen random pattern, and decreases rapidly by the pattern selection to the zigzag state. By the pattern collapse it increases with

$$S_p \propto (a - a_c)^{\beta'}. \tag{7}$$

The exponent β' coincides with β within a numerical error as is expected, since $Q(k)$'s for $k > 1$ contributes to the increase of entropy.

Dynamical aspects of patterns are characterized by the transition matrix of patterns or (more simply) by the pattern dynamical entropy(⇒Appendix). In the frozen pattern, there is no transition among patterns. Thus the pattern dynamical entropy vanishes, even if the static entropy is large. At the onset of defect turbulence, the entropy increases with the form

$$S_d \propto (a - a_c)^{\delta}, \tag{8}$$

with the exponent $\delta \approx \beta$. Since this exponent includes the dynamical aspect also, it is probable that $\delta \neq \beta$.

A difference between the static and dynamic entropies shows how the patterns are spatiotemporally complex in the sense of mutual information (⇒Appendix). The difference is large in the defect turbulence (transition regime).

In the defect turbulence region, the transition matrix may not be a good quantifier, since the temporal correlation has a long-time-tail and is not represented by a Markovian dynamics. The lifetime distribution of the zigzag pattern does not show an exponential decay but has a Pareto-Zipf form, that is,

$$W_1(n) = n^{-\psi}. \tag{9}$$

The life-time distributions for the other sizes of domains have a normal exponential form $W_k(n) = exp(-n/t_{lf}(k))$, where $t_{lf}(k)$ is the lifetime of the domain of size k.

The Pareto-Zipf form is typically observed in the dynamics with the flicker-like noise (see e.g., [57]). Since the power spectra can be written as the sum of the temporal Lorentzian form as

$$P(k,\omega) \sim \int dn W_j(n)/(\omega^2 + 1/n^2) \tag{10}$$

for the corresponding wavenumber k and the domain j, we can expect that the exponent ψ for domain size j and the low-frequency exponent α for the power spectra of the corresponding wavenumber ($k = 1/(2j)$) are related by

$$\psi + \alpha = 3. \tag{11}$$

From our numerical data, this equality is roughly satisfied [6].

[6] for a detailed account of a small discrepancy, see [3].

(IV)Pattern competition intermittency

If the coupling is larger, selected domain sizes are not unique, as can be seen in Fig. 3b). With the increase of nonlinearity, there appear intermittent bursts, which have a nonstationary nature (see Fig.2e),3d)). The burst arises from the mismatching of phases and has much larger structures than a chaotic defect.

The "pattern competition intermittency" here is characterized by the existence of more than one stable patterns and intermittent transition among patterns. Although each pattern remains to be stable by itself, the mismatch of phases of oscillations between domains creates some bursts, which move around in space-time and destroy the pattern. The long-range correlation in this phase is also remarkable.

The present intermittency has many aspects in common with the above defect turbulence. There are two different points: First, the number of selected patterns is more than one here. Secondly, the burst has much larger structures than a defect. This intermittency and defect turbulence belong to the spatiotemporal intermittency (STI), investigated extensively in recent years.

Spatiotemporal Intermittency

Here we briefly look back on STI. STI was first studied as the spatial extension [1,5] (see also [20]) of Pomeau-Manneville's intermittency [59]. The intermittency in the present model is related with the crisis in high dimensional systems (see also [10]). In the STI, local dynamics has topological chaos but a non-chaotic attractor. Through the coupling of the local dynamics, this unobservable topological chaos appears as the observable chaotic burst.

As for the phase transition problem, STI is quite close to the problem of directed percolation. (see also [60]). In some models, however, the critical exponents do not belong to the universality class of the directed percolation[20].

One interesting feature in the intermittency phase is the existence of long-lived transients. If a system size is small, the turbulent pattern can disappear after long transients, and the system finally hits a globally non-frustrated structure, and intermittent bursts disappear. Some of the spatiotemporal intermittency belong to a class of the transient turbulence, as is discussed in [23].

Selective Flicker Noise at the intermittency

In the pattern competition intermittency, we have again observed the selective flicker noise.

At $\epsilon = .3$ and $a = (1.72 \sim 1.76)$, for example, the dynamical form factor $P(k,\omega)$ near $k = 0$ and $k = 1/2$ obey the normal Lorentzian form, while the flicker-like noise is seen around $k = (1/6 \sim 1/3)$, which is fitted by $P(k,\omega) = \omega^{-\alpha}$. $(\alpha \sim (1.5 \sim 2.0))$. At these parameter regions, we have seen the pattern competition intermittency with the selection of domain sizes =2 and 3. The exponent α is ≈ 1.5, for $a = 1.73$.

Quantifiers for pattern dynamics

Pattern dynamical quantifiers show a similar critical behavior as in the defect turbulence.

In the frozen regime, there are various possible patterns which give a large pattern entropy and a vanishing dynamical entropy. As a is increased, the pattern selection process occurs and leads to the decrease of the pattern entropy.

In the intermittency region, the distribution $Q(k)$ consists of the domains of selected patterns (size=2 and 3 for $\epsilon = .3$) and the distribution function by the burst, which has the tail of $exp(-const. \times k)$. Through the collapse of pattern, we can see the critical behavior of the disorder parameter $(1 - \sum_p Q(p))$ (where the sum is over the possible selected patterns, e.g., $p = 2, 3$ for $\epsilon = .3$) and of the entropies, in a similar manner as in the defect turbulence.[7]

In the intermittency region, the lifetime distributions of domains of selected sizes (2 and 3 for $\epsilon = .3$) again shows the power-law. For example, $W_2(n)$ and $W_3(n)$ are fitted by $W(n) \propto n^{-\psi}$ with $\psi \approx 1.75$, for $a = 1.72$ and $\epsilon = .3$, while the lifetime distribution $W_4(n)$) obeys the usual exponential decay. Here again, the exponents ψ and α roughly obey the relation $\psi + \alpha = 3$.

2.6 Fully developed turbulence

As the nonlinearity is increased further, the ordered structure in the space-time diagram can hardly be observed (see Fig. 3e)f)). We call this state "fully developed turbulence" in our lattice system, since this state can be well approximated by the direct product state of local chaos with a rapid decay of spatial correlation. The state can be represented by smooth statistical and dynamical quantifiers. The change of $Q(k)$ with the nonlinearity a is smooth. Pattern distribution function obeys $Q(k) = exp(-const. \times k)$ for large k, which implies that the patterns are generated randomly. This means that the pattern dynamics is well approximated as a Markov process of 1-0 sequence. The difference between the dynamical and static entropies is small, indicating that the transition among patterns is no more restricted and occurs randomly. As for the conflict between the two tendencies (ordering by diffusion and inhomogenizaion by local chaos), the chaotic part exceeds in the ordering process in this phase.

The motion here is roughly approximated by the "heat bath" picture. Take a subsystem, and our dynamics in the subsystem is well approximated by the local deterministic chaos by the variables within the subsystem and the random perturbation from boundaries, which is created by the chaotic motion from other areas. This picture leads to the "self-consistent" Perron-Frobenius theory as is discussed in §6.

2.7 Summing Up...

The main results in §2 are briefly summarized in Table I.

We note that our transition sequence has similarity with the phases in gas, liquid, crystal, and glass. The fully developed turbulent phase can correspond to the gas, while the pattern selection is analogous with the ordered pattern in a crystal. In a frozen random phase, the randomness in a large domain or in a defect is frozen in space, which reminds us of the frozen randomness in the glass. As the nonlinearity is increased, the local chaos in our system is amplified ("temperature is increased"), which leads to the "melting" of frozen patterns successively.

[7] detailed quantitative studies are left for future.

Before closing the present section, we note the tongue structure of pattern selection in the phase diagram (Fig. 1). This kind of tongue-like structure of spatial locking structure is seen in a static pattern of spatially modulated systems (see e.g., [47,48]) and in the resonance structure in low dimensional dynamical systems (Arnold tongue; see e.g., [2]). The pattern selection in our model corresponds to the resonance structure in space. Our result shows that the spatial locking is also seen in the problem of the pattern selection in spatiotemporal chaos, if the selected pattern is fixed in space.

Table I. Features of phases characterized by quantifiers .

Quantifiers	Temporal power spectra	Spatial power spectra	Pattern Distribution $Q(k)$	Pattern Entropy S_p	Dynamical Entropy S_d	Lyapunov Spectra	KS Entorpy
Frozen Random Pattern	Peaks+ Broad Band Noise	$exp(-cst. \times k)$	$Q(k) \neq 0$ for many k's	Large (almost constant)	0	Small stepwise structure	Positive
Pattern Selection	Peaks+ Broad Band Noise	Few peaks (+ Noise)	few $Q(k)$'s $\neq 0$ $Q(\ell)=0$ for $\ell > \ell_c$	Small ↘	0	Large stepwise structure	Small ↘ (or 0)
(Intermittency)	Broad peaks+ Noise	Broad peaks+ Noise	few peaks and $exp(-cst. \times k)$	$(a-a_c)^{\beta'}$	$(a-a_c)^{\delta}$	positive parts grow	start to ↗
Fully Developed Turbulence	Broad Band Noise	$exp(-cst. \times k^2)$	$Q(k) \propto exp(-cst. \times k)$	Large ↗	Large $S_d \approx S_p$	Smooth form	Large ↗

3 Pattern Dynamics in a 2-dimensional Lattice

Extension of our diffusively coupled map lattice to a two-dimensional space is quite straightforward. Here we consider the simplest case, i.e., the nearest-neighbor coupling on a square lattice (see also [16]). The model is given by

$$x_{n+1}(i,j) = (1-\epsilon)f(x_n(i,j)) +$$

$$\epsilon/4[f(x_n(i+1,j)) + f(x_n(i-1,j)) + f(x_n(i,j+1)) + f(x_n(i,j-1))] \qquad (12)$$

where n is a discrete time step and i,j denotes a 2-dimensional lattice point ($i,j = 1,2,\cdots,N$=system size) with a periodic boundary condition. Here the mapping function $f(x)$ is again chosen to be the logistic map $f(x) = 1 - ax^2$.

Examples of snapshots are shown in Fig. 11.

For small ϵ, we have again observed the transition sequence from (i) frozen random state to (ii) pattern selection and to (iii) fully developed turbulence via spatiotemporal intermittency.

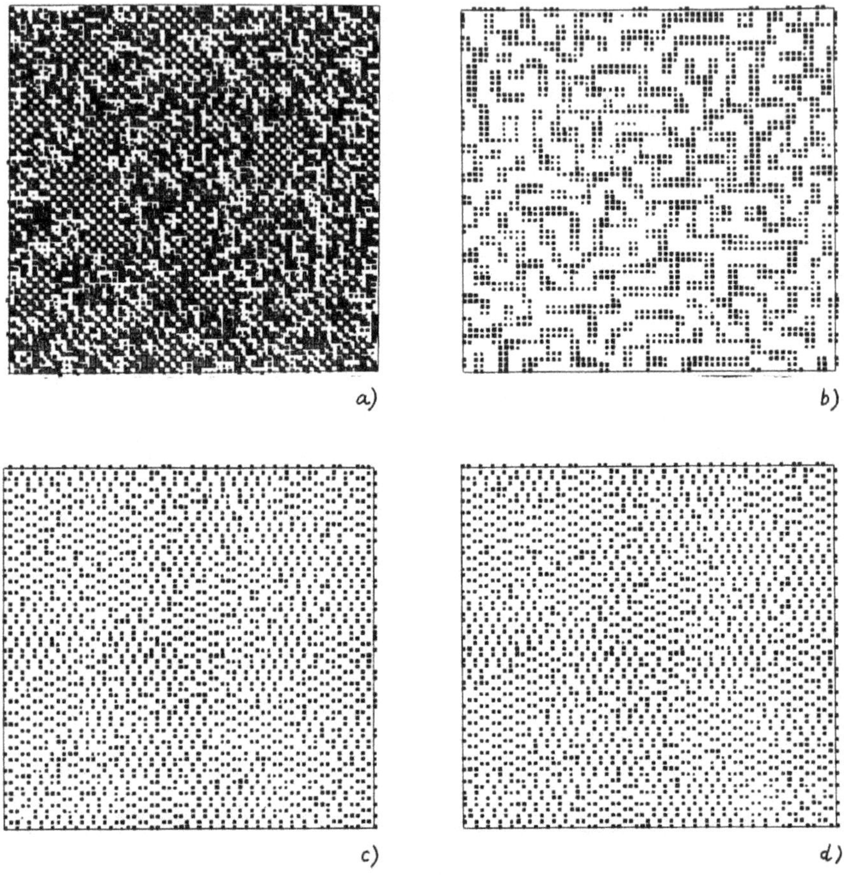

Fig.11: Snapshot pattern for the 2-dimensional logistic lattice. On a lattice point (i, j), a square with a length proportional to $(x_n(i, j) - 0.2)$ is painted if $x_n(i, j) > .2$. Otherwise it is left blank. Size = 64x64.

(a)a=1.903, $\epsilon = .1$, at time step 6000

(b)a=1.8, $\epsilon = .35$, at time step 5000

(c)a=1.94, $\epsilon = .2$, at time step 15000 , (d) time step 40000.

3.1 Checkerboard pattern selection, chaotic string, and spatiotemporal intermittency

For $\epsilon = .1$, a frozen random pattern is observed for $a < 1.75$ (Fig. 6a). A checkerboard pattern is selected for $1.75 \leq a \leq 1.9$ (see Fig. 12) . After some transients, a single checkerboard pattern covers the whole space if the size is even. The selection process is regarded as the pattern formation, since two antiphased checkerboard domains are separated by a string, which moves chaotically in time and moves around space, and disappears by the collisions.

The zigzag pattern formation (Fig.12) is quite similar to the process seen in the ordering process of magnetic systems, or of simulations in the kinetic Ising model or time-dependent Ginzburg-Landau equation (see also §8). We can see the pattern formation process by the Brownian motion of a chaotic string. Here we call the motion as Brownian, since the motion there obeys the normal diffusive behavior triggered by a random walk, as is seen in the Brownian motion of defects in §2. The chaotic motion of a string is again quantitatively checked by Lyapunov exponents. In the same manner as in the Brownian motion of defects, we can estimate the KS entropy of a chaotic string. In fact, we have (additional) positive Lyapunov exponents in the presence of a string, which cannot be seen in the complete checker-board pattern.

Question: Does the chaotic motion at the domain boundary give novel features in the pattern formation? Does it belong to the same universality class as the kinetic Ising model?

At $a = a_c \approx 1.901$, the checkerboard pattern collapses spontaneously. Defects are created spontaneously from the checkerboard pattern. For $a \approx 1.901$, these defects are not percolated. As a is increased further, they propagate and interact with each other. The interaction causes the spatiotemporal intermittency. The spatiotemporal pattern there is understood as the intermittent transition between checkerboards and random patterns (see Fig. 11a) for a snapshot).

The lifetime of checkerboard pattern has a power-law distribution near $a \approx a_c$, which leads to the selective-flicker-like noise again. In fact, we have calculated the dynamical form factor $P(k_x, k_y, \omega)$, a power of the Fourier transform in space and time. As in the 1-d case, it shows the selective flicker noise for the wavenumber of the checkerboard pattern, i.e.,

$$P(k_x = 1/2, k_y = 1/2, \omega) = \omega^{-\alpha} \tag{13}$$

($\alpha \approx 1.9$) while neither the spectrum $P(k_x = 1/2, k_y \approx 0, \omega)$ nor $P(k_x \approx 0, k_y \approx, \omega)$ does show the divergence of low-frequency parts (see Fig. 13 a)b)c)).

Recently Nasuno, Sano, and Sawada [62] have performed a beautiful experiment on the collapse of a grid pattern in the electric convection of liquid crystal. They found the intermittent collapse of the grid pattern and the selective flicker noise for the wavenumber for the grid pattern (with $\alpha \approx 1.9$).

3.2 Selection of larger patterns

2x1 pattern selection

If the coupling ϵ is increased, we have seen the pattern selection with larger wavelengths. If $\epsilon \approx .2$, the selected pattern is a 2x1 unit (see Fig. 11c)d)). In the selection of 2x1 unit, the transient time necessary for the pattern formation is much longer than

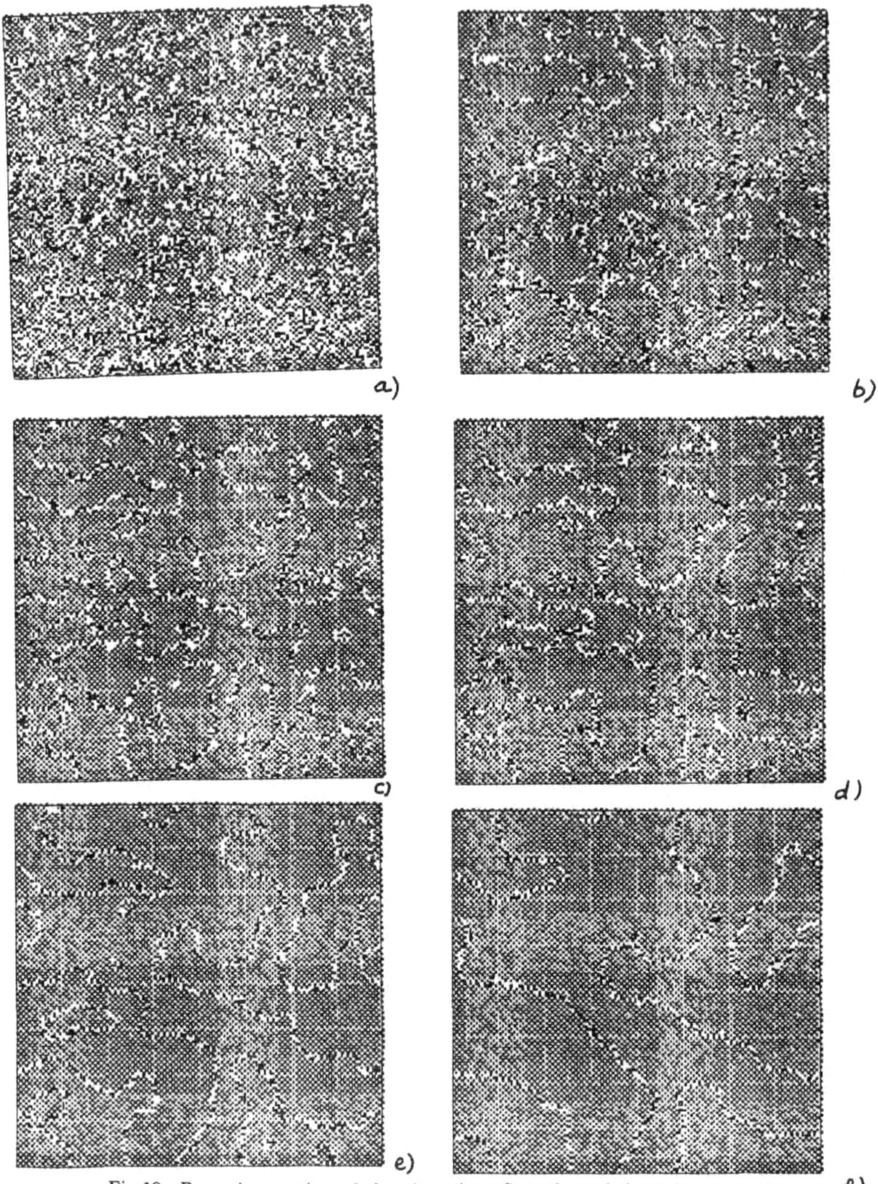

Fig.12: Brownian motion of chaotic string: Snapshot of the 2-dimensional logistic lattice. lattice size=128x128: a=1.8, $\epsilon = .1$

a) time step = 64 b) 192 c) 320 d) 480 e) 640 f) 1952

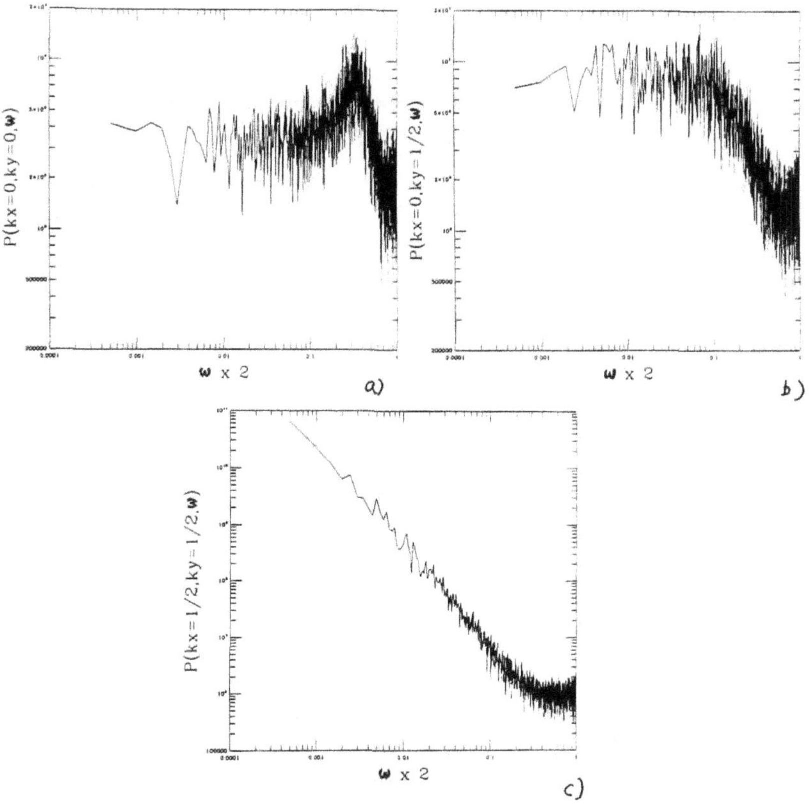

Fig.13: Log-log plot of $P(k_x, k_y, \omega)$ as a function of ω. a=1.904, $\epsilon = .1$. The spectrum is calculated from 10 sequential samples of 4096 steps after 2000 transients for the logistic lattice with the size of 64x64. (a) $k_x = k_y = 0$ (b)$k_x = 0, k_y = 1/2$ (c) $k_x = k_y = 1/2$.

that for a checkerboard pattern. This formation process is very slow because it is the competition process among 4 types of attractors (two types of phases of oscillations, and horizontal and perpendicular roll structures). This kind of pattern was first noted in a phase transition in a 2-dimensional stochastic cellular automaton [61].

Here again, chaos is suppressed by the pattern selection process, which is quantitatively characterized by the decrease of KS entropy.

As the nonlinearity is increased, a collapse of the pattern by the spatiotemporal intermittency again occurs. The existence of 4-fold degeneracy strongly suggests the first order transition, which is true in the 2-dimensional stochastic cellular automaton [61].

If the coupling ϵ is between .1 and .2 (e.g., .15), the competition between the checkerboard and 2x1 is seen, which leads to the intermittent collapse of the two patterns (pattern competition intermittency), or leads to a chaotic string with 2x1 structure in the zigzag domain. These are again similar to the observation in a 1-d lattice. In the experiment by Nasuno et al., these behaviors are again found [62].

labyrinth

We have not yet observed a clear pattern selection with a larger wavelength (e.g., 2x2-domain) in the nearest-neighbor coupling model. For larger coupling we have seen a partially ordered roll-like pattern ("labyrinth") (see Fig.11b).

3.3 Absence of a frozen pattern for stronger coupling and extension of Pierls' argument

If the coupling is further increased ($\epsilon > .4$), neither a frozen random pattern nor a pattern selection is observed. A domain is unstable and its boundary moves in time till a single domain covers the whole space.

Change of Lyapunov exponents with the parameter a is smooth, and the Lyapunov spectra have a smooth shape for all a [4]. These results are consistent with the absence of pattern selection.

The absence of frozen pattern is understood as the extension of Pierls' argument on the absence of ordered phases. The diffusive coupling tries to destroy the domain boundary between two frozen patterns. On the other hand, the motion in each domain is more stable than the boundary. To destroy a domain, a state has to pass through unstable states (recall that $x(i,j)$ has to take a value around the unstable fixed point of a logistic map), which requires some barrier. This effect leads to the preservation of a domain and the pinning of a boundary. The former strength is estimated by the surface tension as $\epsilon M^{(1-1/d)}$ for a domain of length M. The latter part is roughly independent of M, if M is large enough. For $d > 1$, a smaller domain has smaller stability. Thus for a lattice with a dimension ≥ 2, it is expected that there is an upperbound on the coupling strength beyond which a frozen domain structure (the frozen random pattern and the pattern selection) loses its stability.

From our numerical results, it is concluded that this threshold coupling lies around $\epsilon \approx .35$ for our 2-dimensional logistic lattice with the nearest-neighbor coupling.

In higher dimensions, it is expected that the frozen pattern is much harder to sustain in a short-ranged model.

4 Open Flow Models

So far we have investigated the case where the coupling is diffusive.

Another interesting version of CML is the one-way coupling model, introduced as a model for the open fluid flow:

$$x_{n+1}(i) = (1-\epsilon)f(x_n(i)) + \epsilon f(x_n(i-1)). \tag{14}$$

As has already been reported [13,14,15], the model exhibits the spatial period-doubling and selective amplification of noise. See Fig.14a) for the spatial period-doubling. After some doublings, our system goes to a turbulent state.

The model (14) is the extreme limit of the asymmetric coupling model

$x_{n+1}(i) = (1-\epsilon)f(x_n(i)) + \epsilon[\alpha f(x_n(i-1)) + (1-\alpha)f(x_n(i+1)]$

An important notion in the open flow model is the distinction with absolute and convective instabilities. If a small perturbation against a reference state grows in a stationary frame, it is called as "absolute instability", while if the perturbation grows only in some frame with a finite velocity, it is called as "convective instability" (see also "Co-moving Lyapunov exponent" in Appendix and [14]). In the above model, if the coupling is large, our system shows only the convective instability. For example, the stability of a homogeneous periodic state in a Galilean frame with a velocity v is calculated by

$$L(v) = (1/p)log(\prod_{i=1}^{p} f'(x_i)) + log(\frac{1-\epsilon}{1-v}) + vlog(\frac{\epsilon(1-v)}{v(1-\epsilon)}) \tag{15}$$

where x_i is the periodic orbit of a single logistic map. If $L(v)$ is positive for some velocity, our periodic state is convectively unstable. If $L(0) > 0$, it is absolutely unstable. For example, the homogeneous fixed point state $x* = (\sqrt{1+4a}-1)/(2a)$ is convectively unstable if $| f'(x*) | = \sqrt{1+4a} - 1 > 1$, and abosolutley unstable if $| (1-\epsilon)f'(x*) | > 1$.

I.S. Aronson, A.V. Gaponov-Grekhov, and M.I. Rabinovich [17] have recently investigated the above spatial period-doubling in open flow, using the renormalization group approach by Feigenbaum [63]. This procedure works if the coupling is very small, where the absolute instability occurs. They have calculated the scaling relation on the spatial interval in the period-doubling. If the coupling is larger, the instability which causes the period-doubling is not absolutely but *convectively* unstable. In this case the existence of a very small noise is essential to form the pattern [13]. Thus the scaling both with a noise and a space should be necessary.

Corresponding to the phase transitions in the diffusive coupling model, we have obtained similar phase transitions in the pattern dynamics, and a phase diagram similar to Fig.1. We have (i) flow of randomly chosen patterns (ii) flow with selected patterns (iii) transmission of defects (iv) spatiotemporal intermittency and (v) fully developed turbulent state (see Fig.14 and 15 for space amplitude plots and spacetime diagram).

The defect turbulence and transmission of defects to downflow with a fluctuating velocity are found which are triggered by its chaotic motion and interactions among defects. A source of defects is seen at the upper flow.

A novel point here is the spatial bifurcation from a homogeneous periodic state to a pattern selection (periodic state with a periodic spatial structure), and then to a turbulent state as a space goes downflow.

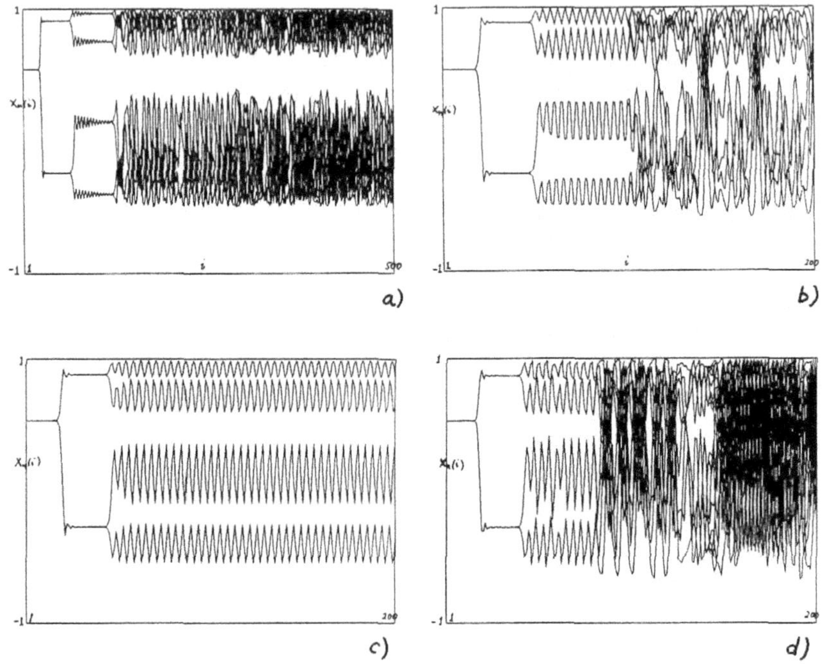

Fig.14: Space-amplitude plot for the one-way coupled logistic lattice (14). Amplitudes $x_n(i)$'s are overlayed for 250 time steps after discarding 10000 transients, starting with a random initial condition. ϵ = 0.5, N=500(a), N=200 (b-d).

(a) a=1.5,(b) a=1.6 (c) a=1.63 (d) a=1.68 .

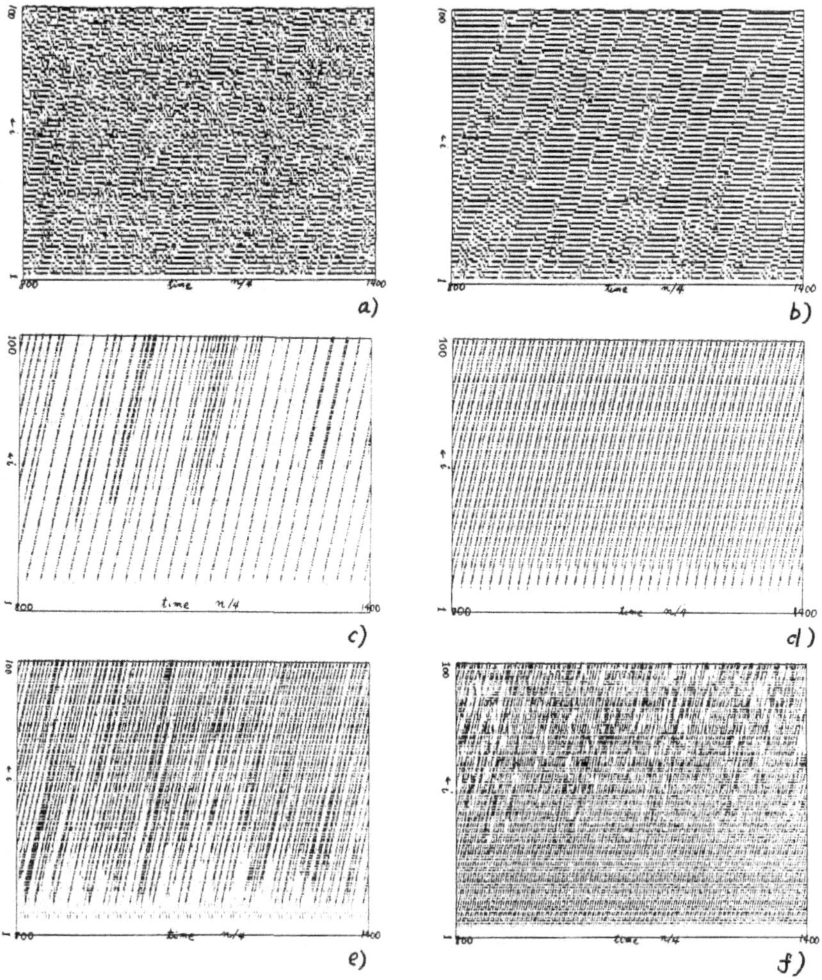

Fig.15: Space-time diagram for the one-way coupled logistic lattice with N=100 and starting with a random initial condition. Every 4th time step is plotted from 0 to 600. For the way of plotting see Fig.3. (a) a=1.58, ϵ = .3 (b) a=1.6, ϵ = .3 (c) a=1.63, ϵ = .3 (d) a=1.75, ϵ = .3 (e) a=1.83, ϵ = .1 (f) a=1.86, ϵ = .1 .

It is rather straightforward to extend the model to a 2-dimensional case with an open flow coupling. The model is given by

$$x_n(i,j) = (1-\epsilon)f(x_n(i,j))+$$

$$(\epsilon/4)(f(x_n(i,j+1)) + f(x_n(i,j-1)) + \alpha f(x_n(i-1,j)) + (1-\alpha)f(x_n(i+1,j)). \quad (16)$$

This class of model has recently been investigated in [18,19], in a possible relation with the spatiotemporal intermittency in the open flow, and leads to some similarity with experiments of turbulence in Benard convection in Libchaber's group [64]. The transport of bursts by the asymmetric coupling in CML corresponds to the transport of hot plumes in the experiment.

5 Information Flow

Chaotic dynamics exhibits the information flow in the bit space [67]. In spatiotemporal chaos, the information flows both in real and bit spaces.

The mutual information in spacetime is useful to see the correlation in spacetime and how the information flows in spacetime. Although there are varieties of ways to see the information flow, we argue the simplest two-point mutual information in space and time, here.

First, take the two-point probability function $P(x_n(i), x_{n+t}(i+m))$, the probability that the lattice site i takes the value $x_n(i)$ at time n and the site $i+m$ takes $x_{n+t}(i+m)$. By this joint probability, the mutual information is defined as [7]

$$I^{st}(t,m;i) = \int logP(x_n(i))dx_n(i) + \int logP(x_{n+t}(i+m))dx_{n+t}(i+m)$$

$$- \int logP(x_n(i), x_{n+t}(i+m))dx_n(i)dx_{n+t}(i+m) \quad (17)$$

Two extreme cases are useful: $I^t(t;i) = I^{st}(t,0;i)$, i.e., the temporal mutual information and $I^s(m;i) = I^{st}(0,m;i)$, i.e., the spatial mutual information.

In the numerical calculation of probability function we have to use a finite bin size, and replace the integral by the summation over bins. If we take the partition number M, the probability function is calculated by the bin $[-1,-1+2/M], [-1+2/M, -1+4/M], \cdots, [-1+2n/M, -1+2(n+1)/M], \cdots, [1-2/M, 1]$. Although changes with the partition number M are another important characterization of a system, we study only the case with a fixed M here.

First we discuss the temporal mutual information flow. In Fig. 16, examples are shown. For the frozen phases (frozen random pattern and pattern selection), the mutual information $I^t(t;i)$ depends on sites (Fig.17a). For the fully developed turbulent phase, the function is independent of sites (Fig.17b), showing the *spatial ergodicity*. The exponential decay is seen. The rate of this decay decreases as the nonlinearity is decreased, till it goes to zero at the intermittency transition. The decay is roughly fitted by the power-law decay in the critical region. In the frozen regime, the decay of information stops at some value (Fig.17a), as far as the precision M is finite, since there is a period-2 peak in the power spectrum. The site dependence of the mutual information clearly shows that the chaotic motion is different by sites.

Spatial mutual information is shown in Fig.18. We again note that the decay is exponential in the fully developed turbulent phase, whose rate of decay decreases as

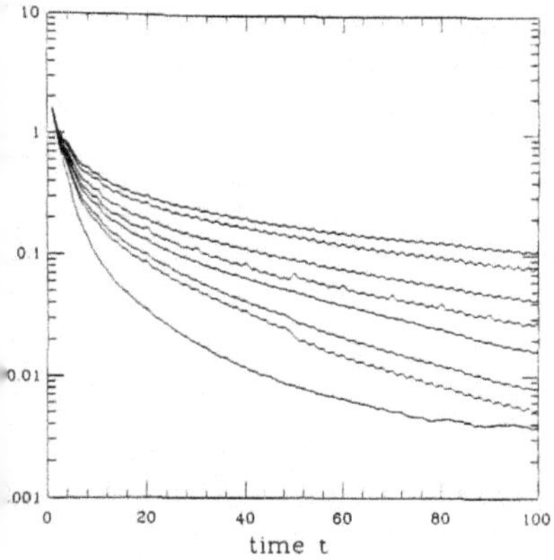

Fig. 16: Temporal mutual information flow for e logistic lattice (1): calculated using the bin mber $M = 64$, and 500000 time step sampling, er discarding 100000 step transients. $N = 100$, $= .3$. a =1.78, 1.79, 1.8, 1.81, 1.82, 1.83, 1.84, d 1.88 from top to bottom. The site dependence small here as is seen in Fig.17b).

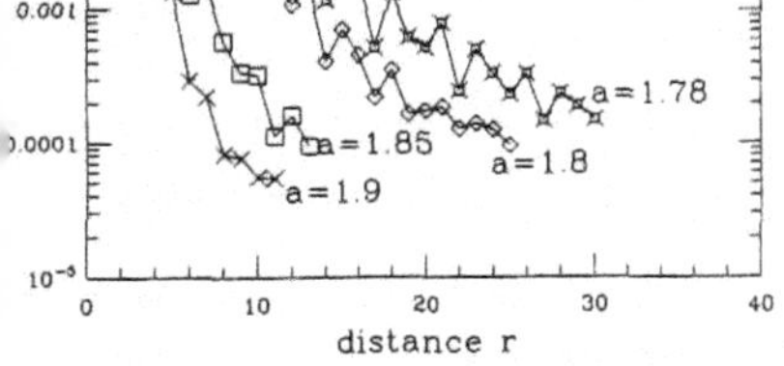

Fig. 18: Spatial mutual information flow for e logistic lattice (1): calculated using the bin mber $M = 64$. Probability is sampled from 00 time steps after 100000 steps, and through- t the whole lattice. $N = 500$, $\epsilon = .3$. $a =$ 72, 1.78, 1.8, 1.85, and 1.90 from top to bottom.

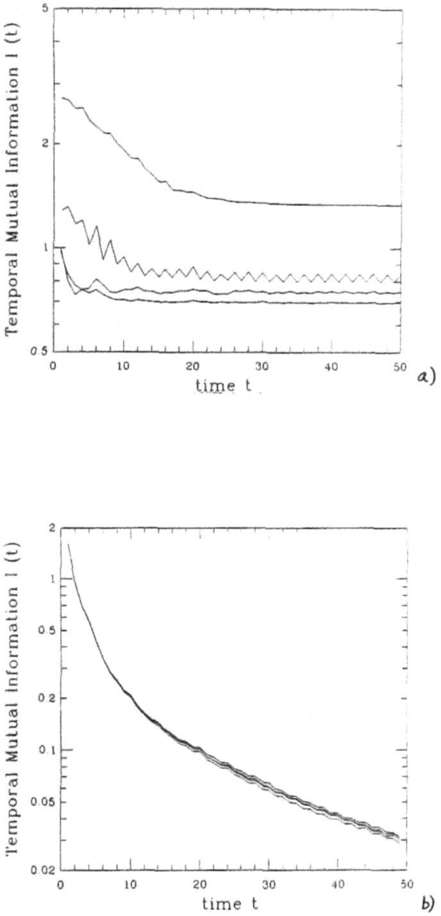

Fig. 17: Temporal mutual information flow calculated in the same manner as in Fig.16. Results from 4 successive lattice points are overlayed. (a) $a = 1.77, \epsilon = .3$ (b) $a = 1.83, \epsilon = .3$.

the parameter approaches the transition point. We also note that a smaller periodic change is added in the mutual information, which comes from the spatial domain structure. Indeed, the wavelength in this smaller variation (Fig. 18) agrees with that of the selected pattern in §2.

The mutual information $I^{st}(t, t/v; i)$ is called as co-moving mutual information flow and gives the measure how the information (or coherence) propagates with the speed v. This measure is especially important in the open-flow model [7].

6 Thermodynamics

In low-dimensional chaos, the statistical mechanics theory has been developed. The essence of statistical mechanics can be seen in Ruelle's textbook[68]. The study of this direction has been developed by I. Shimada, K. Tomita, T. Kai,[70] Y. Oono, Y. Takahashi,[69] and so on, before it has become suddenly popularized in recent days.

According to these works, the chaotic dynamics can be mapped to the statistical mechanics of a 1-dimensional spin system.

Then, it is natural to expect that the thermodynamics of our k-dimensional CML can be mapped onto the statistical mechanics of $k+1$-dimensional spin systems. The interaction, of course, is strongly anisotropic, since the interaction in "spatial" directions is short ranged and bidirectional (e.g., nearest neighbor), but the interaction in "temporal" direction is unidirectional and has a more complicated longer ranged form.

Quite recently Bunimovich and Sinai have rigorously constructed the statistical mechanics corresponding to CML for a special case [24].

Another approach in a different direction has recently been carried out by the author, where the self-consistent approximation on the Perron-Frobenius operator has been formulated. Here it is briefly explained.[25]

The Perron-Frobenius (PF) operator has been a powerful aid in the study of the statistical mechanics of low-dimensional chaos [68,69,71]. The operator has first been extended to spatially extended systems by [72], where the local structure theory of cellular automata is constructed. The operator for the entire dynamical system acts on the measure on N-dimensional space $\rho(x(1), \cdots, x(N))$. The operator is written as

$$H^{PF}\rho(x(1), \cdots, x(N)) = \sum_{y(i)=preimages} \frac{\rho(y(1), \cdots, y(N))}{J(y(0), \cdots, y(N-1))} \tag{18}$$

where the sum takes over all possible sets of $(y(i))$, preimages of $x(i)$ (i.e., $y(i) \to x(i)$ by the map (1)) and $J(y(0), \cdots, y(N-1))$ is the Jacobian of the CML transformation (1).

In our model, the preimages are easily calculated, since it consists of two separated procedures, i.e., $y(i) \to x'(i) = f(y(i))$ and the spatial average by $x(i) = (1-\epsilon)x'(i) + (\epsilon/2)(x'(i+1) + x'(i-1)) \equiv \sum D_{i\ell}x'(\ell)$. The inverses of these two processes are just given by $f^{-1}(x)$ (for the logistic map $f^{-1}(x)$ is given by $\pm\sqrt{(1-x)/a}$) and the inverse matrix D^{-1}.

Since the treatment of N-dimensional distribution is practically impossible, we use a projection to a low-dimensional space. There are many possible ways of the projection. We use the simplest projection here, that is a projection to a k-dimensional space $x(1), \cdots, x(k))$, by intergating out other variables. To get a closed equation within a k-dimensional space, we need a truncation of probability function. It is a traditional problem in statistical mechanics, and various elaborated methods have been developed. Here we use the simplest approximation, that is, the self-consistent approximation. Following this method and after some calculations [25] we have

$$H^{SPF}\rho(x(1),x(2),\cdots,x(k)) = (detD'(k))^{-1}\int\int\sum_{y(1),\cdots,y(s)} dy(0)dy(k+1)$$

$$\frac{\rho(y(1),\cdots,y(k))P(y(2),y(3),\cdots,y(k)|y(k+1))P(y(1),y(2),\cdots,y(k-1)|y(0))}{\prod_{j=1}^{k}|f'(y(j))|} \tag{19}$$

where the conditional probability P is given by

$$P(y(2),y(3),\cdots,y(k)|y(k+1)) = \rho(y(2),y(3),\cdots,y(k),y(k+1))/\rho(y(2),y(3),\cdots,y(k)) \tag{20}$$

and the preimages $(y(1),y(2),\cdots,y(k))$ are given by the solution of

$$y(j) = f^{-1}(D'^{-1}_{j,\ell}(k)x_\ell - (\epsilon/2)(f(y(0))\delta_{j,1} + f(y(k+1))\delta_{j,k})\ . \tag{21}$$

The matrix $D'(k)$ is the diffusion matrix D_{ij} of size k without a periodic boundary (i.e., $(1-\epsilon)\delta_{i,j} + (\epsilon/2)(\delta_{i+1,j} + \delta_{i-1,j})$).

We have applied this self-consistent method with $k = 1$ and $k = 2$ to the problem of spatiotemporal intermittency transition from a laminar regime to a turbulent regime. As for the transition point, the above approximation gives a reasonable result (within 10 % error).

7 Mean-field Type Model

The above self-consistent equation is *different* from the mean-field theory. The mean-field-type model (Fushimi-Temperley-type model) is constructed as a model with couplings to all other elements. The model corresponding to our diffusively coupled map lattice is written as

$$x_{n+1}(i) = (1-\epsilon)f(x_n(i)) + (\epsilon/N)\sum_{j=1}^{N} f(x_n(j))\ . \tag{22}$$

The above model has a remarkably rich behavior (partly similar but much richer than Sherrington-Kirkpatrick (SK) model[83] as a mean-field model for spin-glass).

The model here exhibits the phase transition among a coherent phase, an ordered phase, a glassy phase, an intermittent phase, and a turbulent phase, as the nonlinearity is increased. Also it turns out that the glassy phase and intermittent phase have attractors of similar tree-like structures as are investigated in the SK model. The important difference here is that our tree structure is dynamically changing. Coding of these attractors and switching among attractors are discussed in [43].

8 Simulating Science with Coupled Map Lattices

Our strategy of studying dynamical phenomena in spatially extended systems by CML is based on the separation of parallel procedures and successive operations of them (§1).

In the previous diffusively coupled model, the independent procedures are local transformation of $x(i)$ by a simple 1-dimensional map (eq.(2)) and the diffusion process (eq.(1)), given by the discretized Laplace equation. These two procedures are separated and successively carried out.

If we are interested in phenomena described by local nonlinear processes with local spatial coupling, it is possible to use the above procedures.[8]

8.1 Pattern formation (spinodal decomposition)

Understanding the pattern formation process is the main topic of the present book.

A typical and interesting problem is spinodal decomposition, [74,49] which is a phenomenon when a system is quenched from a disordered state to an ordered state. Traditionally this problem is studied by the time-dependent Ginzburg-Landau equation or the kinetic Ising model with Monte Carlo method.

Oono and Puri have proposed a CML for this problem, which is one of the best success among the applications of CML [27]. Their model is based on the coarse graining in a cell with a large number of sites from the kinetic Ising model. Then the local dynamics of a cell leads to a map with two stable fixed points, e.g.,

$$f(x) = \tanh(\beta x) \quad . \tag{23}$$

The same diffusive coupling form as in the previous sections can be used since the phase transition dynamics includes the term which tries to make two neighboring regions to order ("ferro-coupling").

In the above model the order parameter $\sum_{i,j} x_n(i,j)$ is not conserved. If the order parameter is conserved, the established method in the kinetic Ising model is the use of Kawasaki's exchange dynamics[49]. Corresponding to this class of dynamics, it is possible to construct a CML with the constraint $\sum x_{n+1}(i,j) = \sum x_n(i,j)$ [27]. The model with this constraint is written as

$$x_{n+1}(i,j) = f(x_n(i,j)) - << f(x_n(i,j)) - x_n(i,j) >> \tag{24}$$

where $<< \ldots >>$ denotes the spatial average with the nearest neighbor.

The models by Oono and Puri give correct scaling behaviors for dynamical form factors (⇒ Appendix) corresponding to the conserved and non-conserved phase transition dynamics, respectively. Indeed, their models give much faster simulators than the conventional Monte Carlo method.

8.2 Other pattern formation problems

A similar modelling is possible in the roll formation in convection, pattern formation with complex order parameter(s), and so on. These extensions are rather straightforward. Vortex interactions in the Ginzburg-Landau equation are easily simulated with the use of vector order parameter $\vec{x}_n(i,j)$ and the similar local dynamics as (23) and the same diffusive coupling [28]. Another application is the complex Ginzburg Landau

[8] "local" here means that the interaction is governed by a term decaying exponentially (or faster) in space. Thus we may hope that our method is applicable to a system of reaction-diffusion type equation (including time-dependent Ginzburg-Landau equation), and most spatially extended dynamical phenomena.

equation with complex parameters [75], in which a turbulent behavior with vortices is observed [29,30].

Formation of a roll pattern is also simulated by CML. So far we have two modellings on the roll formation. One is the use of $f(x)$ which is an odd function and has a maximum at $x = x_c$ and goes to zero as x lim ∞. An example is

$$f(x) = G((3\sqrt{3}/2(x/x_0 - (x/x_0)^3), G(y) = y/\sqrt{a^2+x^2} \quad (25)$$

with same diffusive coupling form as in (1)[29] (see Fig.19). The other method is the use of a "lateral" inhibition in the next-nearest-neighbor coupling[31]:

$$x_{n+1}(i,j) = (1 - n_1\epsilon_1 + n_2\epsilon_2)f(x_n(i,j)) + \epsilon_1 \sum_{n.n} f(x_n(i',j')) - \epsilon_2 \sum_{n.n.n} f(x_n(i'',j'')) \quad (26)$$

where $n.n.$ is the sum over nearest neighbors whose number is n_1, while $n.n.n.$ is the sum over next nearest neighbors whose number is n_2. By choosing the couplings suitably, we can have a model of a roll formation.

It is rather straightforward to apply these methods with different classes of ordered patterns and order parameters.

8.3 Crystal growth

Another possible application lies in the nucleation and crystal growth [76]. To take into account a thermal field, it is possible to introduce the following model:

$$x_n(i,j) = (1-\epsilon)f(t_n(i,j), x_n(i,j)) + (\epsilon/4)[f(t_n(i,j+1), x_n(i,j+1)) +$$

$$f(t_n(i,j-1), x_n(i,j-1)) + f(t_n(i-1,j), x_n(i-1,j)) + f(t_n(i+1,j), x_n(i+1,j))] \quad (27)$$

$$t'_n(i,j) = t_n(i,j) + c(f(t_n(i,j), x_n(i,j)) - x_n(i,j)) \quad (28)$$

$$t_{n+1}(i,j) = 1/n_{nghbr} \sum_{nghbr} t'_n(i,j) \quad (29)$$

with $f(t,x) = tanh(\beta(x-t))$.

Here x denotes the order parameter corresponding to the density ($x = 1$ is solid, $x = -1$ is gas). The other variable t corresponds to the temperature field. Threshold on the solidification process is clearly seen in the form of $f(t,x)$. The equation $t \to t'$ represents the effect of latent heat by the solidification. Eq. (29) comes from the diffusion of heat, where $\sum_{nghbr}$ shows the summation over given neighbors, and n_{nghbr} is their number. A simple examples of a growth of a "solid" region is shown in Fig.20. The growth consists of the repetition of the process in the figure.

8.4 Excitable media

Another possible application of CML is a model with an excitable state and a relaxation from it. Possible examples can be seen in reaction diffusion within excitable media, and may also be relevant to some problems of neural response.

A simple 1-dimensional map with an excitable state is introduced by Nagumo and Sato[77], in connection with Harmon's experiment on the firing of an artificial neuron [78]. They have used the map

$$f(x) = b \times (x - H(x)) + c \quad (30)$$

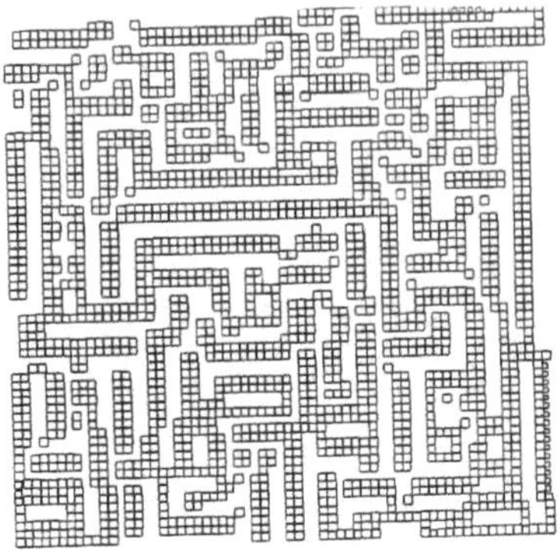

Fig. 19: Roll pattern formation. Snapshot of the diffusively coupled map lattice (25). Size = 64x64. $a = 1/4$, $x_0 = 0.8$, and $w = 0.8$, at the time step 4000. Nearest-neigbor average for diffusion. Squares are depicted only at the points (i, j) with $x(i, j) > 0$ whose length are proportional to $x(i, j)$.

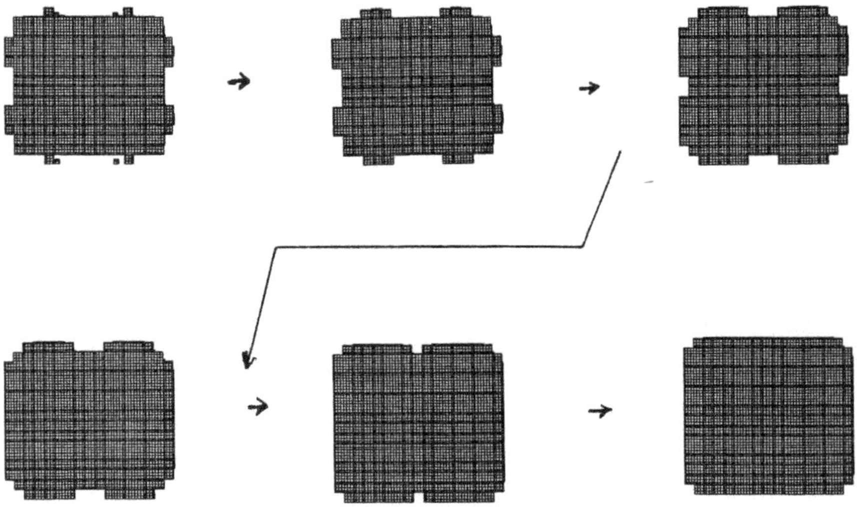

Fig. 20: Crystal growth model (27)-(29). Snapshot of $x(i, j)$, with $\beta = 3.5$, $c = -.55$, starting from a 3-site-seed initial condition. Nearest-neighbor average for diffusion. Only the points with $x(i, j) > 0$ are painted by square whose length is proportional to $x(i, j)$.

where $H(x)$ is the Heaviside's step function ($H(x) = 1$ for $x > 0$ and $H(x) = 0$ for $x < 0$). Here, $x > 0$ corresponds to the " fired state". The constant term "c" comes from an external stimulus applied on a single neuron [77].

We can easily construct a CML corresponding to the above model. If we follow the interpretation of the term "c", we can replace this constant term by stimuli from other lattice points. As a simple model, we assume that a constant pulse is emitted if $x > 0$ ("fired"). Then we obtain the following CML

$$x_{n+1}(i) = b \times (x_n(i) - H(x_n(i))) + d \times (H(x_n(i+1)) + H(x_n(i-1)) \tag{31}$$

(for a 1-dimensional lattice) or

$$x_{n+1}(i,j) = b \times (x_n(i,j) - H(x_n(i,j)))+$$

$$d \times (H(x_n(i+1,j)) + H(x_n(i-1,j)) + H(x_n(i,j+1)) + H(x_n(i,j-1))) \tag{32}$$

for a 2-dimensional lattice.

In the 1-d lattice, we can see the propagation of pulses. At a transition parameter region we can see a soliton turbulence induced by the interaction among pulses [35].

In the 2-dimensional system, we have observed the wave pattern similar to the experiment of BZ reaction. See Fig. 21 for examples.

It will be of use to study a system with more realistic couplings and local functions.

8.5 Dripping handrail and boiling chaos

Another related model is a system with a threshold dynamics and diffusion. As an example of a 1-d lattice, let us take a dripping handrail [5,23]. Water comes from the rain onto a handrail and the amount of water increases with a constant rate up to a threshold. If the amount exceeds the threshold, the water drops from the handrail. These two processes give the local dynamics. There is diffusion along the handrail, which can be modelled by the diffusive coupling. Thus we have $x_{n+1}(i) = (1-\epsilon)f(x_n(i)) + (\epsilon/2)(f(x_n(i-1)) + f(x_n(i+1)))$ with

$$f(x) = x + c(mod\ 1). \tag{33}$$

The spatiotemporal chaos here is of a different nature from that discussed in §2-7, since the model cannot have a state with positive Lyapunov exponents as an attractor. The model exhibits the supertransients [23]. The duration of transients increases with a system size, and a quasistationary chaotic behavior is seen in this transient interval. An extension to a 2-dimensional system is straightforward [32]. The 2-dimensional model may be regarded as a simple model for boiling, since the boiling process consists of the increase of temperature up to a threshold and diffusion.

8.6 Designing fluid dynamics with CML?

One difference between Navier-Stokes-type equations and our diffusively coupled map lattice lies in that the nonlinearity in the former arises from the coupling term as is typically seen in the convective term in the Navier-Stokes equation, while in the diffusively coupled model, the coupling term acts as a smoothing effect. Here we

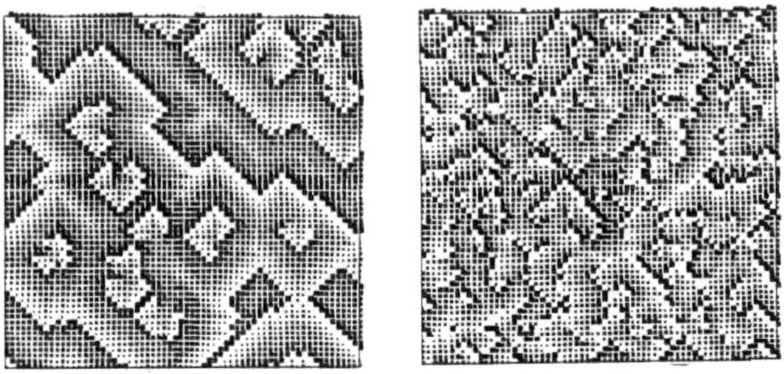

a) b)

Fig. 21: Snapshot for a CML model for the excitable media model (32). $b = .77$. Time step $=1000$, starting from a random initial condition. A length of a side of a square in the figure is proportional to $x(i,j) + .5$. If $x(i,j) < -.5$, the corresponding site is left blank. Size=64x64. (a) $d = 0.06$ (b) $d = 0.1$

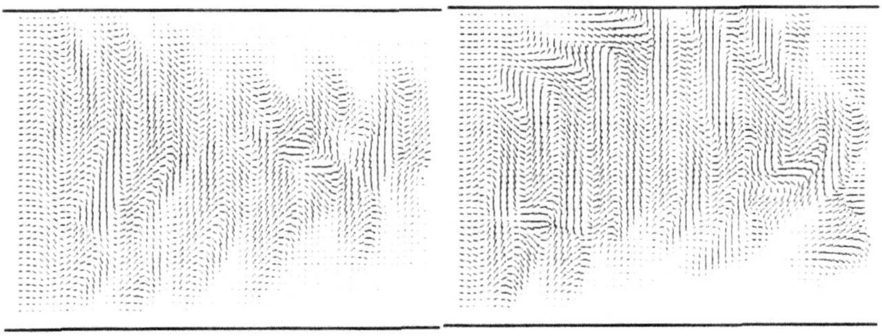

Fig. 22 a) b)

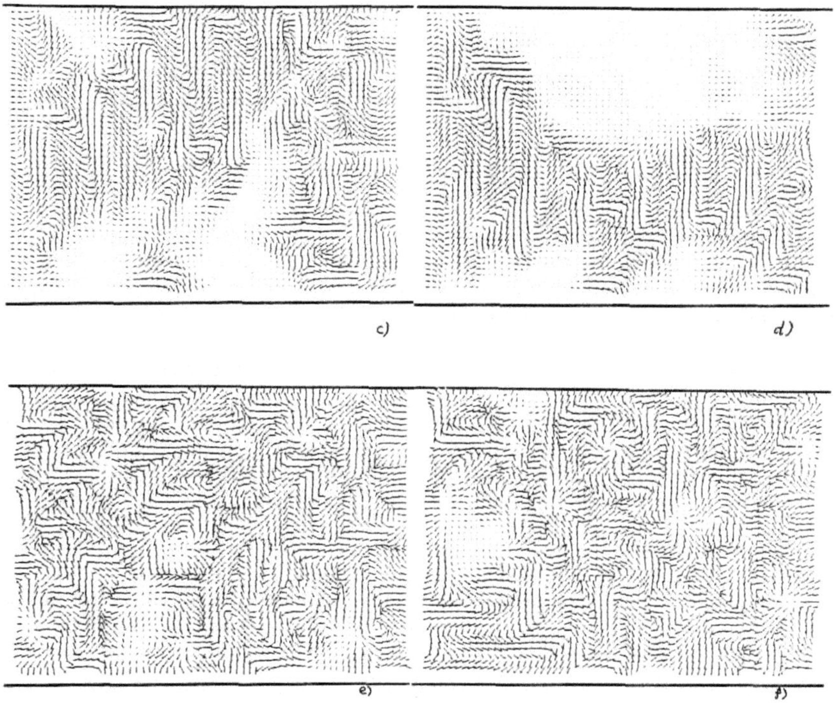

c) d) e) f)

Fig. 22: Snapshot for the 2-dimensional convective coupling model (37)-(39): Left boundary is fixed at $(x, y) = (v_L, 0)$ and right boundary is free. Top and bottom are rigid, i.e., $(x, y) = (0, 0)$. Lattice size is 63x48.

(a) $v_L = 2.3$, $c = 130$, time step =2081

(b) $v_L = 2.5$, $c = 130$, time step =705

(c) $v_L = 3.0$, $c = 200$, time step =225 (d) time step =353

(e) $v_L = 2.2$, $c = 300$, time step =1345 (f) time step =3585 consider a model with nonlinear coupling, which has some similarity with a turbulent behavior in Navier-Stokes-type equations.

soliton turbulence in a convective coupling model

As a simple example for a convective coupling model, let us take the following procedures.

(1) Convective coupling (corresponding to the $-(v\nabla)v$ term in the Navier-Stokes equation)

(2) Diffusive coupling (corresponding to the $\nu\nabla^2 v$ term in the Navier-Stokes)

(3) Cut-off for high velocities: (this term is rather artificial but necessary in order to remove a possible divergent behavior). A simple example is given by $x'(i) = f(|x(i)|) * x(i)$ with a monotonically decreasing function $f(x)$ with $f(0) = 1$ and $f(\infty) = 0$. Here we take $f(x) = exp(-x^2/c)$.

Combining these three procedures, our coupled map lattice is given by

$$(I)x'(i) = x_n(i) + (x_n(i-1) - x_n(i+1))x_n(i) \tag{34}$$

$$(II)x"(i) = (1-\epsilon)x'(i) + \epsilon/2(x'(i-1) + x'(i+1)) \tag{35}$$

$$(III)x_{n+1}(i) = exp(-x"(i)^2/c)x"(i)\ . \tag{36}$$

The evolution $x_n(i) \to x_{n+1}(i,j)$ consists of the successive operations of the procedures (I),(II), and (III).

This model includes two parameters $1/c$ and ϵ, corresponding to the damping and diffusion. As c is increased, the system starts to explore a region with larger nonlinearity. With the increase of c, our attractor changes as follows; (i)"dead" state (i.e., $x(i) = 0$ for all i) $\to$ (ii)soliton turbulence $\to$ (iii) developed turbulent state.

The soliton turbulence is a turbulent behavior generated by the transmission of "solitons" and their interaction and creation. The interaction consists of the emission of solitons from a chaotic nucleus, and collisions which result in pair-annihilation, absorption, passing-through with a phase shift, and a generation of a nucleus. There is sensitive dependence on the phases of collisions.

The soliton turbulence is first found in the coupled circle lattice [11,5] and is also seen in a class of cellular automata [50,52].

2-dimensional flow

The above model is straightforwardly extended to a flow in a two-dimensional system, by taking a two-dimensional vector $\vec{v}(i,j) = (x(i,j), y(i,j))$ instead of the scalar quantity on a 2-dimensional lattice (i,j). The model consists of the following successive procedures.

$$(I)'x'(i,j) = x_n(i,j)+[(x_n(i-1,j)-x_n(i+1,j))x_n(i,j)+(y_n(i,j-1)-x_n(i,j+1))x_n(i)]/2 \tag{37}$$

$$(I)'y'(i,j) = y_n(i,j)+[(x_n(i-1,j)-x_n(i+1,j))y_n(i,j)+(y_n(i,j-1)-x_n(i,j+1))y_n(i)]/2 \tag{38}$$

$$(II)'x"(i,j) = \text{average of } x'(i,j) = \text{for 13 neighbors (} ((i\pm1,j\pm1),(i\pm2,j),(i,j\pm2))$$

$$(III)'x_{n+1}(i,j) = x"(i,j)exp(-(x"(i,j)^2 + y"(i,j)^2)/c)\ . \tag{39}$$

(Procedures for $y(i,j)$ are similar.)

Examples of snapshots are shown in Fig.22, where the boundary condition is fixed at the left end and free at the right end, while it is fixed to zero at the top and bottom. We can see the formation of shocks, vortices, sinks, and sources (since we have not imposed the "incompressibility" condition in our model).

PROBLEM: We have to admit that the model (I)'-(III)' is still a premature attempt towards a modelling of hydrodynamics, compared with the expanding field of lattice gas automata [46]. Drawbacks in our model lie in the existence of artificial cut-off term (III) and a very rough treatment of the conservation law. In fact, our model has a conservation law only in the limit of small diffusion ($\epsilon \to 0$) and $c \to \infty$. One possible refinement of our model is to introduce another quantity corresponding to the density and to take into account of a discrete version of the equation of continuity. Or, another choice is to start from the coarse-graining of microscopic dynamics without any direct consideration of the Navier-Stokes equation.

Without the cut-off term (III)', an interesting behavior is observed only at a critical point in the parameter space. Beyond this point, our model shows a divergent behavior, while below it, the system goes to a laminar flow. It will be important to search for a mechanism to compel to stay near the critical point by a conservation law, as has been discussed in the model of sandbox by [79].

8.7 Neural networks

In the real neuron, states are not discrete. In that sense the discrete-state model of a neural network (e.g., McCllough-Pitts neuron) is an over-simplification. Of course, the model may capture some essential features of the neural dynamics, but may miss some important points, too. In some experiments, a single neuron [80,81] and a small ensemble of neurons [82] are known to exhibit a chaotic behavior. In the standpoint of statistical mechanics, it must be much better to start from a higher-level dynamics than the neural networks with very simple elements, in order to understand a higher-level dynamics in the brain.

Thus it is natural and important to ask the following question: What happens if we use more complex elements (with a chaotic response) with oversimplified couplings instead, as a different limit of simplification from the neural dynamics? The coupling here is not short-ranged.

The global coupling model in §7 may be of importance as the first endeavor towards this direction. The results there [43] are rather promising. We can have a huge number of coded attractors. Some attractors have a tree-like structure with a hierarchical code and have similar complexity with the SK model in spinglass [83], which is the basis of one of the most popular models in neural networks [84,85,86]. Upon the tree structure one can construct a hierarchical dynamics. Through simple inputs on coded attractors, we can switch from them to attractors with different clusterings and tune the strength of chaos.

It is of course interesting to study a CML with local dynamics with an excitable state and a refractory period, and with a coupling which gives the transmission of the firing. A possible example is the use of the model discussed in §8.4, and with a different topology ([35], see also [36]).

8.8 Josephson junction, CDW, ... and coupled circle lattice

The problem of coupled nonlinear oscillators (pendula) can be frequently seen in the solid-state physics. Typical examples include a Josephson junction and a charge density wave (CDW).

Since a forced pendulum can be well modelled by the circle map $f(x) = x + (K/2\pi)sin(2\pi x) + \Omega$ (see e.g., [2]), it is natural to use a coupled circle map lattice for the above problems.

Some preliminary results in the coupled circle map lattice give a soliton turbulence [11,5] and a vortex motion in a 2-d lattice[11]. See [33,34] for applications of these models to problems of CDW. In [42], a problem of a coupled Josephson junction is investigated with the use of a globally coupled circle map, where a transition from a coherent state to a desynchronized state is studied.

Another approach towards this problem is seen in the use of "oscillator lattice", that is a coupled ODE with the use of a differential equation of an oscillator as a local dynamics. [87,88].

8.9 Other applications

Recall that the logistic map has frequently been investigated as a problem of population dynamics. Then it is natural to ask the role of spatial interaction among the populations. Coupled logistic lattices and some other models will be of importance for the study of this problem[37].

The oscillatory behavior of immune response also includes a large number of nonlinear interactions. A possible model for immune networks with a CML is discussed in [90].

8.10 Experimental observation of spatiotemporal chaos

We comment briefly on the possible relation of our logistic lattice with experiments. First, we have to admit that the direct relation with an experiment is not attainable at the present. We do not have to be disappointed with this, however, since no direct relation with the logistic map and a Rayleigh-Benard convection is at hand in the level of equation of motion, but the map explains some aspects of the latter quite well. If the universality class to which our model belongs has a large applicability in the spatiotemporal chaos, which we believe (since our model is so simple and general like the Ising model or the logistic map), we might hope that the same phenomena and quantitative aspects as in our CML are observed in experiments.

We also note that the "universality" is not necessarily defined quantitatively (like critical exponents which we are not interested in). It will be important to define the universality in a qualitative level. For examples, if we have experimentally observed the transition sequence of pattern dynamics similar to our model, we may say that the experiment belongs to the same qualitative universality class as our model.

What we hope to be observed is (1) the transition sequence of the frozen chaotic pattern, the pattern selection, the intermittency (Brownian motion of defect and the defect turbulence), and the fully developed turbulence. If this kind of phenomena is found, the search for (2) selective flicker noise at the transition regime is recommended as a quantitative check. Also, it is desirable to re-investigate (3) the motion of defects in some patterns, to check if the motion is chaotic, and if the Brownian motion is associated with the chaos.

For a qualitative comparison, some visualization techniques introduced in the study of CML, such as the space-time diagram, spatial return maps, co-moving return maps and so on [5] may be useful. For these qualitative and quantitative studies, multiple-point observations will be essential. Through these observations, we can get the graphics and various quantifiers comparable with those investigated in the present paper.

What are possible candidates of the experiments? We have seen some examples of spatiotemporal chaos in recent experiments [89], such as chemical turbulence [91, 92], Benard convection with large aspect ratio [93,94,95,96,97], Taylor-Couette flow with large aspect ratio [98], Faraday experiment [99] electrical convection of liquid crystal [100,101], solid state experiments such as Josephson junction array, electron-hole plasma [102], and charge density wave [103]. For open flow experiments, we have seen many examples of spatiotemporal chaos, such as the pipeflow [104], boundary layer [105], and air jet [106], and the dendritic growth of a crystal [107].

In these experiments, we can see some relations with our observations in the present paper. Spatial bifurcation by the frozen random state is observed in a Benard convection experiment [97]. Brownian motion of defects has been seen in the convection of liquid crystals. Localized chaotic motion (turbator) has been observed in Taylor-Couette flow [98]. Pattern competition is observed in the Faraday experiment and in liquid crystal experiment [99,101]. Flow of turbulent spots can be frequently seen in open flow experiments and in a dendritic growth [107].

Spatiotemporal intermittency has been observed by Ciliberto and Bigazzi [108] on the Benard convection on an annulus. The experiment exhibits the same phenomena as our pattern competition intermittency. This correspondence is supported by the power-law distribution of laminar domains, and the existence of selected wavenumbers and of localized chaos in a subcritical region.

Another important experiment has been carried out by Nasuno et al.[62], in the electrical convection of a liquid crystal. As has been discussed in §3, their experiment shows a remarkable coincidence with our spatiotemporal intermittency.

If we are concerned with an experiment on a phenomenon which includes a local chaotic mechanism and a spatial short-ranged coupling, we may hope, at a level of qualitative "universality', that our prediction on the phases and the quantitative aspects is valid to the experiment.

to conclude ...?

Are all macroscopic dynamical behavior with many degrees of freedom modelled and understood by CML? I hope that the answer is no, but so far there is no counter-example....

Acknowledgement

I would like to thank Professor Kyozi Kawasaki for his continual encouragement, valuable comments, and interests on CML since I had started it several years ago. I am also grateful to him for his giving a chance to talk in the workshop on pattern dynamics in Kyoto and to write in the present book. I would also like to thank Dr. T. Ikegami for useful discussions.

9 Appendix: Quantifiers in Spatiotemporal Chaos (in alphabetical order)

Co-moving Lyapunov spectra

The notion of a co-moving Lyapunov exponent is useful, if a disturbance grows with a characteristic velocity or if macroscopic structures propagate. The co-moving Lyapunov exponent is especially important when a system is convectively unstable, as is seen in the one-way coupled logistic lattice (see §4).

The definition of the convective instability is as follows: Let us take a small perturbation δx_n^i around a state x_n^i. If $\lim_{\delta x_0 \to 0} \lim_{n \to \infty} (\delta x_n^i / \delta x_0) \to \infty$ for an arbitrary fixed lattice site i, it is called absolutely unstable. This definition agrees with the familiar definition of the conventional dynamical system theory. In a case of open flow, even if the system is not absolutely unstable, it can be unstable in some moving frame. If $\lim\ (\delta x_n^{[i+vn]} / \delta x_0) \to \infty$, for a given velocity v, it is called as convective unstable. (Here, [q] is the maximum integer that is smaller than q.) If a system is stable in all arbitrary moving frames, it is absolutely stable. The notion of convective instability was originally introduced in the field of plasma physics and fluid mechanics, where the stability around a stationary state (fixed point or periodic orbit in the terminology of a dynamical system) has been investigated.

The convective (in)stability around any arbitrary state is analyzed by the notion of co-moving Lyapunov exponent. It is defined in the following way: First, the frame is transformed from the stationary laboratory frame to the moving one. In our lattice system the transformation is given by

$$i\prime = i - [vn] \tag{40}$$

where i and $i\prime$ are spatial lattice points (in the stationary and moving frame, respectively) and n is the discrete time step. The Lyapunov exponent in the moving frame with the velocity v is calculated from the Jacobi matrix $J_n = \delta x_n^{i\prime+[v(n+1)]} / \delta x_n^{j\prime+[v_n]}$. A logarithm of the largest eigenvalue of the product of Jacobi matrices divided by the time steps, gives the co-moving Lyapunov exponent with the velocity ".

When a system is only convectively unstable, the conventional Lyapunov exponent is negative, while it takes a positive value for $v_T < v < v_L$. It goes negative again for $v > v_L$, meaning that the disturbance cannot propagate with the velocity larger than v_L. Here the subscripts T and L stand for trailing and leading, and these are appropriate since these speeds describe the trailing and leading edge propagation of amplified perturbations, respectively.

Co-moving mutual information flow

see §5 and [7].

Dimension density

The fractal dimension of an attractor in the phase space is an important quantity. Roughly speaking, its integer part gives the effective degrees of freedom of the attractor. Whereas the Lyapunov dimension is introduced by the Kaplan-Yorke formula;

$$D_L = j + (\sum_{i=1}^{j} \lambda_i / |\lambda_{j+1}|) \tag{41}$$

with j such that $\sum_{i=1}^{j} \lambda_i > 0$ and $\sum_{i=1}^{j+1} \lambda_i < 0$. The Lyapunov dimension gives the upperbound to the fractal dimension of a dynamical system.

In the lattice system, a quantity scaled by size is more relevant. The Lyapunov dimension density is defined by $d_L = D_L/N$. See [38] for a possible estimate of dimension density.

These definitions are introduced in the same manner as in the introduction of "intensive" quantities from "extensive" quantities in thermodynamics.

Dynamical form factor (Spatiotemporal power spectra)

Power spectra in time and space are defined by

$$P(k,\omega) = << |(1/N) \sum_{j=1}^{N} x_n(j) e^{2\pi i (jk - n\omega)}|^2 >>, \tag{42}$$

where the bracket $<< .. >>$ denotes the long-time average. These are useful in the characterization of space-time patterns.

Kolmogorov-Sinai entropy density

Originally, Kolmogorov-Sinai (KS) entropy means the increasing rate of a variety of possible symbol sequences generated by the dynamics, with the increase of a partition precision. It can be numerically computed by the sum of all positive Lyapunov exponents (⇒*Lyapunov Spectra*). It roughly gives the information loss in the phase space.

In a spatially extended system, the KS entropy is proportional to the system size, when the spatial correlation length is finite. Thus the KS entropy density is defined as the *intensive* thermodynamic quantity:

$$h_\mu = (1/N) \sum_{i=1}^{\lambda_i > 0} \lambda_i. \tag{43}$$

Lyapunov spectra

The Lyapunov exponents characterize how a set of independent small deviations in a tangent space is amplified (or reduced). In a N-dimensional dynamical system, there exist N independent directions, and corresponding N eigenvalues, which form the spectrum.

In our coupled map lattice system, Lyapunov spectra can be defined in the following by using the product of Jacobi matrix of the map, i.e.,

$$\lim_{n\to\infty} (1/n) J_n J_{n-1} J_{n-2} J_1 \quad . \tag{44}$$

The logarithm of eigenvalues of the above product divided by time steps n with the limit of $n \to \infty$ gives the Lyapunov exponents. The exponents λ_i, ordered from the largest to the smallest, gives a spectrum of the system.

If the largest Lyapunov exponent takes a positive value, two nearby orbits separate exponentially, and the system exhibits the orbital instability, or in other words, sensitive dependence on initial conditions. If the largest one is negative, the orbit is stable and the system is attracted to a periodic state. If the largest one is zero, the motion is quasiperiodic. If there are k vanishing exponents, the motion is on a $(k+1)$-torus.

Corresponding eigenvectors to each eigenvalues of the matrix (44) are called Lyapunov vectors.

Mutual information in spacetime

see §5.

Pattern dynamics quantifiers

(a)Pattern distribution Q(j) and pattern entropy

Since the spacetime patch $x_n(i)$ has too much information, it is useful to reduce the information by a coarse-grained measurement of the continuous variables into some discrete states. Here we digitize a spatial pattern as a symbol patch. We first assign $\sigma_n(i) = 0, 1, 2, \cdots, K-1$ to $x_n(i)$ according to the following rule

$$\begin{aligned}
\sigma_n(i) &= 0 \quad && if \quad x_n(i) < X_1 \\
\sigma_n(i) &= 1 \quad && if \quad X_1 < x_n(i) < X_2 \\
&\cdots \\
&\cdots \\
\sigma_n(i) &= K-1 \quad && if \quad X_{K-1} < x_n(i) \ .
\end{aligned} \tag{45}$$

The simplest digitization K=2 is carried out by the symbolization $\sigma_n(i) = 0$ for $x_n(i) < x*$ and $\sigma_n(i) = 1$ for $x_n(i) > x*$. In the domain structure we study in §2, each domain is separated by the separatrix of the unstable fixed point of the logistic map $x* = (\sqrt{1+4a} - 1)/(2a)$. Thus a snapshot pattern can be represented by a symbol sequence such as $0111 \cdots 001$.

Order parameter for the pattern selection is constructed by the domain distribution. A domain in §2 is defined as the maximal continual sequence in which $\sigma_n(i)$ takes the same value. Thus the probability distribution $Q(j)$ of the spatial length of the same symbol ($11 \cdots 11$ or $00 \cdots 00$) is chosen as the order parameter(s) for our pattern dynamics. $Q(j)$ is defined by the following procedure: From the spatial sequence $x_n(i)$ at a given time n, one obtains the minimum length in which $\sigma_n(k)$ takes the same symbol ($k = i-h, i-h+1, \cdots, i, \cdots i+m$; the length j is $h+m+1$). The length j gives the domain size of the lattice point i at each time step n. From the spatiotemporal sampling through the entire lattice and many time steps, the probability that a lattice point belongs to the domain size j is obtained. For example, $Q(1) = 1$ and $Q(k) = 0$ for $k \neq 1$ for the complete zigzag pattern.

A static complexity of a pattern is measured by the following (static) pattern entropy:

$$S_p = -\sum_j Q(j) log Q(j). \tag{46}$$

(b)Pattern transition matrix and pattern dynamical entropy

The above quantifiers are not sufficient to study the dynamics of a pattern. We are required to construct the transition matrix of successive pattern evolution. The transition matrix is calculated as follows: First, compute the domain size to which a given lattice site i belongs at a time n (assume that the size is m). After a given time step, compute the domain size again at the lattice point (assume that the size is k). We call this as the transition m $\rightarrow$ k. By taking the spatiotemporal samplings, we obtain the probability of the transition

$$T(m \rightarrow k) = \text{transition from the domain with size m to that with size } k\,. \tag{47}$$

To see the dynamical complexity, a dynamical entropy of the transition is defined as follows:

$$S_d = -\sum_{i,j} Q(i)T(i \rightarrow j) log T(i \rightarrow j), \tag{48}$$

which is first introduced in the study of attractors of CA by the author [50].

The meaning of the quantity is as follows; Assume that we know that the lattice point belongs to the domain of some size. After a given time step, we again observe the size of domain to which the lattice point belongs. The mutual information gain through the observation is given by S_d.

As is easily shown, the relation $S_p \geq S_d$ holds. If a pattern is spatially complex but temporally frozen, $S_d = 0$ and $S_p > 0$. In the turbulent regime, both S_d and S_p are positive. If the transition occurs without a memory of the previous pattern (i.e., $T(i \rightarrow j) = p(j)$, independent of i), the equality $S_p = S_d$ holds.

If the pattern has a long-time memory and cannot be represented by the Markovian dynamics, the above transition matrix is no more useful. In such case, a distribution of the life-time is important to characterize the power-law-type behavior.

Power spectra in space or time

The spatial power spectra are defined by

$$S(k) = << |(1/N)\sum_{j=1}^{N} x_n(j)e^{2\pi ikj}|^2 >>; \tag{49}$$

where $<< *** >>$ shows the long time average.

Temporal power spectra are the power of Fourier transform of a time series of $x(i)$:

$$P(\omega) = << |(\sum_{n=t_0}^{t_1} x_n(j)e^{2\pi in\omega}|^2 >> \tag{50}$$

These are traditional characterizations of space/time data.

References

[1] K. Kaneko, Prog. Theor. Phys. 72 (1984) 480, 74 (1985) 1033 in *Dynamical Problems in Soliton Systems* (Springer, 1985, ed. S. Takeno) 272-277

[2] K. Kaneko, Ph. D. Thesis *Collapse of Tori and Genesis of Chaos in Dissipative Systems*, 1983 (enlarged version is published by World Sci. Pub., 1986)

[3] K. Kaneko, Physica 34D (1989) 1

[4] K. Kaneko, Physica 35D (1989) in press

[5] J. P. Crutchfield and K. Kaneko,"Phenomenology of Spatiotemporal Chaos", in *Directions in Chaos* (World Scientific, 1987)

[6] I. Waller and R. Kapral, Phys. Rev. 30A (1984) 2047; R. Kapral, Phys. Rev. 31A (1985) 3868

[7] K. Kaneko, Physica 23D (1986) 436

[8] K. Kaneko, Phys. Lett. 125 A (1987) 25;Eur. Phys. Lett., 6 (1988) 193

[9] T. Yamada and H. Fujisaka, Prog. Theor. Phys. 72 (1984) 885; 74 (1985) 918:

[10] J. D. Keeler and J. D. Farmer, Physica 23D (1986) 413

[11] K. Kaneko"Phenomenology and Characterization of Spatio-temporal Chaos" in *Dynamical Systems and Singular Phenomena* (World Sci. Pub., 1987, ed. G. Ikegami)

[12] F. Kaspar and H.G. Schuster, Phys. Lett. 113A (1986) 451; Phys. Rev. A36 (1987) 842:

[13] K. Kaneko, Phys. Lett. 111A (1985) 321

[14] R.J. Deissler and K. Kaneko, Phys. Lett. 119A (1987) 397

[15] R. J. Deissler, Phys. Lett. 120A (1987) 334

[16] G.L. Oppo and R. Kapral, Phys. Rev. A33 (1986) 4219; Phys. Rev. A 36 (1987) 5820; Physica 23 D (1986) 455: R. Kapral, G.L. Oppo and D.B Brown, Physica 147 A (1987) 77

[17] I.S. Aronson, A.V. Gaponov-Grekhov, and M.I. Rabinovich, Physica 33D (1988) 1:

[18] for open flow models see also:

J.Brindley and R.M. Everson, Phys. Lett. A 134 (1989) 229:

D. Rand and T. Bohr, preprint (1988):

[19] M. H. Jensen, Phys. Rev. Lett. 62 (1989) 1361:

[20] see also H. Chate and P. Manneville, C.R. Acad. Sc. 304 (1987) 609; Phys. Rev. Lett. 58 (1987) 112

[21] H. Chate and P. Manneville, Physica 32D (1988) 409; Phys.Rev. A38 (1988) 4351; Europhys. Lett. 6 (1988) 591

[22] J.P. Crutchfield and K. Kaneko, "Space-time Information Theory", in preparation

[23] J.P. Crutchfield and K. Kaneko, Phys. Rev. Lett. 60 (1988) 2715 and in preparation

[24] L.A. Bunimovich and Ya.G. Sinai, Nonlinearity 1 (1989) 491:

[25] K. Kaneko,"Self-consistent Perron Frobenius Operator for Spatiotemporal Chaos" (LA-UR-89-1088), Phys. Lett. A, in press

[26] M. H. Jensen et al., private communication

[27] Y.Oono and S. Puri, Phys. Rev. Lett. 58 (1986)836; Phys. Rev. 38 A (1988) 1542

[28] H. Nishimori and T. Nukii, J. Phys. Soc. Jpn. 58 (1989) 563

[29] K. Kaneko, unpublished (1987)

[30] T. Bohr and M.H. Jensen, private communication

[31] J. Crutchfield, private communication (1988)

[32] R. Shaw,"boiling chaos" (movie presented at the conf. on Spatiotemporal Chaos, Los Alamos 1986):

[33] P. Alstrom and R. K. Ritala, Phys. Rev. 35A (1987) 300: H. Daido, Prog.Theor. Phys. 75 (1986) 1460:

[34] S. Coppersmith, Phys. Rev. 38 A (1988) 375

[35] K. Kaneko, talk at "the First International Conference on Form "(1986) (Tsukuba)

[36] K. Aoki and N. Mugibayashi, Phys. Lett. 128A (1988) 349:

[37] M. Rotenberg, Physica 30D (1988) 192:

[38] G. Mayer-Kress and K. Kaneko, J. Stat. Phys. 54 (1989) 1489

[39] P. Grassberger, preprint (1987)"Lumped and Distributed dynamical systems"

[40] for an application of CML to Hamiltonian systems (symplectic map lattices) see
R. Levi et al., J. Stat. Phys. 46 (1987) 147
K. Kaneko and T. Konishi, J. Phys. Soc. Jpn. 56 (1987)2993: preprint (1988)
H. Kanz and P. Grassberger, J. of Phys. A 21 (1988) L127:

[41] for CML see also;
T. Bohr et al., Phys.Rev. Lett. 58 (1987) 2155:
N.H. Packard, private communications (1986):
G. Grinstein, preprint(1988):
L.A. Bunimovich, A. Lambert, and R. Lima, preprint (1989)
T. Bohr and O.B. Christensen, preprint (1989)

[42] P.Hadley and K. Wiesenfeld, Phys. Rev. Lett. 62 (1989) 1335

[43] K. Kaneko, preprint (1989, LA-UR-89-698) K. Kaneko, preprint "Clustering, Coding, Switching, and Hierarchical Ordering in Network of Chaotic Elements"

[44] *Theory and Applications of Cellular Automata*, ed. S. Wolfram, (World Sci. Pub., 1986)

[45] J.E. Hopcroft and J.D. Ullman, *Introduction to Automata Theory, Languages, and Computation*, (Addison Wesley, 1979)

[46] U. Frisch, B. Hasslacher, and Y. Pomeau, Phys. Rev. Lett. 56 (1986) 1505; see also Complex Systems, (1987) No.4

[47] S. Aubry, in *Solitons and Condensed Matter Physics*, ed. A.R. Bishop and T. Schneider (Springer, N.Y., 1978)

[48] P. Bak, Rep. Prog. Phys. 45 (1982) 587

[49] K.Kawasaki, in *Phase Transitions and Critical Phenomena*, vol.2, ed. C. Domb and M.S. Green (Academic Press, London, 1972);

[50] K. Kaneko, in *Theory and Applications of Cellular Automata* pp. 367-399 (World Scientific, 1986, ed. by S. Wolfram)

[51] K. Kaneko, Prog. Theor. Phys. 69 (1983) 1477

[52] Y. Aizawa, I. Nishikawa, and K. Kaneko, in preparation

[53] P. Grassberger, Physica 10D (1984) 52

[54] C. Grebogi, E. Ott, and J. A. Yorke, Phys. Rev. Lett. 48 (1982) 1507; 57 (1986) 1284; Physica 7D (1983) 181

[55] C. Grebogi, E. Ott, and J. A. Yorke, to appear in Phys. Rev.A

[56] This high-dimensional crisis picture may be related with the homoclinic excursions in high-dimensional systems: A.C. Newell, D.A. Rand, and D. Russel, preprint

[57] Y. Aizawa, C. Murakami, and T. Kohyama, Prog. Theor. Phys. 79 (1984) 96

[58] I. Procaccia and H. Schuster, Phys. Rev. A28 (1983) 1210, B.C. So, N. Yoshitake, H. Okamoto and H. Mori, J. Stat. Phys. 36 (1984) 367

[59] Y. Pomeau and P. Manneville, Comm. Math. Phys. 74 (1980) 189

[60] S.P. Obukhov, Physica 101A (1980) 145: see Y. Pomeau, Physica 23D (1986) 3, for a suggestion on the relation with intermittency.

[61] K. Kaneko and Y. Akutsu, J. Phys. A19 (1986) L69-74

[62] S. Nasuno, M. Sano, and Y. Sawada, in *Cooperative Dynamics in Complex Systems* (Springer, 1989, ed. H. Takayama) and private communications

[63] M.J. Feigenbaum, J. Stat. Phys. 21 (1979) 669

[64] F. Heslot, B. Castanig, and A. Libchaber, Phys. Rev. A 36 (1987) 5870

[65] L.D. Landau and E.M. Lifshitz, *Fluid Mechanics* (Pergamon, London, 1959) Chapt.3

[66] D. Ruelle and F. Takens, Comm. Math. Phys. 20 (1971) 167, 23 (1971) 343

[67] R. Shaw, Z. Naturforshung, 36a (1981) 80; *The Dripping Faucet as a Model Chaotic System*, (Aerial Press, Santa Cruz, 1984)

[68] R. Bowen and D. Ruelle, Inventiones Math. 29 (1975) 181; D. Ruelle, *Thermodynamic Formalism*, Addison Wesley, Reading, MA (1978)

[69] Y. Oono, Prog. Theor. Phys. 60 (1978) 1944; Y. Oono and Y. Takahashi, Prog. Theor. Phys 63 (1980) 1804.

[70] T. Kai and K. Tomita, Prog. Theor. Phys. 64 (1980) 1532

[71] H. Mori, B-C. So and T. Ose, Prog. Theor. Phys (1981) 1266

[72] H. A Gutowitz, J.D. Victor, and B.W. Knight, Physica 29D (1987) 18; H. A. Gutowitz and J. D. Victor, J. Stat. Phys. 54 (1989) 495

[73] Y. Oono and C. Yeung J. Stat. Phys. 48 (1987) 593; see also, Y. Oono and M. Kohmoto, Phys. Rev. Lett. 55 (1985) 2927

[74] J. S. Langer, M. Bar-On, and H.P. Miller, Phys. Rev. A 11 (1975) 1417

[75] Y. Kuramoto and T. Tsuzuki, Prog. Theor. Phys. 55 (1976) 356

[76] see for the approach with CA and temperature field, N. H. Packard, in *Theory and Applications of Cellular Automata* (World Scientific, 1986, ed. by S. Wolfram)

[77] J. Nagumo and S. Sato, Kybernetik 10 (1972) 155

[78] L.D. Harmon, Kybernetik 1 (1961) 89

[79] P. Bak, C. Tang, and K. Wiesenfeld, Phys. Rev. Lett., 59 (1987) 381

[80] H. Hayashi et al., Phys. Lett. 88 A (1982) 265

[81] K. Aihara and G. Matsumoto, in *Chaos* (ed. A.V. Holden, Princeton, 1986)

[82] W. Freeman and C. A. Skarda, Brain Res. Rev. 10 (1985) 147; W. Freeman, Brain Res. Rev. 11 (1986) 259; Biol. Cybernetics 55 (1987) 1;

[83] M. Mezard, G. Parisi, and M.A. Virasoro eds., *Spin Glass Theory and Beyond* (World Sci. Pub. 1987)

[84] L.N. Cooper, in *Proc. of Nobel Symposium on Collective Properties of Physical Systems*, B. Lundruist and S. Lundruist eds. (1973, Academic Press)

[85] W.A. Little and G.L. Shaw, Behavioral Biology 14 (1975) 115

[86] J. J. Hopfield, Proc. Natl. Acad. Sci. 79 (1982) 2554

[87] Y. Kuramoto, *Chemical Oscillations, Waves, and Turbulence*, Springer, 1984

[88] G.B. Ermentrout and N. Kopell, SIAM J. Math. Anal. 15 (1984) 215; N. Kopell and G.B. Ermentrout , in *Nonlinearity in Biology and Medicine*, (eds A.S. Perelson, B. Goldstein, M. Dembo,and J.A. Jacquez, Elsevier, 1988)

[89] *Spatiotemporal Coherence and Chaos in Physical Systems*, (North Holland, ed. A. R. Bishop, G. Gruner, and B. Nicholaenko, 1986)

[90] T. Ikegami and K. Kaneko, to be published

[91] A. T. Winfree, in *Oscillation and Traveling Waves in Chemical Systems*, ed. R.J. Fields and M. Burger (John Wiley and Sons, 1985): see also Y. Kuramoto, *Chemical Oscillations, Waves and Turbulence*, (Springer, N.Y., 1984) for theoretical stage.

[92] H. Yamazaki, Y. Oono and K. Hirakawa, J. Phys. Soc. Jpn. 44 (1978) 335

[93] G. Ahlers and R. P. Behringer, Prog. Theor. Phys. 64 (1978) 186

[94] E. Moses and V. Steinberg, Phys. Rev. Lett.57(1986) 2018

[95] S. Ciliberto and M. A. Rubio, Phys. Rev. Lett. 58 (1987) 2652

[96] R.W. Walden, P. Kolodner, A. Passner, and C.M. Surko, Phys. Rev. Lett. 55 (1985) 496

[97] S. Sato, M. Sano and Y. Sawada, private communication

[98] R. J. Donnelly et al., Phys. Rev. Lett. 44(1980) 987: G. W. Baxter and C. D. Andereck, Phys. Rev. Lett 57(1986) 3046

[99] S. Ciliberto and J.P. Gollub, Phys. Rev. Lett. 52 (1986) 922

[100] S. Kai and K. Hirakawa, Prog. Theor. Phys. Suppl. 64 (1978) 212

[101] M. Lowe, J. P. Gollub and T. C. Lubensky, Phys. Rev. Lett. 51(1983) 786

[102] R. V. Buskirk and C. D. Jeffries , Phys. Rev. A 31 (1985) 887

[103] A. Zettl, Physica 23D (1986) 155

[104] see e.g., K. R. Sreenivasan, in *Frontiers of Fluid Mechanics*, (Springer, 1985, eds. by S.H. Davis and J.L. Lumley)

[105] see e.g., J.J. Riley and M. Gad-el-Hak, in *Frontiers of Fluid Mechanics*, (Springer, 1985, eds. by S.H. Davis and J.L. Lumley)

[106] M. Bonetti, R. Meynart, J.P. Boon, and D. Olivari, Phys. Rev. Lett. 55 (1985) 492

[107] Y. Sawada, Physica 140A (1986) 134

[108] S. Ciliberto and P. Bigazzi, Phys. Rev. Lett. 60 (1988) 286

PATTERN DYNAMICS
IN EXCITABLE REACTION-DIFFUSION MEDIA

Takao Ohta and Masayasu Mimura*

Department of Physics, Ochanomizu University, Tokyo, 112 Japan
*Department of Mathematics, Hiroshima University, Hiroshima, 730 Japan

ABSTRACT

Recent studies of pattern formation and pattern evolution far from equilibrium by means of the interfacial approach are reviewed. The stability of spatially localized solutions exhibited in excitable reaction-diffusion system is investigated and is compared with computer simulations. Applications to the phase field model for a crystal growth and to a one-dimensional model system are also presented.

1. INTRODUCTION

A reaction-diffusion approach has been used extensively to study pattern formation and pattern evolution in the systems far from equilibrium.[1)~4)] Despite its apparently simple structure a set of reaction-diffusion equation exhibits a surprising variety of spatio-temporal orders. Typical examples are a limit cycle, traveling waves and chaos.[5)] These are now found generically in nature. Substantial parts of our understanding of the pattern formation far from equilibrium have indeed been put forward for these two decades by the study of various reaction-diffusion equations.

Numerical simulations of reaction-diffusion models are very useful to study pattern formation far from equilibrium.[6),7)] However in order to extract the essential properties one needs to develop an analytical theory. Unfortunately rigorous treatment of nonlinear reaction-diffusion equations is formidable except for few very special cases. Attempts have, therefore, been made to construct an approximate but reasonable formulation of the dynamical theory of patterns. Among them most successful are probably the phase dynamics[8)~15)] and the interface dynamics.[16)~18)] The basic idea of these approaches is that there is a particular variable associated with patterns, which governs the slow motion of large scale deformation of the patterns. It should be noted that emergence of any regular patterns out of a uniform system violates the translational invariance of the system so that there exists a Goldstone mode. Thus the time-evolution of a slightly deformed pattern can be described reasonably well by the Goldstone variable. The phase dynamics was indeed originally introduced in and applied to the periodic solutions in reaction-diffusion equations,[8),9)] where the large length scale deformations can be described by allowing the space-time dependence in the phase variable.

The interface dynamics is also concerned with the relevant slow variable in a system which consists of two states separated by an interface. However there is an essential difference from the phase dynamics. One might think that the interface dynamics is a special case of the phase dynamics of a periodic pattern with an infinite period. This is correct in a very restricted sense. In the phase dynamic description the length scale of the order of spatial period is coarse-grained whereas in the interface dynamics the smallest length scale of the problem is the interface width. Thus the interface dynamics contains more information than the phase dynamics. Sometimes the word "extended phase dynamics" is used emphasizing that it is formulated to include not only the original phase dynamics but also the interface dynamics.[19]

In this article we review mainly our recent investigations of pattern formation and pattern evolution in excitable reaction-diffusion media by means of the interface dynamics. As a prototype of excitable media we have studied the Bonhoffer-van der Pol equation which is a coupled set of equations for two variables u and v

$$\varepsilon\tau\partial_t u = \varepsilon^2\nabla^2 u + f(u) - v \qquad (1.1a)$$

$$\partial_t v = D\nabla^2 v + u - \gamma v, \qquad (1.1b)$$

where ε, τ, γ, and D are positive constants. f(u) is assumed to be a cubic-like nonlinearity. The simplest choice is

$$f(u)=u(1-u)(u-a) \qquad (1.1c)$$

with $0<a<1/2$. It is often useful for analytical study to approximate f(u) by a piecewise-linear form

$$f(u)=\begin{cases} -u & u<a \\ 1-u & u>a \end{cases} \qquad (1.2)$$

This was first utilized by McKean[20] to study propagating pulse in the FitzHugh-Nagumo equation[21),22] which is a special case of Eqs. (1.1) with $D=\gamma=0$.

The nullcline of Eqs. (1.1) is shown in Fig. 1-1. When the parameter γ is large, there are three equilibria, under which the system is bistable where two of the equilibria are stable whereas

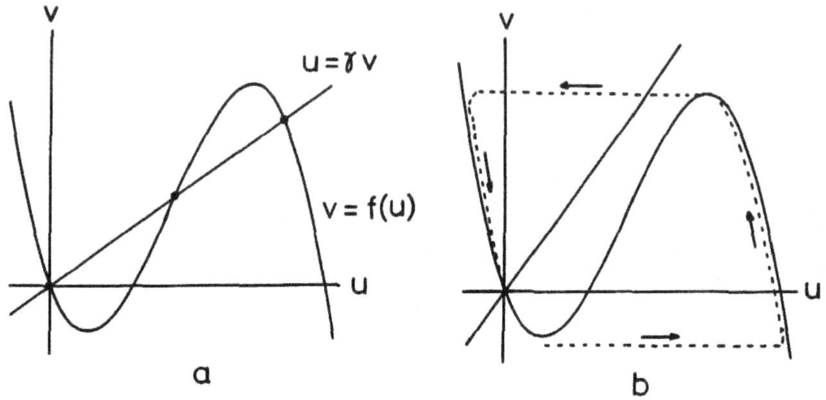

Fig. 1-1. Nullcline corresponding to bistable (a) and monostable (b) states.

the middle one is unstable. We call the stable state (u, v)=(0, 0) a rest state and the other an excited state. When γ is small, on the other hand, the state (0, 0) is the only stable equilibrium state, which is called a rest state. The system is, then, mono-stable. Usually the case shown in Fig. 1-1b is called excitable. In this situation the rest state is very sensitive to perturbation. If the perturbation is small the disturbance is rapidly damping out. However if the amplitude of the perturbation is large beyond a threshold value, the system responds very differently. As shown in Fig. 1-1b by the dotted line u grows far away from the origin and then as v increases u decreases gradually and finally the system returns to the origin. Since u represents the degree of excitation and v inhibits the growth of u, we call u and v an activator and its inhibitor, respectively. When the diffusion terms are present as in

Eqs. (1.1), the situation is drastically changed. That is, the system does not necessarily return to the original uniform rest state but enters into other non-uniform states such as propagating pulses or spatially localized patterns as in Fig. 1-2.

Equations (1.1) should be compared with the simplest nonlinear reaction-diffusion model equation:

$$\partial_t S = \varepsilon^2 \nabla^2 S + S - S^3 - h, \tag{1.3}$$

where ε and h are constant. Note that if v is constant in Eq. (1.1a) it reduces to the form of (1.3). Equation (1.3) is called the time-dependent Ginzburg-Landau equation in the field of phase transitions.[23] Even this simple equation exhibits nontrivial behavior of the solutions. In fact, if (1.3) is bistable for some values of h, we have two kinds of domains where S has the different stationary values and which are separated by a sharp layer or interface for small values of ε. Recently several authors have developed the interface dynamics in the study of this domain growth problem.[17], [24], [25], [68]

The reason as to why we are concerned with the excitable model (1.1) is three fold. First of all Eqs. (1.1) does not have any Lyapounov functionals or free energy functionals. This is one of the most important features of the systems far from equilibrium which traces back to the violation of the Onsager's reciprocal relation and equivalently to the absence of the detailed balance in a general nonequilibrium open system.[26],[27] An immediate consequence of this is the possibility of temporal oscillation in purely dissipative systems. It is instructive to see that the TDGL equation (1.3) can be written in terms of the free energy functional H[S] as

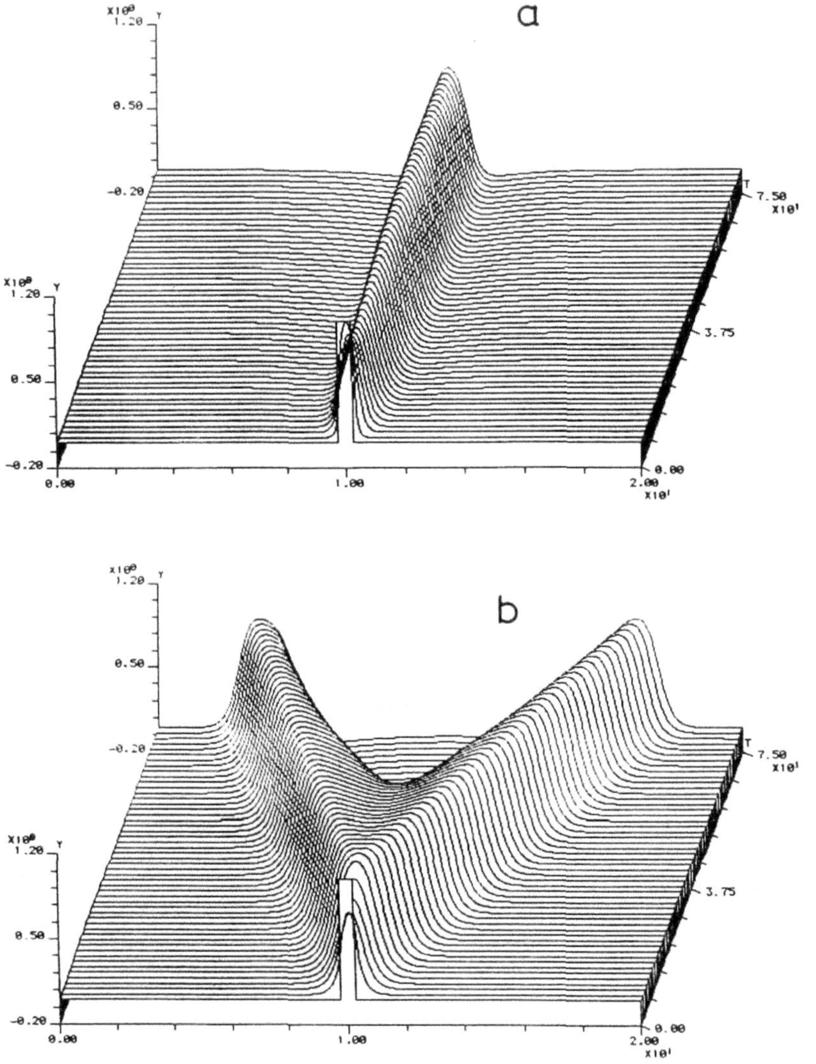

Fig. 1-2. Time-evolution in one dimension (a) u-profile of localized pattern for ε=0.1412, τ=1, γ=5 and D=10 (b) u-profile of propagating pulses for ε=0.1412, τ=0.15, γ=5 and D=1.

$$\partial_t S = -\frac{\partial H}{\partial S}, \tag{1.4a}$$

$$H(S) = \int d\mathbf{r}\left[\frac{1}{2}(\nabla S)^2 - \frac{S^2}{2} + \frac{S^4}{4}\right]. \tag{1.4b}$$

Equations (1.1) cannot be expressed in the form of (1.4). If a system has a Lyapounov functional it simply tends to decay to the state of the "energy" minimum asymptotically. Hence no oscillatory behavior is expected in a dissipative system. The TDGL equation was introduced in the dynamical theory of phase transitions where relaxation of a non-equilibrium state to the final equilibrium is a main concern.[23] Hence the existence of the free energy functional is not serious restriction. When a system has a Lyapounov functional the interface or phase dynamics can be constructed systematically. One may use the concept of energy balance and virtual work under deformations of the pattern.[28],[29] However the problem is more difficult in non-equilibrium systems due to the absence of such functionals. Attempt of constructing a general theory is still under developing.

We here point out another very important property. That is, there exist generally no spatially localized stable patterns in a reaction-diffusion system if it possesses a Lyapunov functional and the system is extended infinitely. This can be proved by an argument similar to Derrick's theorem for a non-dissipative system with a Lagrangian.[30] This states that if a field-theoretic model has a Lagrangian and the interaction is short ranged, no stable localized solution can exist in two and more spatial dimensions. Let us consider a Lagrangian L (S) which consists of two parts:

$$L\{S\} = \int \mathbf{dr}\ \{V_1 + V_2\},\qquad (1.5a)$$

where

$$V_1 = (1/2)(\nabla S)^2 \qquad (1.5b)$$

and V_2 is the the local part which satisfies $V_2=0$ for the minimum value of S. Suppose that $S(\mathbf{r})$ is the localized solution obtained by variation of $L\{S\}$. We introduce a one-parameter family of the solution $S(\mathbf{r}, \lambda) = S(\lambda\mathbf{r})$. By changing λ the Lagrangian is transformed as

$$L(\lambda) = \lambda^{2-d}V_1 + \lambda^{-d}V_2\,. \qquad (1.6)$$

Since $S(\mathbf{r}, \lambda)$ with $\lambda=1$ is the stationary solution, we should have the relation

$$dH/d\lambda_{\ \lambda=1} = 0 \qquad (1.7)$$

or

$$(d-2)V_1 + dV_2 = 0. \qquad (1.8)$$

For $d > 2$ (1.8) holds only for $V_1 = V_2 = 0$. Thus no non-uniform solution exists. For d=2 we have $V_2 = 0$, which is the minimum value so that we may apply the variational principle only to V_1. Then we trivially have $V_1=0$. Thus for $d \geq 2$ we may conclude that stable localized solutions do not exist for a system with the short range interaction.

In a dissipative system a Lyapounov functional plays a role of Lagrangian. Thus the absence of a Lyapounov functional is a necessary condition for existence of a steady localized solution in the spatial dimension greater than one. A similar argument for the existence of localized solutions can be seen in Ref. 31)

Another feature of Eqs. (1.1) is that when $\varepsilon^2 \ll D$ and $\tau < 1$ the time-evolution of the variable $v(\mathbf{r},t)$ turns out to be very slow. It becomes even slower than the motion of an interface separating two regions of the rest and the excited states. This provides us with an interesting example of the coupling between the Goldstone variable associated with the interface (or the phase) and the extra slow degrees of freedom.[32)~34)] However no systematic theory has been available so far.

The third reason why we are interested in the excitable model (1.1) is of course its realizability in real chemical and biological systems. In fact Eqs. (1.1) which possesses both bistability and monostability serves as a simple and fundamental model for the Belousov-Zhabotinskii reaction, pulse propagation on nerve axion and others. This will be discussed further in section 2.

This review article is organized as follows: After presenting several experimental phenomena which are described by Eqs. (1.1) in the next section we solve (1.1) in section 3 to obtain spatially localized steady solutions. The localized excited domain is separated form the surrounding rest state by a sharp interface. In section 4 we derive the interface equation of motion for a gently distorted pattern by using a singular perturbation method. The linear stability analysis of the interface equation of motion is performed in section 5. It is shown that the steady localized solutions lose their stability by changing the parameters a, τ and γ in Eqs. (1.1). We obtain the stability diagram. The results are compared with the numerical simulations. Sections 6 and 7 deal with the related problems. In section 6 we derive the interface equation of motion for a phase field model of crystal growth demonstrating that the present formulation can be applied to wide class of problems in non-equilibrium open systems. In section 7 we study the kink-antikink system in one dimension. The short range interaction which appears between a kink-antikink pair is analyzed. The final section (section 8) is devoted to summary and discussion.

2. EXCITABLE SYSTEMS

In this section we discuss some of the excitable non-equilibrium open systems described by the Bonhoffer-van der Pol type equation (1.1).

Among them the most famous is the Belousov-Zhabotinskii reaction.[35,36] This is the oxidation of malonic acid (or citric acid as found originally by Belousov) by bromate in the presence of cerium ions (or ferroin ions) as catalyst. Reference 37) describes a useful model approach. The overall reaction proceeds as follows:

$$2BrO_3^- + CH_2(COOH)_2 + 2H^+$$
$$\rightarrow 2BrCH(COOH)_2 + 3CO_2 + 4H_2O.$$

In the course of this reaction the cerium ions oscillate between Ce^{3+} and Ce^{4+} states. Later in a thin layer of the oscillating solutions a concentric wave pattern (target pattern) was found, which propagates outward from leading centers formed spontaneously.[38] Formation of spiral waves was also discovered.[39),40] This is of particular interest since the spiral waves can be observed even in a non-oscillatory medium. The most recent studies of the Belousov-Zhabotinskii reaction can be found in the special issue of Journal Physical Chemistry.[41)]

The details of the reaction steps have been investigated by Field, Koros and Noyes.[42)] The major reaction process consists of the following two steps:

$$2Ce^{+3} + BrO_3^- + HBrO_2 + 3H^+$$

$$\rightarrow 2Ce^{+4} + 2HBrO_2 + H_2O, \qquad (R1)$$

$$HBrO_2 + Br^- + H^+ \rightarrow 2HOBr. \qquad (R2)$$

The reaction (R1) is the autocatalytic reaction of $HBrO_2$ which increase rapidly. On the other hand (R2) causes a rapid decrease of $HBrO_2$. The oscillation occurs by the switching of these two processes. Suppose that there is a sufficient number of Br^- ions initially. By the reaction (R2) Br^- decreases. Furthermore there is another reaction which consumes Br^-.

$$BrO_3^- + Br^- + 2H^+ \rightarrow HBrO_2 + HOBr. \qquad (R3)$$

Hence when the concentration of Br^- ions becomes very low by the reactions (R2) and (R3) the reaction (R1) turns on and the rapid increase of $HBrO_2$ and Ce^{+4} occurs. These increases, then, trigger the following two reactions:

$$2HBrO_2 \rightarrow BrO_3^- + HOBr + H^+, \qquad (R4)$$

$$4Ce^{+4} + BrCH(COOH)_2 + 2H_2O$$

$$\rightarrow 4Ce^{+3} + Br^{-} + HCOOH + 2CO_2 + 5H^{+}. \qquad \text{(R5)}$$

The reaction (R5) produces Br^- so that the reaction (R2) becomes active again.

Noyes et. al.[42] have shown that the above expectation indeed occurs by constructing a model equation for the reaction processes. Let us put $X=[HBrO_2]$, $Y=[Br^-]$ and $Z=[Ce^{+4}]$ where [ ] means the concentration of each chemical agent. Other chemical reagents are assumed to exist sufficiently and to be constant. By suitable scaling of space, time and the concentrations the above five reactions lead to the coupled set of equations[37] which is called the Oregonator;

$$\varepsilon\dot{X} = X + aY - X^2 - XY \qquad (2.1a)$$

$$\varepsilon'\dot{Y} = aY + bZ - XY \qquad (2.1b)$$

$$\dot{Z} = X - Z. \qquad (2.1c)$$

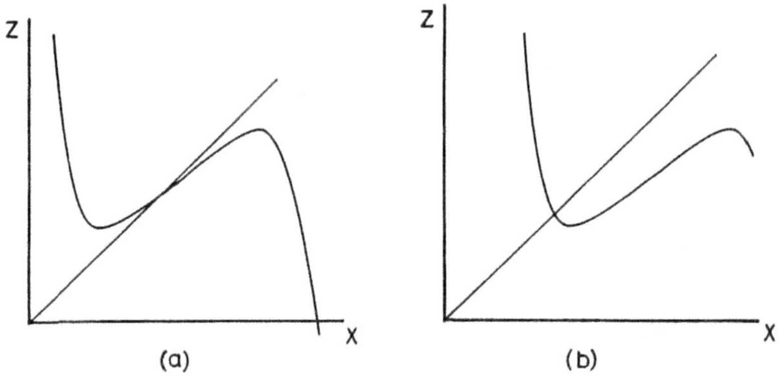

Fig. 2-1. The nullcline of $\dot{X}=0$ and $\dot{Z}=0$. (a) oscillatory case and (b) excitable case.

The behavior of the solution can be analyzed qualitatively as follows: In typical experimental situations $\varepsilon' \ll \varepsilon \ll 1$, $a \ll 1$ and $b \approx 1$. Hence we may simplify (2.1) by putting $\dot{Y}=0$. The nullclines of $\dot{X}=0$ and $\dot{Z}=0$ are, then, given as shown schematically in Figs. 2-1a or 2-1b depending on the values of the parameters a and b. In the case of Fig. 2-1a a relaxational oscillation emerges. In the case of Fig. 2-1b, on the other hand, the system is excitable. If we add the diffusion terms the system is essentially described by the Bonhoffer-van der Pol type equation (1.1). Recently there have been a number of theoretical studies of the spiral waves and its three-dimensional version, the scroll waves, in the excitable situation by means of the interfacial approach.[18] The target patterns and the spiral waves in the oscillatory regime had also been investigated.[43]

The next example of the excitable system is the band pattern formation on the surface of *Characean* internodal cells under illumination.[44]~[47] Alternating acidic and alkaline zones appear along the length of the cell. The spatial distribution of pH is detected by using phenol red. It has been shown that the distribution of the electric membrane potential and the surface electric potential are the same as that of pH. In most experimental conditions this pattern is symmetric with respect to the circumferential direction of the cell so that it is regarded as a periodic band pattern. Existence of such a pattern is believed to play an important role on the growth and the structural formation of the living cells.[46]

This pattern formation is caused by a positive feedback of ion transport across the membrane. Here we construct a simple model equation[48] by taking into account the experimental facts. We approximate the shape of the internodal cells of *Chara* by a cylinder and consider a two-dimensional surface of the cylinder. Let us denote the concentration of H^+ ions inside (outside) the cylinder by u_i (u_e). There are two kinds of H^+ current across the membrane (the surface of the cylinder). One is the passive flow J_p which is proportional to the difference $u_i - u_e$:

$$J_p = -p(u_e - u_i), \tag{2.2}$$

where p is a positive constant. The sign of J_p has been chosen such that it is positive when the current is directed outward the surface.

The other is the active transport which is caused by the ATPase existing at the membrane. Even when u_i is smaller than u_e there is a finite current J_a by the locally active H^+ pumps. We assume J_a as

$$J_a = f(u_e - u_i, v), \tag{2.3}$$

where the function f(x, v) takes the form shown in Fig. 2-2. The parameter v stands for a threshold for the active transport. From (2.2) and (2.3) and assuming that the diffusion constants of u_i and u_e along the surface are the same, which is denoted by D, we obtain the equation for $u = u_e - u_i$ as

$$\partial_t u = D\, \Delta u + 2[f(u,v) - pu], \tag{2.4}$$

where Δ is the two-dimensional Laplacian.

The threshold value v depends on the energy which makes the H^+ pumps active and, therefore, depends on the ATP concentration A. We may expect that the threshold v is small if the magnitude of A is large and *vice versa* . Thus we simply put

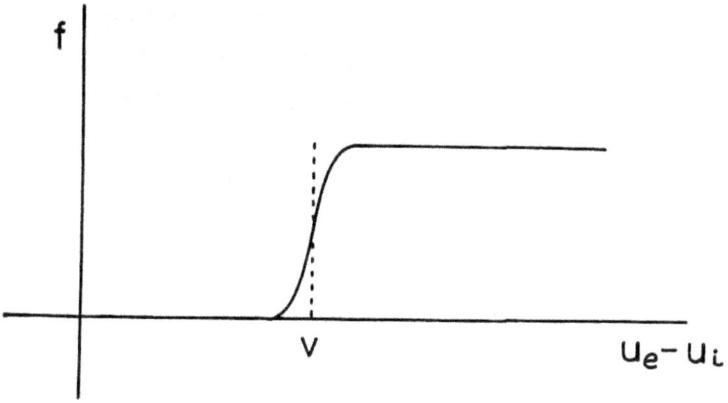

Fig. 2-2. The form of $f(u_e-u_i)$.

$$\partial_t(v+A) = 0. \tag{2.5}$$

Since ATP is transformed from ADP by light and it is consumed by pumping the H^+ ions it obeys the equation

$$\partial_t A = cI - bf(u,v), \tag{2.6}$$

where c and b are positive constants. I is the intensity of light. Putting (2.5) and (2.6) together and adding a diffusion term of v we obtain

$$\partial_t v = d\,\Delta v - cI + bf(u,v), \tag{2.7}$$

where d is a diffusion constant.

In this way we obtain the coupled set of equations (2.4) and (2.7). The nullclines of these equations are shown in Fig. 2-3 by changing the light intensity I. Thus it is found that this system is essentially equivalent with the excitable system (1.1).

The third example is pattern formation arising in mathematical ecology. It is observed that plankton often displays spatial heterogeneity in a homogeneous environment (for instance, a lake).[78),79)] Such phenomenon is called patchiness in ecology. Okubo[80)] suggested the prey-predator species model where plant plankton is a prey and zoo plankton is its predator and showed the possibility of patchiness when the combination of prey autocatalysis with high dispersal ability of predator.

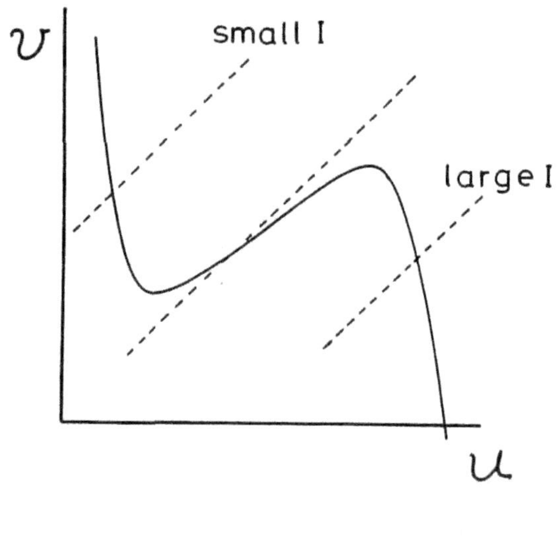

Fig. 2-3. The functions pu=f(u,v) (solid line) and cI= bf(u,v) (broken line) for several values of I.

Let us derive the modeling by using Lotka-Volterra model equations including diffusion effect. Let $u_1(\mathbf{r}, t)$ and $u_2(\mathbf{r}, t)$ be the population densities of a plant plankton (prey) and a zoo plankton (predator), respectively for time t and position **r**. The Lotoka-Volterra diffusion model is described by

$$\partial_t u_1 = d_1 \nabla u_1 + f_1(u_1, u_2)u_1 \tag{2.8a}$$

$$\partial_t u_2 = d_2 \nabla u_2 + f_2(u_1, u_2)u_2 , \tag{2.8b}$$

where d_1 and d_2 are the diffusion rates and f_1 and f_2 are the growth rates of the two populations. Among many ecologically possible models, one example is as follows:[81)]

$$f_1 = r\left(1 - \frac{u_1}{K}\right) - \frac{ku_2}{u_1 + D} . \tag{2.9}$$

Here the first term is known as the Verhulst-Pearl logistic form with the intrinsic growth rate r and the carrying capacity K. The second term is as the Holling predation rate with a constant k and the critical value of the prey population D beyond which the predator's attack capability begins to saturate. The case when D = 0 implies that the predator has a *constant* predation rate. We note that when $K > D$, f_1 has the humped nonlinearity with the maximum at $u_1 = (K - D)/2$ so that, $f_1(u_1, u_2)u_1 = 0$ is of cubic like nonlinearity. On the other hand,

$$f_2 = s\left(1 - \frac{u_2}{\gamma u_1}\right) , \tag{2.10}$$

which is given by Leslie. Here the growth of the predator population is similar to Verhulst-Pearl logistic form where K is replaced by γu_1. All parameters in f_1 and f_2 are assumed to be positive constants. For suitable choice of K, D and γ, we have two cases as in Fig. 2-4.

Finally we should point out here that an apparently similar spatial structure has been observed in electric glow discharges where current filament exhibit a periodic structure.[49] This phenomenon has indeed been interpreted by the model equation essentially same as (1.1).[50]

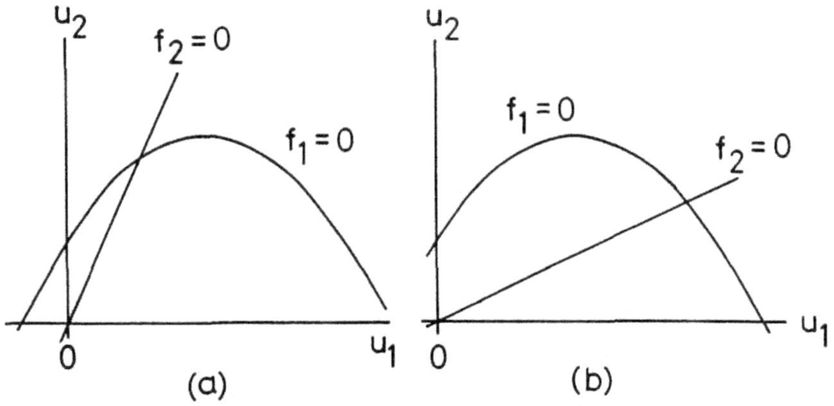

Fig. 2-4. The functions $f_1(u_1, u_2)u_1 = 0$ and $f_2(u_1, u_2)u_2 = 0$ (a) oscillatory case and (b) excitable case.

3. LOCALIZED PATTERNS

One of the most interesting features of the coupled set of Eqs. (1.1) is that it exhibits stable localized patterns. So far almost all studies of (1.1) are concerned with the limit where D is very small or at most of the order of ε.[18] In this case Eqs. (1.1) has a propagating pulse solution in the monostable region. What we are interested in here is the opposite limit: $\varepsilon^2 \ll D$.[51] The presence of the diffusion of the inhibitor v will change the behavior of the solution drastically. Suppose that initially u is positive finite in some finite area of space and is zero otherwise. By the reaction term in Eq. (1.1b) v is, then, generated where u is finite. Once v is produced, this, in turn, tends to diminish u by the reaction term in Eq.(1.1a). Hence a motionless non-uniform solution is unstable in the monostable region. However when the diffusion of v is present the produced v can flow out of the domain. Therefore if this diffusion balances with the production the localized domain can neither shrink nor expand. Thus we may expect a stable spatially localized domain.

The above expectation was first confirmed by Koga and Kuramoto.[52] They studied Eqs.(1.1) with a simplification of the nonlinearity in f(u) as (1.2), which will also be used in the present paper. A one-dimensional excited localized domain was indeed shown to be stable in some parameter regimes. What is remarkable in their results is that the localized domain undergoes sustained oscillation like a breathing motion for small values of τ. The existence and the stability of a localized excited domain in one dimension have also been investigated rigorously by Nishiura and Mimura for general cubic-like nonlinearity.[53]

Here we will show an extension of the study by Koga and Kuramoto to higher dimensions.[51] As was mentioned in Introduction, existence of a higher dimensional localized domain is a feature pertinent to a system having no Lyapounov functional. Furthermore as will be seen later higher dimensional localized patterns provide several inherently new instabilities which was not found in one dimension. We do not treat an assembly of localized domains but investigate the stability and the time-evolution of a single domain. The very interesting problem of the mutual interaction between adjacent domains can, in principle, be solved by the present approach. It will, however, be left for a future study.

Hereafter in this section we use the piecewise-linear form (1.2) for f(u) to derive spherically symmetric localized solutions. In the steady state Eqs. (1.1) becomes as follows:

$$\varepsilon^2\left[\frac{d^2}{dr^2}+\frac{d-1}{r}\frac{d}{dr}\right]u - u + H(u-a) - v = 0 \tag{3.1a}$$

$$\left[\frac{d^2}{dr^2}+\frac{d-1}{r}\frac{d}{dr}\right]v + u - \gamma v = 0 \quad , \tag{3.1b}$$

where d is the dimensionality of space. H(x) is the Heaviside step function. The boundary conditions are imposed as

$$\frac{du}{dr} = \frac{dv}{dr} = 0, \qquad \text{at } r=o \tag{3.2}$$

$$\lim_{r\to\infty} u = \lim_{r\to\infty} v = 0 \tag{3.3}$$

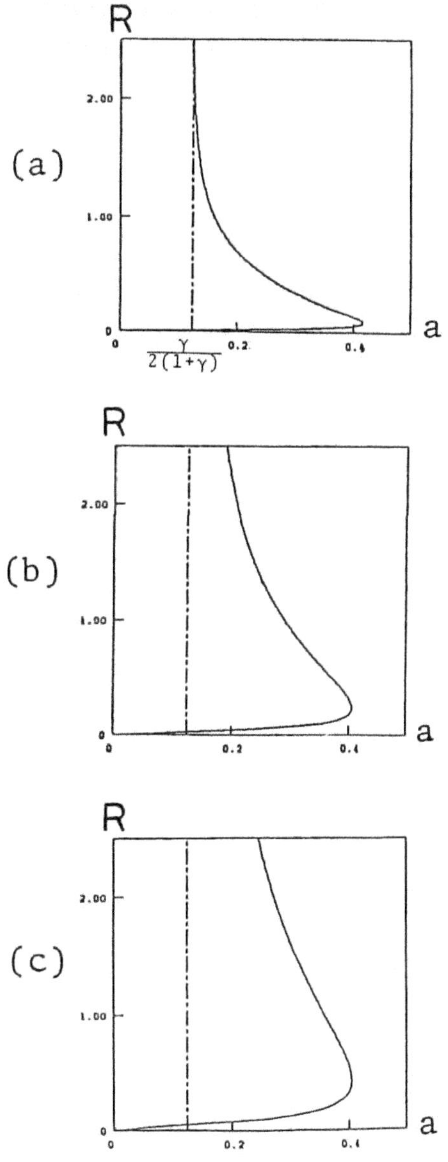

Fig. 3-1. Equilibrium radius R as a function of a for d=1 (a), 2 (b) and 3 (c).

and u and v and their first derivatives are continuous at r=R. Here R is the radius of the localized domain defined through the relation

$$u(R)=a\,. \qquad (3.4)$$

In one dimension the localized solution takes a band-shaped domain with the domain width 2R. In two and three dimensions the solutions of Eqs.(3.1) give us a disk-shaped and a spherical domain, respectively.

Equations (3.1) can be solved exactly as was performed in Ref. 51). In Fig. 3-1 we plot the radius R as a function of a for d=1, 2, and 3. It is noted that R is a multivalued function in the interval between a_1 and a_2. The value a_1 is given by $a_1 = \gamma/(2(1+\gamma))$ where the excited and the rest states can coexist with each other in equilibrium. In other words at $a=a_1$ the Maxwell construction holdsas shown in Fig. 3-2 where the shaded regions have equal area. In order to see further the meaning of the two solutions of R we take the limit $\varepsilon \to 0$ in Eqs.(3.1). Then a_2 is shown to move

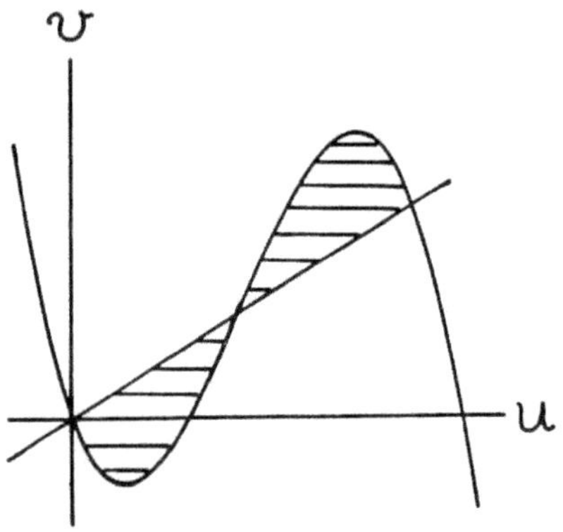

Fig. 3-2. The case where Maxwell rule holds.

toward a=1/2 so that the lower branch (smaller value of R) vanishes in this limit. Thus only the upper branch survives. These facts strongly suggest that only the upper branch is physically meaningful. We will explore this point in another aspect in the next section by the interface dynamics. In fact it will be shown that the lower branch is unstable for disturbance of uniform contraction and expansion while the upper branch is the equilibrium stable radius of domains.

Before closing this section we show an example of stable localized solutions. In Fig. 3-3 the solution for a small value of ε in one dimension is displayed together with the limiting solution for $\varepsilon \to 0$ for comparison.

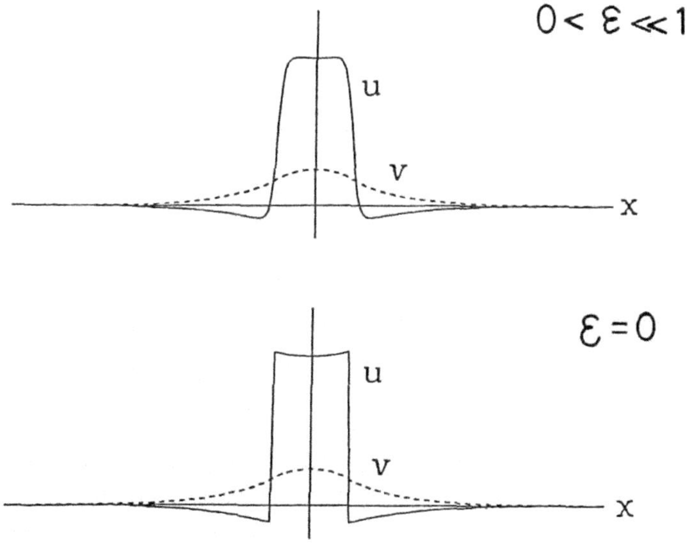

Fig. 3-3. Spatial profiles of the localized solution.

4. INTERFACE EQUATION OF MOTION

As can be seen from Fig. 3-3 in the previous section there exists a well-defined interface in the limit $\varepsilon \ll R$ separating the excited and the rest domains. This suggests that in order to study the time-evolution of the pattern it is more efficient to analyze the motion of the interface rather than to solve the original Eqs. (1.1). Here we construct the interface dynamics for the model system Eqs. (1.1).

Before going into this problem we illustrate the basic idea of the interface dynamics by using the simpler equation (1.3). It is well known that (1.3) for $h \neq 0$ possesses a propagating kink solution in one dimension:

$$S(x,t)=S_0(x-ct), \tag{4.1}$$

where the velocity c depends on h. Substitution (4.1) into (1.3) yields with $z=x-ct$

$$-c\frac{d}{dz}S_0(z) = \varepsilon^2 \frac{d^2}{dz^2}S_0(z)+ S_0(1-S_0^2) - h\,. \tag{4.2}$$

By supposing temporally that z is "time" and S_0 is the "position" this equation can be regarded as the Newton equation of a ball moving under the potential $(1/2)S_0^2-(1/4)S_0^4-hS_0$ with the friction proportional to the "velocity" dS_0/dz. As shown in Fig. 4-1 the ball sitting initially near point A falls down and approaches to the vicinity of point B asymptotically if the value of the "friction" coefficient c is chosen appropriately. This clearly indicates the existence of a kink solution of the original equation (1.3).

Having prepared the existence of kink in one dimension we next

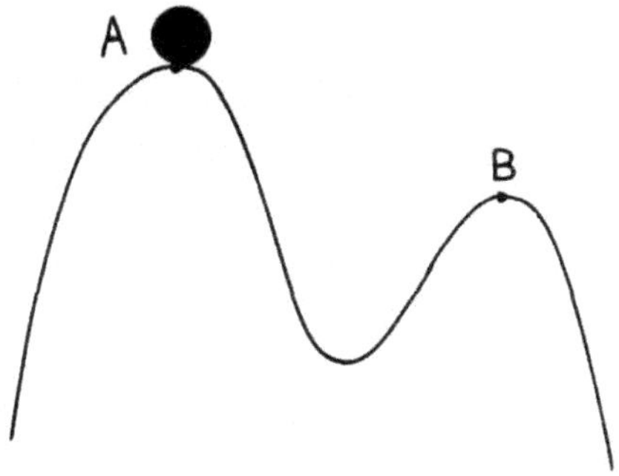

Fig. 4-1. Motion of a ball.

apply the above argument to higher dimensions. Let us assume a spherically symmetric domain which is expanding with a time-dependent radius R(t) and put the solution as

$$S(\mathbf{r},t)=\hat{S}_0(r-R(t)). \tag{4.3}$$

From (1.3) we have

$$-\frac{dR}{dt}\frac{d}{dz}\widehat{S_0}-\varepsilon^2\frac{(d-1)}{r}\frac{d}{dz}\widehat{S_0}=\varepsilon^2\frac{d^2}{dz^2}\widehat{S_0}+\widehat{S_0}(1-\widehat{S_0}^2)-h \tag{4.4}$$

When R(t) is sufficiently larger than ε we may approximate $\hat{S}_0$ by the one-dimensional solution S_0. Furthermore since $d\hat{S}_0/dz$ is finite only near r=R(t) we may replace (d-1)/r by (d-1)/R(t). Hence we obtain by comparing the left hand sides of (4.2) and (4.4)

$$\frac{dR}{dt}=-\frac{\varepsilon^2(d-1)}{R}+c(h) \tag{4.5}$$

The stationary solution of (4.5) is given by $R(t) = \varepsilon^2(d-1)/c(h)$. However as is seen from Fig. 4-2 this solution is clearly unstable. Thus there is no steady localized spherically symmetric solution to the TDGL equation. This is an example that a system which has a Lyapounov functional and the interaction is short range does not exhibit any stable localized solutions in two or more dimensions as was stated in section 1.

Although (4.5) was obtained for a particular form of domain generalization to an arbitrary domain shape is straightforward. The velocity $dR(t)/dt$ is replaced by the normal component $V(t)$ of the interface velocity. The mean curvature $K(t)$ should be used in stead of $-(d-1)/R(t)$. Therefore we have

$$V(t)=\varepsilon^2K(t)+c(h), \qquad (4.6)$$

where the sign of $K(t)$ is defined such that when the unit normal on the interface is directed to the opposite direction to the center of the curvature, $K(t)$ should be chosen to be negative and *vice versa* . It is also noted that the sign of $c(h)$ depends on whether the domain concerned is thermodynamically stable or not compared to the surrounding region.

The interface equation of motion (4.6) for TDGL model has been

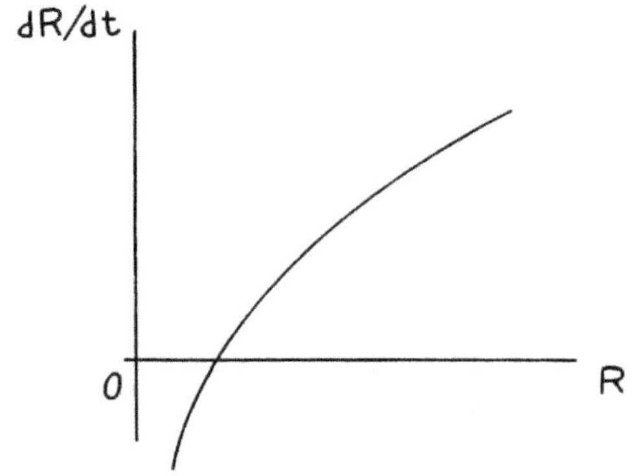

Fig. 4-2. dR/dt vs. R .

derived by several authors[17),28)] and applied to the ordering dynamics of quenched thermodynamically unstable systems.[25)] Equation (4.6) has also been used in pattern formation far from equilibrium such as spiral waves in the Belousov-Zhabotinskii reaction.[18)]

Now we apply the above method to the excitable system (1.1).[54),55)] In the situation $\varepsilon^2 \ll D$ the activator $u(\mathbf{r},t)$ changes very rapidly at the interface while the inhibitor $v(\mathbf{r},t)$ is smooth over the large length scale. This is clearly exhibited in Fig. 3-3. Thus in the limit $\varepsilon \to 0$ we may approximate $v(\mathbf{r},t)$ in Eq. (1.1a) by the value $v_I(t)$ at the interface:

$$\varepsilon\tau\partial_t u = \varepsilon^2\nabla^2 u + f(u) - v_I. \tag{4.7}$$

This is equivalent with the TDGL equation (1.3) with only modification that the external field h is now time-dependent. Thus the interface equation of motion is of the same form as (4.6)

$$\tau\varepsilon\, V(t) = \varepsilon^2\, K(t) + c(v_I,\, a), \tag{4.8a}$$

where the dependence on v_I and a in the last term is shown explicitly. If we take the piecewise-linear form of f(u), the velocity is given by[56)]

$$c(v_I,a) = 2(1/2-a-v_I)[(a+v_I)(1-a-v_I)]^{-1/2}. \tag{4.8b}$$

Since Eq. (4.8a) is not closed, we need the value of $v(\mathbf{r},t)$ at the interface. This requires to solve Eq. (1.1b) under a given configuration of the interface. When ε is small, we may substitute

$$v(\mathbf{r},t)=f(u)$$

or

$$u(\mathbf{r},t)=h_{\pm}(v) \tag{4.9}$$

into Eq. (1.1b) where $h_{\pm}$ (v) defines the two branches of u as a function of v as in Fig. 4-3. Then Eq. (1.1b) becomes

$$\partial_t v = D\nabla^2 \mathbf{v} + h_{\pm}(v) - \gamma v \tag{4.10}$$

We should assign h_+ in the excited domain and h_- in the rest domain. Thus we have the coupled set of equations (4.8) and (4.10) for the time-evolution of the interface. It is noted that we have not used the piecewise-linear form of f(u) in the derivation of these equations except for the explicit form of $c(v_l,a)$.

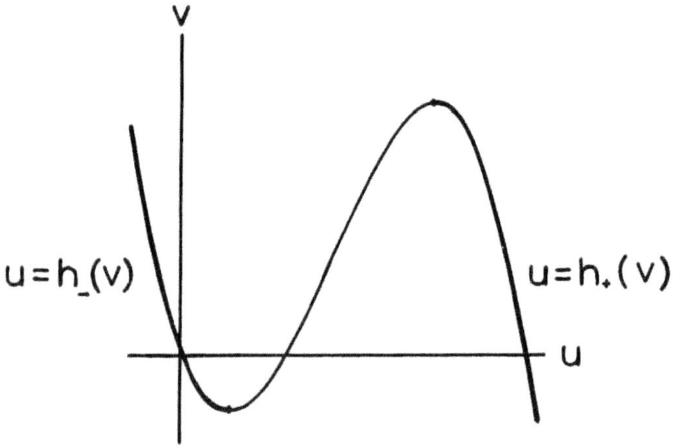

Fig. 4-3. Two branches h_+ and h_- .

5. STABILITY OF LOCALIZED PATTERNS

In this section we solve the interface equation of motion (4.8) and (4.10) for small deviations around the spherically symmetric localized solutions.[51] The piecewise-linear form of f(u) will be used fully throughout this section. The diffusion constant D is put to be unity by rescaling the unit of length.

First we consider a band-shaped domain. Let $\zeta^{+}(y,t)$ ($\zeta^{-}(y,t)$) be the deformation of the right (left) interface as shown in Fig. 5-1. Our task is to derive the time-evolution equations for $\zeta^{\pm}$ from (4.8) and (4.10), but it is very difficult in general. Here we restrict ourselves to linear order. Even in this case the resulting equation is not trivial since the slow variable $v(\mathbf{r},t)$ causes a non-local interaction between two points distant on the interfaces.

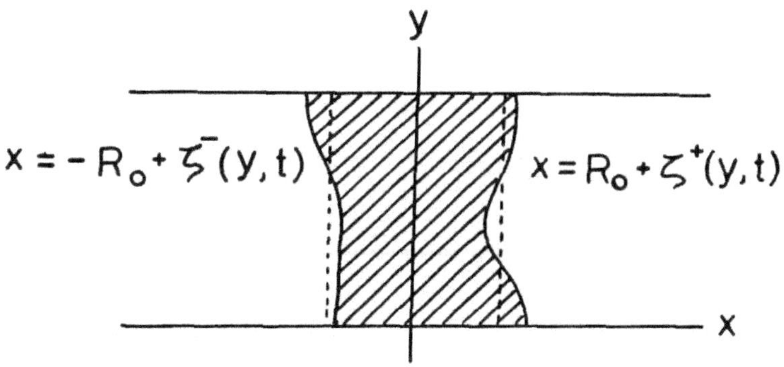

Fig. 5-1. Deformations ζ_{+} and ζ_{-} .

When f(u) is piecewise linear Eq. (4.9) can be written in terms of $\zeta^{\pm}$ as

$$u(x,y,t) = -v(x,y,t) + H(R_0+\zeta^{+}(y,t)-x)H(x+R_0-\zeta^{-}(y,t)), \tag{5.1}$$

where $2R_0$ is the equilibrium width of the band structure. Equation (4.10) is, then, solved formally for given $\zeta^{\pm}$. The value v_I at the interfaces is thus obtained as

$$\begin{aligned} v_I^{\pm}(y,t) &= v(\pm R_0 + \zeta^{\pm}(y,t),\ y,t) \\ &= v_0 + v_1^{\pm}(y,t) + v_2^{\pm}(y,t) + O({\zeta^{\pm}}^2), \end{aligned} \tag{5.2a}$$

where

$$v_0 = v(R_0), \tag{5.2b}$$

$$v_1^{\pm}(y,t) = \frac{dv}{dR}\bigg|_{R=\pm R_0} \zeta^{\pm}(y,t), \tag{5.2c}$$

$$v_2^{\pm}(y,t) = \int \frac{dq}{(2\pi)^2} \int_0^t ds\, e^{-s(1+\gamma+q^2)-iq_xR_0-iq_yy}$$

$$\times \int dy'\, e^{iq_yy'}\, [\, \zeta^{+}(y',t-s)\, e^{iq_xR_0} - \zeta^{-}(y',t-s)\, e^{-iq_xR_0}]. \tag{5.2d}$$

The superscript + (−) means that the quantity is defined on the right (left) interface. Substituting (5.2) into (4.8b) yields

$$\tau c(v_I) = \tau c(v_0) + \tau c'(v_0)\,(v_1{}^{\pm} + v_2{}^{\pm}) + O(\zeta^2), \tag{5.3}$$

where $c(v_0)=0$ and $\tau c'(v_0)=-4$. Noting the facts that $V= -\partial_t\zeta^{\pm} + O(\zeta^2)$ and $K= -\partial_y{}^2\zeta^{\pm} + O(\zeta^2)$ we can finally write (4.8a) in terms of $\zeta^{\pm}(y,t)$. By introducing the Fourier components

$$\zeta_k{}^{\pm}(t)= \int dy\; \zeta^{\pm}(y,t)\; e^{iky}, \tag{5.4}$$

it takes the following form up to $O(\zeta)$

$$\tau\frac{d}{dt}\zeta_k(t) = -\varepsilon k^2\zeta_k(t) + \Gamma\zeta_k(t) - \int_0^t ds\; \phi_k(t)\,\zeta_k(t-s) \quad , \tag{5.5a}$$

where

$$\zeta_k(t) = \begin{cases} \zeta_k^{+}(t) \\ \zeta_k^{-}(t) \end{cases} , \tag{5.5b}$$

$$\Gamma = \frac{2}{(1+\gamma)^{1/2}}\,(1 - e^{-2R_0(1+\gamma)^{1/2}}), \tag{5.5c}$$

$$\phi_k(s) = \frac{2}{\sqrt{\pi}}\, e^{-s(1+\gamma+k^2)}\frac{1}{\sqrt{s}} \begin{pmatrix} 1 & -e^{-R_0^2/s} \\ -e^{-R_0^2/s} & 1 \end{pmatrix} . \tag{5.5d}$$

The non-local and time-delayed interaction mediated by $v(\mathbf{r},t)$ is contained in the last term in (5.5a). Note here that the right and the left interfaces can interact each other since the off-diagonal part in $\phi_k(s)$ is finite.

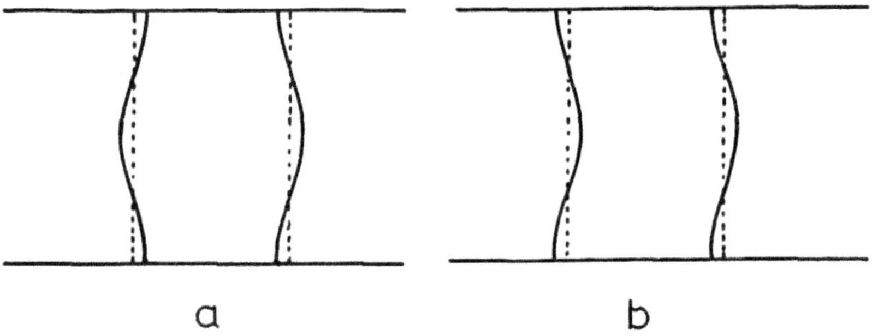

Fig. 5-2. Varicose (a) and zig-zag (b) modes .

We now consider two fundamental deformations. One is the symmetric deformation with respect to the band center. $\zeta_k{}^+ = -\zeta_k{}^- = \zeta_k$. The other is antisymmetric such that $\zeta_k{}^+ = \zeta_k{}^- = \zeta_k$. If the band pattern loses stability for these deformations we call such instabilities a varicose and a zig-zag instability, respectively. These are shown in Fig. 5-2.

In terms of ζ_k Eqs. (5.5) can be written as

$$\tau\frac{d}{dt}\zeta_k(t) = -\varepsilon k^2\zeta_k(t) + \Gamma\zeta_k(t) - \int_0^t ds\, \phi^{\pm}{}_k(t)\,\zeta_k(t-s) \quad , \qquad (5.6a)$$

where

$$\phi_k^{\pm}(s) = \frac{2}{\sqrt{s\pi}}\,(1 \pm e^{-R_0^2/s})\; e^{-s(1+\gamma+k^2)}. \qquad (5.6b)$$

The superscript + (−) means the symmetric (antisymmetric) case. Substituting

$$\zeta_k(t)=ce^{zt} \qquad (5.7)$$

with c a constant into Eq. (5.6a), we have the following eigenvalue equation:

$$\tau z = -\epsilon k^2 + \frac{2}{\sqrt{1+\gamma}}\{1 - e^{-2R_0(1+\gamma)^{1/2}}\}$$

$$- \frac{2}{\sqrt{1+\gamma+k^2+z}}\{1 \pm e^{-2R_0(1+\gamma+k^2+z)^{1/2}}\}. \qquad (5.8)$$

If the real part of z is positive the band-shaped domain is unstable. Note here that the k=0 mode is always neutral (z=0) for the antisymmetric deformation. It corresponds to a uniform translation of the band-shaped domain.

Equation (5.8) indicates that the curvature effect expressed by the first term always tends to stabilize the band-shaped domain while the second and the third terms contain the possibility of destabilization. These terms vanish in the limit $\gamma \to \infty$ where $v(\mathbf{r},t)$ does not exist as is seen from Eq. (1.1b). This implies that the coupling with $v(\mathbf{r},t)$ is indispensable for the instability.

First we consider the case where z is real and changes its sign from negative to positive by changing the external parameters. Solving Eq. (5.8) numerically for ε=0.05 and γ=1/3, we obtain the stability diagram as in Fig. 5-3. The region indicated by z (v) is a region where the band-shaped domain is unstable for the antisymmetric (symmetric) deformation. Thus if we increase the equilibrium band width R_0 by, e.g. changing the parameter a the band-shaped domain undergoes first the zig-zag instablility at $R_0=R_{02}$. In Fig. 5-3 we have plotted the line $R_0=R_{01}$ which is the critical value for the existence of a band-shaped solution. Between this line and the broken line there is a region where the varicose instability occurs. These two lines cannot be obtained from Eq. (5.8) since it is correct only for $\varepsilon \ll R_0$. Actually we have solved rigorously Eq. (1.1) with the piecewise-linear form of f(u) to get the behaviors near $R_0 \approx R_{01}$.

In order to verify the stability diagram in Fig. 5-3 numerical simulations of Eqs. (1.1) is performed with the cubic nonlinearity

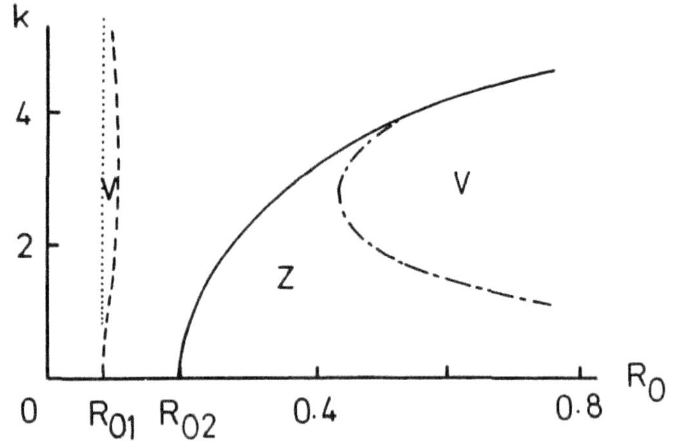

Fig. 5-3. Stability diagram for a band-shaped domain.

(1.2).[51] The system was a rectangular form with the Neumann boundary condition. Figure 5-4a shows that when the band width is small the band-shaped domain is stable for both the symmetric and the antisymmteric deformations. However for larger band width the band-shaped domain becomes unstable first for the antisymmetric deformation and then for the both deformations as shown in Fig. 5-4b and 5-4c. These observations are completely consistent with the theoretical results obtained above.

Now we discuss the reason as to why the band-shaped domain is more unstable by the antisymmetric deformation rather than by the symmetric one. Deformations give rise to a local unbalance of $v(\mathbf{r},t)$ due to the reaction term in Eq. (1.1b). In the + regions in Fig. 5-5 $v(\mathbf{r},t)$ increases locally whereas it decreases in the - regions. This produces the diffusion current of $v(\mathbf{r},t)$ from the + regions to the - regions. If the diffusion does not take place rapidly then the value of $u(\mathbf{r},t)$ decreases (increases) in the + (-) regions because of the reaction in Eq. (1.1a). This acts as a suppression of the deformations. However if the diffusion is strong enough the deformed domain does not have any restoring effects (except for the curvature) so that deformations are enhanced. Comparing the two Figures in Fig. 5-5 we find that the local increase of $v(\mathbf{r},t)$ for the antisymmetric deformation can be smeared out more efficiently than that for the symmetric deformation since the + region in Fig. 5-5a is surrounded by three nearest neighbor - regions whereas in Fig. 5-5b only by two - regions. Thus we note that the zig-zag instability is more preferable than the varicose instability.

Next we show that the band-shaped domain exhibits an oscillatory (breathing) instability for small values of τ. When τ is very small the characteristic time of relaxation of $v(\mathbf{r},t)$ turns out to be very slow compared to the motion of the interface so that an

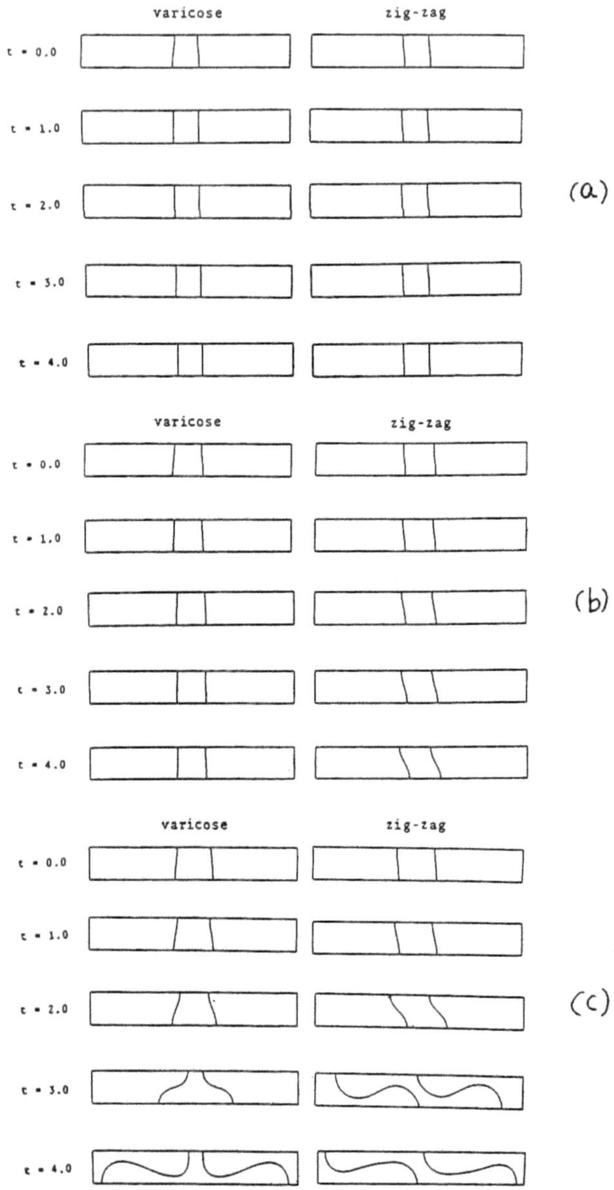

Fig. 5-4. Time-evolution of band-shaped domains.

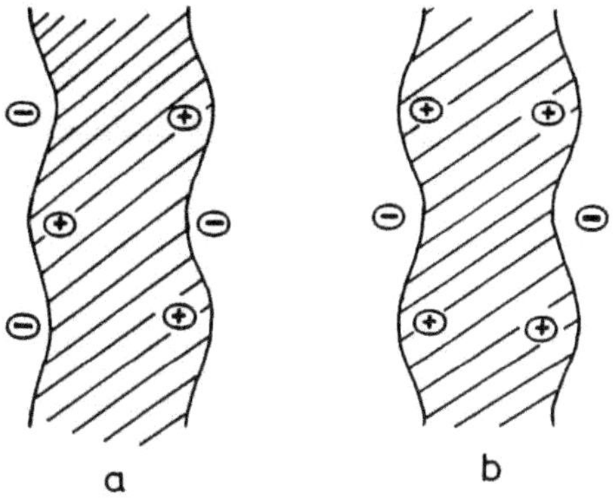

Fig. 5-5. Local unbalance of v associated with the deformations .

overshooting of the interface motion is expected to occur. In fact Eq. (5.8) is shown to exhibit a Hopf bifurcation for $\tau \ll 1$. Figure 5-6 was obtained numerically from Eq. (5.8). It shows that the oscillatory varicose instability emerges first by decreasing τ and the first unstable mode is k=0. This is a sharp contrast to the static instability where the zig-zag instability dominates the varicose instability.

The above fact can also be understood by Fig. 5-5. In order for the overshooting to occur the diffusion of v($\mathbf{r}$,t) have to be very weak. This is realized more easily by the symmetric deformations. Furthermore the most inefficient deformation for the diffusion is the uniform mode with k=0.

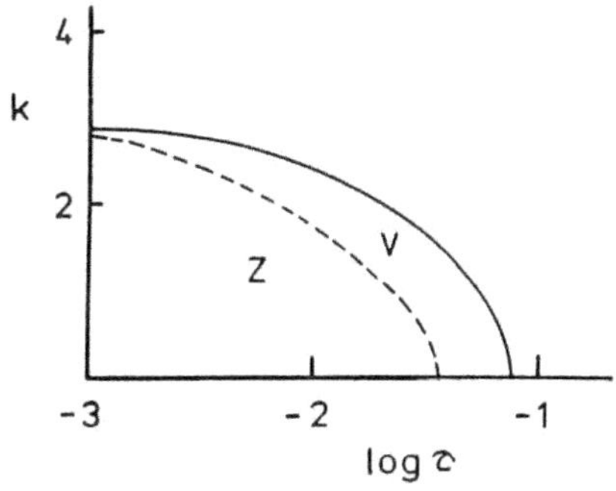

Fig. 5-6. Stability diagram for oscillatory instability.

This breathing instability with k=0 mode was also confirmed by numerical simulations. Koga and Kuramoto[52)] had obtained the breathing instability in a strictly one-dimensional system. It is interesting to see that since the k=0 mode is most unstable their result is not altered even when the deformations along the y-axis is allowed.

The method described above can be applied to a disk-shaped and a spherical domain too. Here we present the result only for the former case. The surface of a deformed disk is represented by the polar coordinate:

$$r=R_0+\zeta(\varphi,t), \tag{5.9}$$

where φ is the azimuthal angle. We assume that $\zeta(\varphi,t)$ is a single-valued function of φ provided that the deformation is sufficiently weak. It is convenient to introduce the Fourier components defined by

$$\zeta_n(t) = \frac{1}{2\pi}\int_0^{2\pi} d\phi\, \zeta(\varphi,t)\, e^{-in\varphi}. \tag{5.10}$$

The linearized equation of motion for $\zeta_n(t)$ is, then, given from (4.8) and (4.10) by

$$\tau\frac{d}{dt}\zeta_n(t) = -\frac{\varepsilon}{R_0^2}\zeta_n(t) + \Gamma\zeta_n(t) - \int_0^t ds\, \phi_n(t)\, \zeta_n(t-s) \quad , \tag{5.11a}$$

where

$$\Gamma = 4R_0 I_1(R_0(1+\gamma)^{1/2})\, K_1(R_0(1+\gamma)^{1/2}), \tag{5.11b}$$

$$\Phi_n(s) = 4R_0 \int_0^\infty dq\, q\, e^{-s(1+\gamma+q^2)}\, (J_n(qR_0))^2. \tag{5.11c}$$

$I_1(x)$ and $K_1(x)$ are the modified Bessel functions and J_n is the Bessel function of the first kind. As a byproduct in the derivation of (5.11) the zeroth order solution for R_0 is also obtained as

$$\tau \frac{dR_0}{dr} = -\frac{\varepsilon(d-1)}{R_0} + \tau c(a, v(R_0)) \tag{5.12}$$

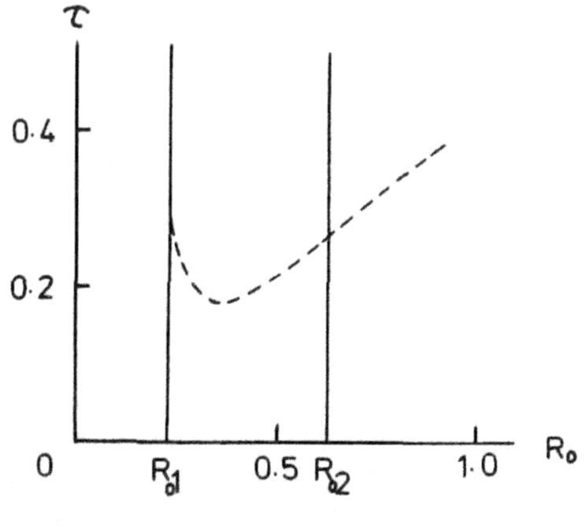

Fig. 5-7. Stability diagram for a disk-shaped domain .

Putting $\zeta_n(t) \propto e^{\lambda t}$ we obtain from (5.11) the eigenvalue equation for λ. The resulting stability diagram is shown in Fig. 5-7. For $R_0 < R_{01}$ a band-shaped domain does not exist. Between R_{01} and R_{02} and when τ is large enough the disk-shaped domain is stable.

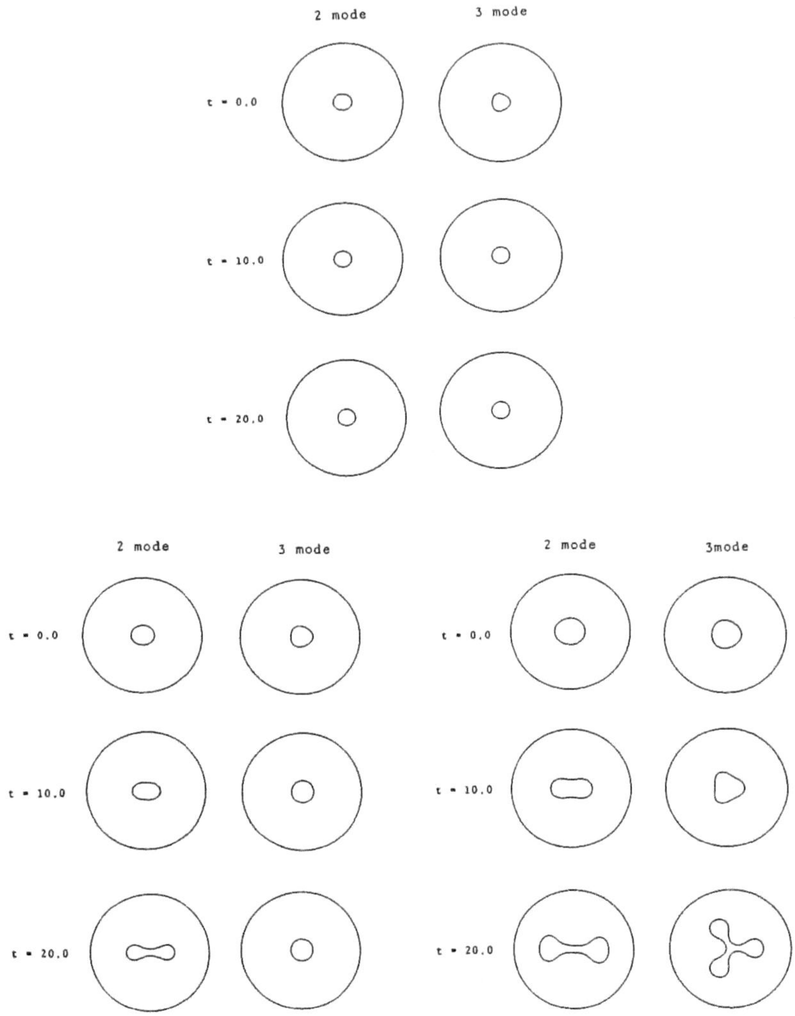

Fig. 5-8. Numerical simulationsof the motion of a disk-shaped domain.

However the domain loses its stability for $R_0 > R_{02}$. The first instability appears for the n=2 mode. The dotted line is the Hopf bifurcation line, below which the disk-shaped domain undergoes a sustained oscillation with the most unstable n=0 mode. The oscillatory instability with larger values of n emerges for smaller τ.

These results were confirmed again by numerical simulations in Fig. 5-8.[51] The disk-shaped domain is indeed stable when its radius is small. However as the radius is increased it becomes unstable first for the n=2 mode distortion and then for the n=2 and 3 distortions. Although it is not shown here the breathing motion of a disk with n=0 has also been found for small values of τ.

Finally we plot in Fig. 5-9 the right hand side of Eq.(5.12) as a function of R_0. This should be compared with Fig. 4-2 for the localized solution of TDGL equation. Equation (5.12) for $a < a_c$ has two equilibrium solutions $R_0 = R_{0u}$ and R_{0s} ($R_{0u} < R_{0s}$). The smaller one R_{0u} is clearly unstable. This fact corresponds to the argument made in section 3 and provides the evidence that only the domain with larger radius is stable.

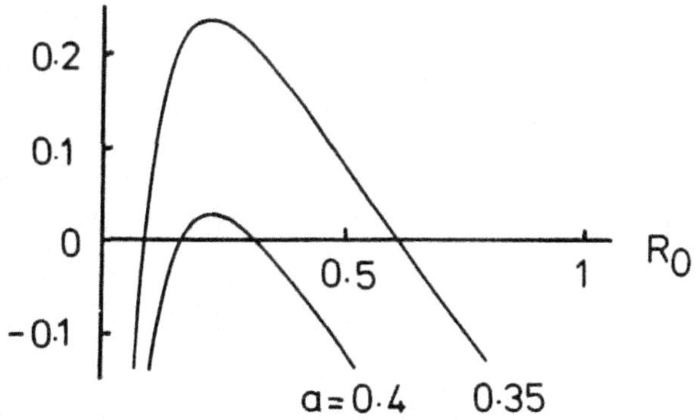

Fig. 5-9. $\tau dR_0/dt$ vs. R_0 .

6. APPLICATION TO CRYSTAL GROWTH

The excitable reaction-diffusion model discussed in the preceding sections has some similarity to the phase field equation for a crystal growth problem.[58)~60)] That is, in the phase field approach $S(\mathbf{r},t)$ represents the phase of the system such that $S(\mathbf{r},t)=1$ (0) expresses the solid (melt) phase. The phase field $S(\mathbf{r},t)$ must be accompanied with the local temperature field $T(\mathbf{r},t)$ since solidification produces latent heat. The model equation constructed in this way is given as follows:

$$\partial_t S = \varepsilon^2 \nabla^2 S + W(S,a(T)) \qquad (6.1a)$$

$$\partial_t T = D\nabla^2 T + L\, \partial_t S, \qquad (6.1b)$$

where $W(S,a)$ has a cubic nonlinearity such as

$$W(S,a(T))=S(1-S)(S-a(T)) \qquad (6.1c)$$

with $0<a(T)<1/2$. The function $a(T)$ is assumed to increase as T increases. The concrete form of $a(T)$ is not necessary at present. The positive constant L in (6.1b) is proportional to the latent heat.

Historically the interfacial approach was first successfully applied to a crystal growth problem.[61)] Since then various morphological instabilities of growing crystals were studied.[62)] More recently substantial progress has been made on the pattern selection in needle crystal and in a dendritic growth.[63)] Although this is a very active current topic we here do not enter the pattern selection problem. What we wish to show is an application of our interfacial approach to a growing crystal modeled by Eqs. (6.1).

One of the features of the phase field description is that one neither assume the steady condition for the diffusion equation for concentration nor employs any thermodynamic conditions such as the Gibbs-Thomson relation as a moving boundary condition.[62] Thus derivation of the interface equation of motion which is free from the quasistatic and the local equilibrium approximations is of fundamental importance. Furthermore since the phase field model has been used for numerical simulations[60] the analytic study by means of the interface equation of motion is necessary to understand morphological instability obtained by simulations.

Let us consider a crystal growing in the direction parallel to the x-axis. The boundary condition is imposed such that $S(\mathbf{r},t)=1$ for $x \to -\infty$ and $S(\mathbf{r},t)=0$ for $x \to \infty$. The thermal diffusion constant D is assumed to be the same in the solid and in the fluid phases. The interface position is denoted by $X(\mathbf{y},t)$ with $\mathbf{y}$, d-1 dimensional vector orthogonal to the x-axis. First let us consider Eq. (6.1a). When $\varepsilon \ll 1$ and D is an ordinary magnitude as we assume hereafter the interface width is very small compared to the characteristic length of the temperature variation. Therefore we may approximate $T(\mathbf{r},t)$ in $a(T)$ by the value $T_I(t)$ just at the interface, which will be determined self-consistently. If we fix T_I at a particular constant value a steady solution of (6.1a) is given by

$$S(\mathbf{r},t)=S_0(x-ct) \tag{6.2}$$

$$-cS_0' = \varepsilon^2 S_0'' + W(S_0,a), \tag{6.3}$$

where

$$c = \varepsilon(1-2a)/\sqrt{2}\,. \tag{6.4}$$

Let $v(a)$ be the constant stationary velocity of a planar interface where T_I is the equilibrium value determined uniquely from Eqs. (6.1). Suppose that this interface is deformed slightly.

Introducing $Z(\mathbf{y},t)$ which denotes a deformation around the planar interface the interface position $X(\mathbf{y},t)$ is written as

$$X(\mathbf{y},t)=v(a)t + Z(\mathbf{y},t). \qquad (6.5)$$

As was done in section 4 when the deformation is weak we approximate $S(\mathbf{r},t)$ by the one-dimensional solution (6.3). Substituting $S(\mathbf{r},t) = S_0(x-X(\mathbf{y},t))$ into (6.1a) and using (6.3) with $a=a(T_I)$ we obtain up to $O(Z)$

$$\partial_t Z(\mathbf{y},t) = \varepsilon^2 \partial_y{}^2 Z + c(a(T_I)) - v(a). \qquad (6.6)$$

Next we have to determine $T_I(t)$ from Eq. (6.1b) under the given configuration $Z(\mathbf{y},t)$. Since this step is quite analogous to the treatment presented in section 4, we do not repeat the derivation. The final result is given up to $O(Z)$ by

$$\partial_t Z_{\mathbf{q}}(t) = -\varepsilon^2 q^2 Z_{\mathbf{q}}$$

$$+ \frac{dv}{dT_I} \frac{L}{\sqrt{4\pi D}} \int_0^t ds \frac{1}{\sqrt{t-s}}\, e^{-v^2(t-s)/4D} \left(\frac{-v^2}{2D}\right) Z_{\mathbf{q}}(t)$$

$$+ \frac{dv}{dT_I} \frac{L}{\sqrt{4\pi D}} \int_0^t ds \frac{1}{\sqrt{t-s}}\, e^{-v^2(t-s)/4D - Dq^2(t-s)} \left(\frac{-v^2}{2D}\right) \left(\frac{\partial Z_{\mathbf{q}}(s)}{\partial s} + \frac{v^2}{2D} Z_{\mathbf{q}}(s)\right).$$

(6.7)

The Fourier component $Z_{\mathbf{q}}(t)$ has been defined as in section 4.

The stability of the moving planar interface can be examined by putting $Z_{\mathbf{q}}(t) \propto e^{\lambda t}$. Equation (6.7) then becomes

$$\lambda = -\varepsilon^2 q^2$$

$$+ (dv/dT_I)L\{-(v/2D) + (v^2 + 4D^2q^2 + 4D\lambda)^{-1/2}(\lambda + (v^2/2D))\}. \tag{6.8}$$

dv/dT_I is determined from (6.4) with $c(a) = v(a)$ and by specifying the form of $a = a(T_I)$. If we simply put $a = T_I$, dv/dT_I is given by

$$dv/dT_I = -\sqrt{2}\,\varepsilon(L + T_\infty), \tag{6.9}$$

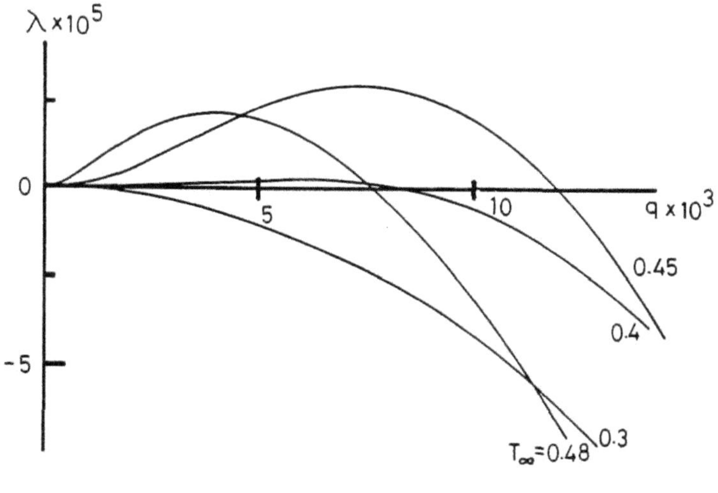

Fig. 6-1. Eigenvalue λ as a function of q for D=10, L=0.01, ε=1 and T_∞=0.3, 0.4. 0.45 and 0.48. The expression (4.8b) with v_I=0 was used for v in this plot. We are indebted to J. Tsunasawa and K. Yamada for their help in solving (6.8) numerically.

where we have used the fact that the steady value of T_l in the absence of deformations is given by $T_l = L + T_\infty$ with T_∞ the temperature at $x=\pm\infty$. Note that the solution $\lambda=q^2=0$ always satisfies Eq. (6.8). This is the translational mode. The eigenvalue λ as a function of q was evaluated numerically. The result is displayed in Fig. 6-1. For T_∞ greater than the particular value the planar interface loses its stability. This is the well-known instability in directional solidification.[64] It is noted here that any oscillatory instabilities do not appear in the model system (6.1).

7. ONE-DIMENSIONAL SYSTEM

In this section we discuss the motion of interface in one-dimensional bistable systems. The interface equation of motion in one-dimension is completely different from those in higher dimensions since the mean curvature is identically zero. We hereafter use the word "kink" instead of interface emphasizing the zero mean curvature.

We start with the TDGL equation with the constant external field ξ:

$$\partial_t u = \varepsilon^2 \partial_x{}^2 u + u(1-u^2) - \xi . \tag{7.1}$$

The non-uniform stable solution of (7.1) takes the forms as shown in Fig. 7-1. When ξ (>0) is finite the kink moves with the velocity

$$V(\xi)=(3/\sqrt{2})\, \varepsilon\, \xi + O(\xi^2). \tag{7.2}$$

This is essentially equivalent with (6.4). We sometimes distinguish the two solutions in Fig. 7-1 by calling a kink (Fig. 7-1a) and an antikink (Fig. 7-1b).

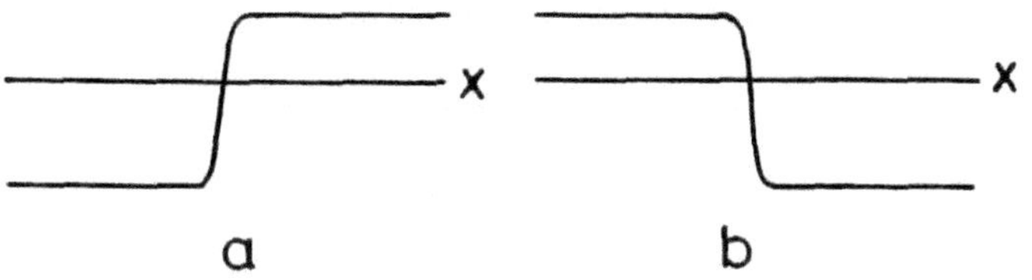

Fig. 7-1. Kink (a) and antikink (b) .

Now let us consider a kink-antikink pair shown in Fig. 7-2 for $\xi=0$. It is readily seen that this solution is linearly unstable. The question is, then, how the domain where $u=1$ evolves with time. This problem has recently been studied by several authors.[65)~68)] See also Ref. 19). What has been found is that there is a short range interaction between the kink and the antikink. In the domain shown in Fig. 7-2 the distance $x_2-x_1=z$ obeys

$$\sigma \, dz/dt = -32 \exp(-\sqrt{2} z/\varepsilon), \tag{7.3}$$

where $\sigma = 2\sqrt{2}/(3\varepsilon)$ is the interfacial energy. Equation (7.3) was derived under the assumption $z \gg \varepsilon$. Thus we know that the kink-antikink pair moves toward each other. Of course the region $z \approx \varepsilon$ cannot be treated correctly by (7.3). But we may expect that the kink-antikink pair annihilates after collision in the dissipative system (7.1). This type of short range interaction is not limited to a kink system but appears in traveling wave pulse solutions in

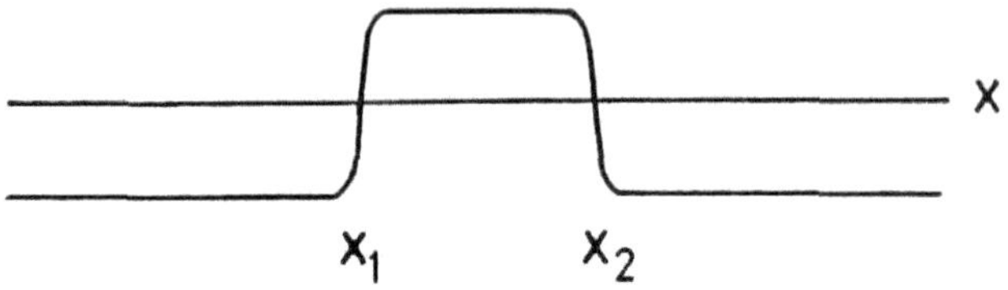

Fig. 7-2. Kink-antikink pair.

more general nonlinear systems. Strictly speaking the short range interaction (7.3) exists in higher dimensions too. However since the curvature effect is more dominant there we have ignored the short range interaction in the analysis presented in sections 4 and 5.

Now we discuss the time-evolution of a kink-antikink system in further detail. As a simple but nontrivial system where the short rang interaction plays an important role we consider the model where ξ in (7.1) is not constant but obeys the following equation

$$\partial_t \xi = \frac{1}{L}\int_0^L u dx - \gamma \xi . \tag{7.4}$$

The system is derived from (1.1) in the situation when the diffusion coefficient D of v is infinity. We consider it for the finite interval (0, L) with the Neumann boundary conditions. In the limit $D \uparrow \infty$, v might be spatially constant, that is, it is a function of t only, say $\xi(t)$. Therefore, integrating (1.1b) with respect to x, we obtain (7.4). Thus the coupled set of equations (7.1) and (7.4) is regarded the limiting system of (1.1) with $\tau = 1/\varepsilon$ as $D \uparrow \infty$.

Suppose that kink and antikink are placed alternatively as in Fig. 7-3 whose positions are denoted by $x_i(t)$ $(i=1,2,\ldots,N)$. We consider the case where the total number N is odd. (The case that N is even can be treated similarly.) If we take account only of the nearest neighbor interaction between the adjacent kink-antikink pair the equation of motion is given for $i=2,\ldots,N-1$ by

$$\frac{d}{dt} x_i = c_0 \{e^{-b(x_{i+1}-x_i)} - e^{b(x_{i-1}-x_i)}\} + V_i(\xi), \tag{7.5a}$$

where

$$c_0 = 16/\sigma$$

$$b = \sqrt{2}/\varepsilon$$

$$V_i(\xi) = (-1)^{i+1}\, V(\xi) \tag{7.5b}$$

and for simplicity we have assumed that the absolute value of ξ is sufficiently small. The Neumann condition is incorporated by putting an extra kink at $x=x_0=-x_1$ and an antikink at $x=x_{N+1}=L+(L-$

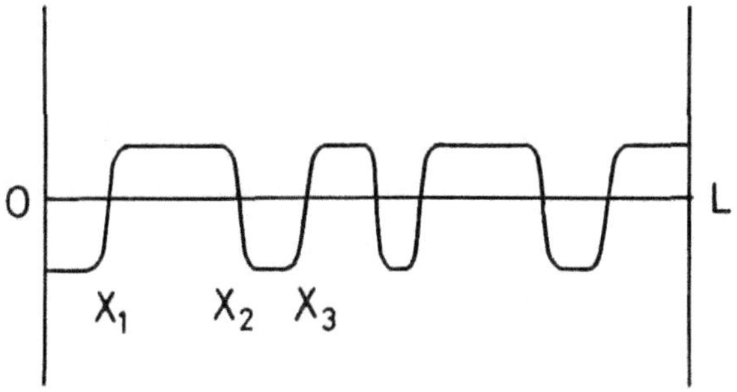

Fig. 7-3. Configuration with kinks and antikinks.

x_N). Then the equations for x_1 and x_N are given respectively by

$$\frac{d}{dt} x_1 = c_0 \{e^{-b(x_2-x_1)} - e^{2bx_1}\} + V_1(\xi), \qquad (7.6a)$$

$$\frac{d}{dt} x_N = c_0 \{e^{-2b(L-x_N)} - e^{b(x_{N-1}-x_N)}\} + V_N(\xi). \qquad (7.6b)$$

In the limit $\varepsilon \to 0$ Eq. (7.4) may be approximated as

$$d\xi/dt = (1/L)\{u_- x_1 + u_+(x_2-x_1) + \ldots + u_+(L-x_N)\} - \gamma\xi, \qquad (7.7)$$

where $u_\pm$ ($u_+ > u_-$) are the steady uniform solutions:

$$u_\pm(1 - u_\pm^2) - \xi = 0. \qquad (7.8)$$

For small values of ξ we have

$$u_\pm = \pm 1 - \xi/2 \qquad (7.9)$$

so that Eq. (7.7) becomes

$$\frac{d\xi}{dt} = \frac{2}{L}\sum_{i=1}^{N} (-1)^i x_i + (1-\frac{\xi}{2}) - \gamma\xi \qquad (7.10)$$

Thus we have the coupled set of equations (7.5), (7.6) and (7.10).

As a special case we examine the linear stability of the configuration where kinks and antikinks are equally spaced. Let us put

$$x_i(t) = (L/N)(i - 1/2) + \delta x_i(t), \qquad (7.11)$$

where $\delta x_i(t)$ is a small deviation. The mirror symmetry at the system boundary x=0 and x=L requires the condition

$$\delta x_0 = -\delta x_1$$

$$\delta x_{N+1} = -\delta x_N . \qquad (7.12)$$

Using (7.11) and (7.12) and the Fourier transform

$$\delta x_i = \sum_{p=0}^{N} x_p \sin \frac{\pi p}{N}(i-\frac{1}{2}) , \qquad (7.13)$$

Eqs. (7.5) and (7.10) can be written respectively as

$$dx_p(t)/dt = 2bc_0e^{-L/N}(1-\cos(\pi p/N))x_p + c_1\delta_{p,N}\xi \qquad (7.14)$$

$$d\xi/dt = -2(N/L)x_N - (\gamma+1/2)\xi \qquad (7.15)$$

with $c_1=(3/\sqrt{2})\,\varepsilon$. Putting x_p, $\xi \propto e^{\lambda t}$ it is readily found that if $N \neq 1$ there are always positive eigenvalues λ given by

$$\lambda = 2bc_0e^{-bL/N}(1 - \cos(\pi p/N)), \qquad 0<p \leq N-1. \qquad (7.16)$$

Thus we note that the most unstable mode is the p=N-1 mode although the value of λ is quite small in magnitude for $\varepsilon \ll L/N$. In Fig. 7-4 the motion of the kinks for the most unstable mode is shown schematically for N=3. On the other hand Eqs. (7.14) and (7.15) tell us that if the system has only one kink (or antikink), i.e., N=1 and the constant γ is not extremely large, all the eigenvalues (or at least the real part of the eigenvalues) are negative. It is stressed that in contrast to (7.16) the absolute value of the negative eigenvalues is an ordinary magnitude. The above result has a simple physical interpretation. Equation (7.5) acts as a kind of conservation law. If x_i $(i=1,\ldots,N)$ are equally spaced, the condition $\xi>0$ (<0) implies

$$\int_0^L u dx < 0 \ (>0). \tag{7.17}$$

Hence we may expect from (7.5) that the equilibrium solution is

$$\xi = \int_0^L u dx = 0. \tag{7.18}$$

Thus the kinks have to move in such a way that the conservation (7.18) is satisfied. The motion for N=3 shown in Fig. 7-4 is consistent with this expectation. Furthermore when N=1 the center of the system is the only stable position of the kink in accord with the conservation.

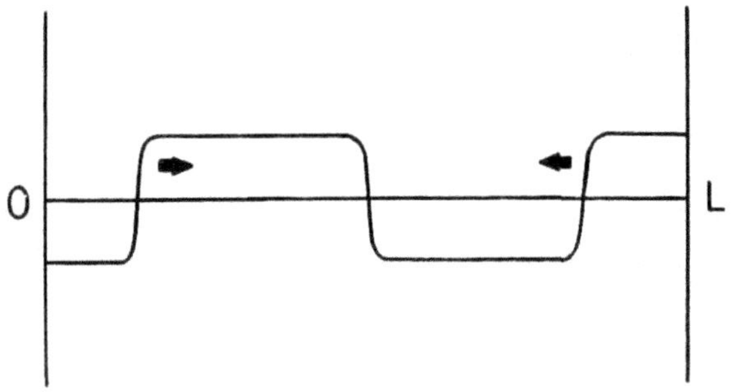

Fig. 7-4 Motion for N=3. The arrows indicate the direction of the velocity .

When γ is sufficiently large, the above argument does not holds. In this case the total value of u integrated over x for $0<x<L$ can be finite although ξ is still very small. In fact, it is shown from (7.14) and (7.15) that one kink solution is linearly unstable for $\gamma > (c_1/2bLc_0)\exp(bL) - 1/2$ and the uniform state is the only stable solution.

8. SUMMARY AND DISCUSSION

We have reviewed mainly our recent studies of pattern evolutions in excitable reaction-diffusion media. The time-evolution of spatially localized motionless patterns was investigated by means of the interface dynamics. The coupling with the extra slow variable $v(\mathbf{r},t)$ was shown to produce a number of instabilities of the patterns. All the predictions were confirmed by numerical simulations.

Here we wish to discuss some outstanding problems.

(1) First of all, our stability analysis is up to linear order of the deviation. It is of great interest to see what is a new stable pattern, if such a solution exists, when a spatially symmetric domain loses its stability. Some of the answers will be obtained at least qualitatively by a standard technique such as the normal form analysis.[69] However if a deformed disk-shaped domain,for instance, shown in Fig. 5-8 breaks up to two disconnected domains this provides us with an interesting problem of self-organization characteristic to the spatially localized pattern.[57]

(2)As was mentioned in section 1 we have not considered the interaction between the localized domains. This interaction arises mediated by the slow variable $v(\mathbf{r},t)$. Dynamics of an assembly of breathing domains has a potential applicability to cooperative biological oscillations. Synchronization of oscillators has recently been studied by a simple model equation.[70)~73)] It is of fundamental importance to derive the interaction in the present model system and to see the relation with the coupled oscillator model.[57)]

(3) It should be noted that the formation of an excited domain has a similarity to the nucleation of a stable domain in a metastable state which occurs, e. g. in the TDGL equation. The difference is that the excited domain cannot grow indefinitely but is blocked at a finite radius. In the ordinary nucleation and growth a statistical theory was developed by Langer[74)] and others[75)] where the theory relies fully on the existence of a thermodynamic functional. This theory should be generalized to the non-variational excitable system with an *external* noise term so that one can treat spontaneous formation of excited domains. This problem is now under study and will be published in the near future.

Finally we make a remark that although we have analyzed the localized domain in the excitable media a localized pattern can exist in wider class of systems. The most well-defined localized patterns observed experimentally are probably the localized traveling waves in the Rayleigh-Benard convection in binary mixtures[76)] and the localized domains of traveling rolls in the electroconvection of a nematic liquid crystal.[77)] An attempt[31)] to show a localized wave has recently been performed based on a complex TDGL equation which is not excitable. As was emphasized in section 1 existence of localized patterns is a characteristic feature far from equilibrium. We hope that in the next few years this fascinating problem will be developed further to understand the pattern formation far from equilibrium.

REFERENCES

1) Nicolis, G. and Prigogine, I., "Self-Organization in Nonequilibrium Systems" (Wiley Interscience, New York, 1977).
2) Fife, P. C., Mathematical Aspects of Reaction and Diffusion Systems", Lecture Notes in Biomathematics 28 (Springer, Berlin,1979).
3) Kuramoto, Y., "Chemical Oscillations, Waves and Turbulence", (Springer, Berlin, 1984).
4) Mimura, M. and Yamaguti, M., Adv. Biophys. **15**, 19 (1982).
5) Field, R. J. and Burger, M., "Oscillations and Traveling Waves in Chemical Systems", (Wily, New York, 1985).
6) Meinhardt, H., "Models of Biological Pattern Formation", (Academic Press, London, 1982).
7) Vidal, C. and Pacault, A. (eds.), "Non-Equilibrium Dynamics in Chemical Systems", (Springer, Berlin, 1984).
8) Ortoleva, P. and Ross, J., J. Chem. Phys. **58**, 5673 (1973).
9) Yamada, T. and Kuramoto, Y., Prog. Theor. Phys. **56**, 681 (1976).
10) Kuramoto, Y., Prog. Theor. Phys. Suppl. No 64, 346 (1978).
11) Pomeau, Y. and Manneville, P., J. de Phys. Lett. **40**, 609 (1979).
12) Cross, M. C. and Newell, A. C., Physica **10D,** 299 (1984).
13) Kuramoto, Y., Prog. Theor. Phys. **71**, 1182 (1984).
14) Ohta, T., Prog. Theor. Phys. **73**, 1377 (1985).
15) Ohta, T. and Kawasaki, K., Physica **27D**, 21 (1987).
16) Sivashinsky, G. I., Acta Astronautica **4**, 1177 (1977) and **6**, 569 (1979).
17) Kawasaki, K. and Ohta, T., Physica **118A**, 175 (1983).
18) Tyson, J. J. and Keener, J. P., Physica **32D**, 327 (1988) and the references cited therein.
19) Kuramoto, Y., in the Proceedings for the International Conference on "Spatial Inhomogeneities and Transient Behavior in Chemical Kinetics", Brussels, 1987 (Manchester University Press,

1988).

20) McKean, H. P., Adv. Math. **4**, 209 (1979).

21) FitzHugh, R., Biophys. J. **1**, 445 (1961).

22) Nagumo, J., Arimoto, S. and Yoshizawa, S., Proc. IRE **50**, 2061 (1962).

23) Gunton, J. D., San Miguel, M. and Sahni, P. S., in Phase Transitions and Critical Phenomena, vol 8, C. Domb and J. L. Lebowitz, eds. (Academic Press, New York, 1983).

24) Allen, S. M. and Cahn, J. W., Acta Metal. **27** 1085 (1979).

25) Ohta, T., Jasnow, D. and Kawasaki, K., Phys. Rev Letters **49**, 1223 (1982).

26) Graham, R. and Haken, H., Z. Physik **213**, 420 (1968).

27) Graham, R. and Haken, H., Z. Physik **237**, 31 (1970).

28) Kawasaki, K., Ann. Phys. **154**, 319 (1984).

29) Ohta, S., Ohta, T. and Kawasaki, K., Physica **128A**, 1 (1984).

30) Coleman, S., "Classical Lumps and Their Quantum Descendants", Lectures delivered at the 1975 International School of Subnuclear Physics "Ettore Majorana".

31) Thual, O. and Fauve, S., J. Phys. France **49**, 1829 (1988).

32) Siggia, E. and Zippelius, A., Phys. Rev. **A26**, 1788 (1982).

33) Siggia, E. and Zippelius, A., Phys. Rev. **A24**, 1036 (1981).

34) Pomeau, Y., Zaleski., S. and Manneville, P., Phys. Rev. **A27**, 2710 (1983).

35) Belousov, B. P., in Collections of Abstracts on Radiation Medicine (Medgiz, Moscow, 1958).

36) Zhabotinskii, A. M., Biophysics **9**, 329 (1964).

37) Tyson, J. J. and Fife, P. C., J. Chem. Phys. **73**, 2224 (1980).

38) Zaikin, A. N. and Zhabotinskii, A. M., Nature **225**, 535 (1970).

39) Winfree, A. T., Science **175**, 634 (1972).

40) Zhabotinskii, A. M. and Zaikin, A. N., J. Theor. Biol. **40**, 45 (1973).

41) J. Phys. Chem. **93**, No. 7 (1989).

42) Field, R. J., Koros, E. and Noyes, R. M., J. Am. Che. Soc. **94**, 8649 (1972).

43) Kuramoto, Y. and Yamada, T., Prog. Theor. Phys. **56**, 724 (1976).

44) Spear, D. J., Barr, J. K. and Barr, C. E., J. Gen. Physiol. **54**, 397 (1969).

45) Lucas, W. J. and Smith, F. A., J. Exp. Bot. **24**, 1 (1973).

46) Toko, K., Chosa, H. and Yamafuji, K., J. Theor. Biol. **114** 127 (1985).

47) Toko, K., Nosaka, M., Fujiyoshi. T., Yamafuji. K. and Ogata, K., Bull. Math. Biol. **50**, 255 (1988).

48) Mimura. M., unpublished work.

49) Radehaus, Ch., Dirksmeyer, T., Willebrand, H. and Purwins, H.-G., Phys. Letters **125**, 92 (1987).

50) Purwins, H.-G., Radehaus, Ch., Dirksmeyer, T., Dohmen, R., Schmeling. R. and Willebrand, H., Phys. Letters **136**, 480 (1989).

51) Ohta, T., Mimura, M. and Kobayashi, R., Physica **34D**, 115 (1989).

52) Koga, S. and Kuramoto, Y., Prog. Theor. Phys. **63**, 106 (1980).

53) Nishiura, Y. and Mimura. M., to appear in SIAM J. Appl. Math.

54) Fife, P. C., SIAM-AMS Proc. **10**, 23 (1976).

55) Ermentrout, G. B., Hastings, S. P. and Troy, W. C., SIAM J. Appl. Math. **44**, 1131 (1984).

56) Rinzel, J. and Terman, SIAM J. Appl. Math. **42**, 1111 (1982).

57) Ikeda, T., Movie of numerical simulations on "Interfacial Dynamics".

58) Caginalp, G., Ann. Phys. **172**, 136 (1986).

59) Collins, J. B. and Levine, H., Phys. Rev. **B31**, 6119 (1985).

60) Kobayashi, R. and Mimura, M., unpublished.

61) Burton, W. K., Cabrera, N. and Frank, F. C., Phil. Trans. Roy. Soc. London **A243**, 290 (1951).

62) Langer, J. S., Rev. Mod. Phys. **52**, 1 (1980).

63) Kessler, D. A., Koplik, J. and Levine, H., Adv. Phys. **37**, 255

(1988).

64) Mullins, W. W. and Sekerka, R. F., J. Appl. Phys. **34**, 323 (1963).

65) Kawasaki, K. and Ohta, T., Physica **116A**, 573 (1982).

66) Coullet, P. and Elphick, C., Phys. Letters **121A**, 233 (1987).

67) Elphick, C., Meron, E. and Spiegel, E. A., Phys. Rev. Lett. **61**, 496 (1988).

68) Carr, J. and Pego, R. L., preprints.

69) Guggenheim, J. and Holmes, P., "Nonlinear Oscillations, Dynamical Systems and Bifurcations of Vector Fields" (Springer, New York, 1983).

70) Winfree, A. T., "The Geometry of Biological Time", Biomath. 8 (Springer, Berlin, 1980).

71) Kuramoto, Y and Nishikawa, I., J. Stat. Phys. **49**, 569 (1987).

72) Daido, H., Phys. Rev. Letters **61**, 231 (1988).

73) Strogatz, S. H. and Mirollo, R. E., Physica **D31**, 143 (1988).

74) Langer, J. S., Ann. Phys. **54**, 258 (1969).

75) Landauer, R. and Swanson, J. A., Phys. Rev. **121**, 1668 (1961).

76) Kolodner, P., Bensimon ,D. and Surko, C. M., Phys. Rev. Letters, **60**, 1723 (1988).

77) Joets, A. and Ribotta, R., Phys. Rev. Letters, **60** (1988) 2164.

78) Cassie, R. M., Oceanogr. Mar. Biol. Ann. Rev. **1** (1963) 223.

79) Steele, J., in "Ecological Stability" M. B. Usher and M. H. Williamson, eds. (Chapman and Hall, London, 1973) 179.

80) Okubo, A., Biomathematics Vol. 10, (Springer, 1980).

81) May, R., "Stability and Complexity in Model Ecosystems" (Prinston University Press, 1973).

STATISTICS AND GEOMETRY OF RANDOM INTERFACE SYSTEMS

Hiroyuki Tomita

College of General Education, Kyoto University, Kyoto

ABSTRACT

Statistical properties of the random interface system which is typical at the late stage of the phase separation process are studied, and used to discuss singularity and symmetry of its correlation function and structure function. Important characteristics for such system are curvature invariants. In order to calculate and illustrate them a statistical model of the random interface generated by stochastic field is introduced. Based on the results of this calculation, a phase transition phenomenon of mean topology of the random interface system is discussed and related to the percolation problem in continuous media. The percolation limit is determined as a scale invariant point.

§ 1. INTRODUCTION

The phase separation process in binary mixtures after quenching from homogeneous phase serves as a typical example of pattern formation at far-from-equilibrium situations.[1] The particular points of pattern formation in this system are the randomness of the pattern and its continual growth due to following two mechanisms included in the dynamics: One is the thermodynamical instability, which causes unstable growth of thermal fluctuations of the concentration of mixed components. Another is the nonlinear saturation effect, i.e. the concentration in each separated phase approaches respective equilibrium value at the state of thermodynamical coexistence. An interesting problem at the early stage of the process is how local heterogeneity arises from initial homogeneity. At the late stage, the evolution of the random pattern formed of almost saturated coexisting domains is concerned with. For example there exists an apparent scaling property[2] in the course of evolution: The pattern at any time in this stage is statistically equivalent with each other if it is scaled with a proper, time-dependent characteristic length. In this paper, however, these dynamical phenomena are not taken up. The main part of the present work is devoted to investigate general statistical aspects of this random pattern system.

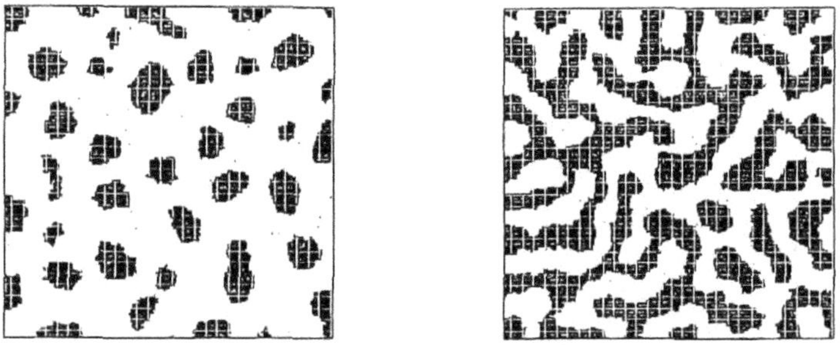

a b

Fig.1 Typical random patterns in phase separating process with volume fraction **a**: 1/4 and **b**: 1/2.

The random, smooth and semi-macroscopic interface picture is useful in investigating the late stage process of phase separation.[3),4] Nonlinearity included in its dynamics is essentially incorporated in the theory by assuming sharp profile of the interface between separated phases. The conventional droplet theory[5] is an example which successfully has employed this picture and has explained nonlinear evolution of a binary system with small volume fraction of the minority phase.

Let us briefly review the conventional notion of the interface separating the coexisting phases by considering a simplified model of a binary alloy system composed with A and B atoms:

The microscopic Hamiltonian[1] of this system is given by the following

Ising model on a N-site lattice;

$$H = (1/4) \sum_{(ij)} \{u_{AA}(1+s_i)(1+s_j) + u_{AB}(1+s_i)(1-s_j) + u_{BA}(1-s_i)(1+s_j) + u_{BB}(1-s_i)(1-s_j)\} \tag{1.1}$$

$$= -J \sum_{(ij)} s_i s_j - h \sum_i s_i + \text{constant} , \tag{1.2}$$

where the variable s_i is defined by

$$s_i = \begin{cases} +1 \text{ if the site } \mathbf{r}_i \text{ is occupied by A atom,} \\ -1 \text{ if the site } \mathbf{r}_i \text{ is occupied by B atom,} \end{cases} \tag{1.3}$$

and (ij) denotes the nearest pair of neighboring sites. The interaction coefficients J and h in Eq.(1.2) are related to the inter-atomic energy constants u_{AA}, u_{BB} and u_{AB} in Eq.(1.1) as

$$J = -(1/4)[u_{AA} + u_{BB} - 2u_{AB}] ,$$

and

$$h = (z/4)[u_{BB} - u_{AA}] ,$$

where z is the coordination number of the lattice. In the case of binary mixtures, we have an additional constraint

$$\overline{s} = (1/N) \sum_i s_i = (N_A - N_B)/N = \text{constant} , \tag{1.4}$$

where N_A and N_B are the numbers of A and B atoms, respectively, and $N_A+N_B=N$.

The interface picture is a macroscopic notion, and we need thermodynamics corresponding to this microscopic model. This is performed by introducing the coarse grained variable, i.e. the local concentration

$$s(\mathbf{r}) = (1/v_L) \Sigma_L s_i , \tag{1.5}$$

and the coarse grained free energy functional

$$F\{s(\mathbf{r})\} = -k_BT \log \mathrm{Tr}' \exp(-H/k_BT) \ , \tag{1.6}$$

where Σ_L is a summation within a semi-macroscopic cell located at $\mathbf{r}$, its size L or volume v_L being chosen properly. Tr' denotes the constrained trace over $s_i = \pm 1$ with the condition $(1/v_L)\Sigma_L s_i = s(\mathbf{r})$ for all $\mathbf{r}$. However, we cannot perform this coarse graining calculation explicitly in general. Instead we usually start with a phenomenological form given by

$$F\{s(\mathbf{r})\} = \int [(K/2)(\nabla s(\mathbf{r}))^2 + f(s(\mathbf{r}))]\, d\mathbf{r}, \tag{1.7}$$

where the first term in the integrand is due to spatial inhomogeneity and the coefficient K is to be related to the surface tension later. The second part is the homogeneous part and usually approximated by

$$f(s) = f_0(s) + hs \ ,$$

where

$$f_0(s) = -(1/2)a\, s^2 + (1/4)g\, s^4 \ . \tag{1.8}$$

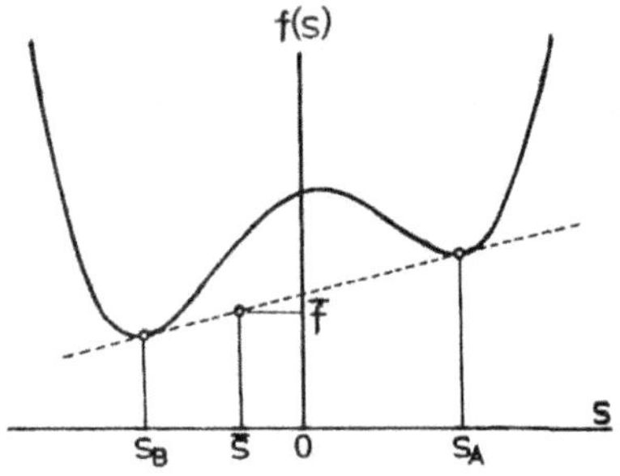

Fig.2 Free energy f(s) for uniform system.

This is the well known Ginzburg-Landau free energy. The same formulation was given by Cahn and Hilliard[8] in the phase separation problem, independently. The coefficients a and g are positive below the Curie temperature T_c, because the free energy f(s) should have a common tangent at two points, say $s=s_A$ and $s=s_B$, which correspond to the equilibrium concentrations of the coexisting A-rich and B-rich phases, respectively. In the present simplified model, s_A and s_B are given by

$$s_A = +\sqrt{a/g} \quad \text{and} \quad s_B = -\sqrt{a/g}\,, \tag{1.9}$$

because using these values one finds the common tangent whose gradient is given by

$$\bar{\mu} = f'(s_A) = f'(s_B) = (f(s_A)-f(s_B))/(s_A-s_B) = -h\ .$$

Define the interface free energy by

$$\Delta F = \int \{[(K/2)(\nabla s(\mathbf{r}))^2 + f(s(\mathbf{r}))] - \bar{f}\}\, d\mathbf{r}\ , \tag{1.10}$$

where $\bar{f} = f(s_B) + \bar{\mu}(\bar{s}-s_B)$ is the equilibrium mean free energy shown in Fig.2 and can be rewritten as

$$\bar{f} = (1/V)\int [f(s_B) + \bar{\mu}(s(\mathbf{r})-s_B)]\, d\mathbf{r}\ . \tag{1.11}$$

Here the new definition of the mean concentration

$$\bar{s} = (1/V)\int s(\mathbf{r})\, d\mathbf{r}\ , \tag{1.12}$$

is used, where V is the volume of the total system. The coexisting state with separating interfaces is attainable by minimizing this excess free energy. This condition yields the following interface profile equation

$$-K\nabla^2 s(\mathbf{r}) + df_0(s(\mathbf{r}))/ds(\mathbf{r}) = 0\ . \tag{1.13}$$

The simplest solution of it is the plane interface whose profile equation is given by

$$-K(d^2s/d\zeta^2) + df_o(s(\zeta))/ds(\zeta) = 0 , \tag{1.14}$$

or equivalently given by an integrated form

$$(K/2)(ds/d\zeta)^2 - f_o(s(\zeta)) = \text{constant} , \tag{1.15}$$

where ζ is the coordinate perpendicular to the plane. Using the explicit expression Eq.(1.8) for $f_o(s)$ and assuming the boundary conditions $ds/d\zeta=0$ at $\zeta=\pm\infty$, $s(-\infty)=s_B$ and $s(+\infty)=s_A$, one finds

$$s(\zeta) = \sqrt{a/g}\tanh(\zeta/2\xi) , \tag{1.16}$$

where

$$\xi = \sqrt{K/2a} , \tag{1.17}$$

which is used as the thickness parameter of the interface as is shown in Fig.3. Note that this ξ is equivalent to the correlation length of fluctuations of $s(\mathbf{r})$.[7] The plane interface solution Eq.(1.16) yields the resultant interface free energy[6),8]

$$\begin{aligned} \Delta F &= K\int (ds/d\zeta)^2\, d\mathbf{r} \\ &= (Ka/2g\xi)\int da \int_{-\infty}^{+\infty}[1-\tanh^2(\zeta/2\xi)]^2\, d\zeta/2\xi \\ &= (2Ka/3g\xi)\int da , \end{aligned} \tag{1.18}$$

where da is the interface area element. The coefficient in the last expression is the surface tension τ itself, which is defined as the interface free energy per unit area, i.e.

$$\tau = 2Ka/3g\xi = (1/6)\times[2a(\Delta s)^2\xi], \tag{1.19}$$

where

$$\Delta s = s_A - s_B = 2\sqrt{a/g} \tag{1.20}$$

is the so-called order parameter gap and

$$2a = f''(s_A) = f''(s_B) .$$

Note that the profile shape given by Eq.(1.16) is expected to be insensitive to the coarse graining cell size L if we choose it properly. It should be much larger than the lattice constant but not much larger than ξ, so that each cell is contained in one of the single phase regions.[1]

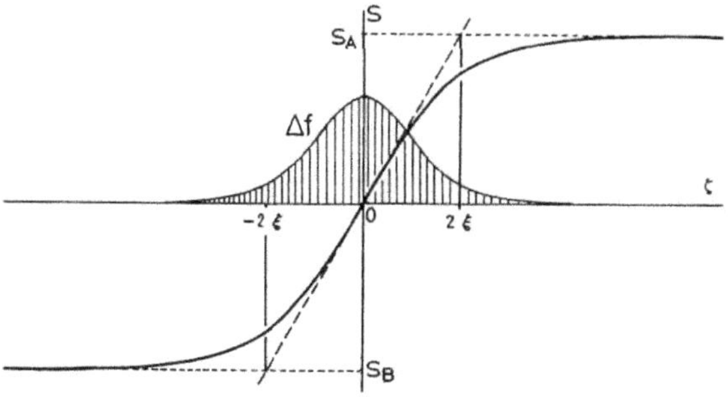

Fig.3 Interface profile and excess free energy Δf.

The semi-macroscopic interface means the intermediate stage, i.e. not the last stage of phase separation: The characteristic length scale of it is not macroscopic yet, but is sufficiently larger than microscopic length scale ξ, the thickness of the interface defined above. At this stage the local profile of the interface is well approximated by Eq.(1.16). A useful length scale to characterize this interface system is the inverse of the interface area density[8] defined by

$$A = (1/V) \int da . \tag{1.21}$$

Then the system considered here is particularized by

$$\xi \ll A^{-1} \ll V^{-1/3} . \tag{1.22}$$

The smoothness of the interface is due to the effect of surface tension. If a singular position such as edge, corner, intersection and contact etc. appears on the interface, it disappears immediately by evaporation, or coalescence. Then one may expect that the curvature is finite almost everywhere on the interface. In § 2, the spatial averages of some invariants constructed with the principal curvatures are shown to be important characteristics as well as the interface density A. It should be noted that, when the smoothing process due to surface tension cannot follow the growing process, one may find a kind of fractal surface,[9] where curvatures cannot be defined anywhere. This problem is also mentioned in § 2 partially.

In addition to above properties, randomness of the interface is assumed. This means that, starting from isotropic and homogeneous initial state, the symmetry is preserved globally at this intermediate stage, though appearance and growth of local heterogeneity are considered in the phase separation problem. Here the notion of random interface does not always mean a complicated and extended object. Randomly scattered spherical droplets system, even mono-dispersive, is a typical example of this category.

General statistical treatment for random interface systems seems to be difficult, since it is ambiguous what ensemble should be chosen at our starting point of the statistical theory. Fortunately, however, the interface concerned with in the present problem has definite foundation: The interface is defined as a curved surface in the Euclidean space, i.e. as a sub-set of points by

$$\{ \mathbf{r} \mid s(\mathbf{r}) = 0 \}, \tag{1.23}$$

where $s(\mathbf{r})$ is the coarse grained concentration at $\mathbf{r}$. This definition brings a random interface system, because the concentration $s(\mathbf{r})$ is fluctuating quantity in general.

Now modeling is very simple: Let $\{u(\mathbf{r})\}$ be a stochastic scalar field in the Euclidean space of arbitrary dimension, its statistics being properly chosen. Then define sub-sets $\mathbf{U}$ and $\partial\,\mathbf{U}$ as follows;

$$\mathbf{U} = \{ \mathbf{r} \mid u(\mathbf{r}) < U \}, \tag{1.24}$$

and

$$\partial \mathbf{U} = \{ \mathbf{r} \mid u(\mathbf{r}) = U \}. \tag{1.25}$$

The sub-sets $\mathbf{U}$ and $\partial \mathbf{U}$ are called an excursion set and its boundary set,[10] respectively. In other words, $\mathbf{U}$ and $\partial \mathbf{U}$ are 'sea' and 'coast' in a random topography $\{u(\mathbf{r})\}$.[11] As is shown in § 3, expectations of the area density and curvature invariants of $\partial \mathbf{U}$ are calculable, at least in principle, by using statistics of $\{u(\mathbf{r})\}$. We have an exactly tractable model, i.e. a Gaussian random field, which plays an important role in the present problem. Many works on the dynamical theory of the ordering processes are in this frame, i.e. the concentration field is approximated by a linearized Gaussian field[12)-15)] and the nonlinearity included in its dynamics is incorporated in a separate way, e.g. by a nonlinear transformation of field variables[12),13)] or by a dynamical saturation effect.[15)]

Note that the excursion set $\mathbf{U}$ is composed with scattered random manifolds. Each manifold has the simplest geometry as a 'pond' or a 'lake', i.e. has closed, oriented surface(s) around itself. Exception is an infinitely extended object, say an 'ocean', which is expected to appear at a certain value of the water level U. This makes a general percolation problem in continuous media. This is discussed in § 4 with relation to the statistical geometry.

§ 2. SINGULARITY AND SYMMETRY OF THE CORRELATION FUNCTION

2-1 The Porod Law and an Additional Sum-Rule for Structure Function

Hereafter let us assume a sharp profile of the interface, i.e. $\xi \to 0$ and use the normalized concentration variables $\{c(\mathbf{r})\}$ defined by

$$c(\mathbf{r}) = \begin{cases} 1 & \text{in A-rich phase region,} \\ 0 & \text{in B-rich phase region,} \end{cases} \tag{2.1}$$

instead of $\{s(\mathbf{r})\}$. A statistically significant quantity is the correlation function

$$g(r) = (1/V) \int c(\mathbf{r}_1)c(\mathbf{r}_1+\mathbf{r})\, d\mathbf{r}_1 - \bar{c}^{\,2}$$

$$= \overline{c(0)c(\mathbf{r})} - \bar{c}^{\,2} , \qquad (2.2)$$

where

$$\bar{c} = (1/V) \int c(\mathbf{r})\, d\mathbf{r} = \phi$$

is the volume fraction (of A-rich phase) and the bar $\bar{}$ denotes the spatial average. Note that global uniformity and isotropy are assumed here as is mentioned in § 1. It is more advantageous to use

$$g_1(r) = \overline{\nabla c(0)\cdot \nabla c(\mathbf{r})}$$

$$= -\nabla^2 g(r) , \qquad (2.3)$$

instead of $g(r)$ itself for the purpose of finding its explicit expression with use of interface informations, because $\nabla c(\mathbf{r})$ has non-zero value only just on the interface. This new vector-field variable is rewritten as

$$\nabla c(\mathbf{r}) = \mathbf{n}(\mathbf{r})\, \delta(\zeta(\mathbf{r})) , \qquad (2.4)$$

where $\mathbf{n}(\mathbf{r})$ is the normal unit vector at $\mathbf{r}$ on the interface and $\zeta(\mathbf{r})$ the normal coordinate in this direction with conditions $\zeta=0$ and $|\nabla\zeta|=1$ at the interface position, and $\delta(\zeta)$ is the ordinary delta function. With use of Eqs.(2.3),(2.4) one finds that

$$4\pi r^2\, g_1(r)\, dr = (1/V) \int da_1 \int{}' (\mathbf{n}_1\cdot\mathbf{n}_2)\, da_2 , \qquad (2.5)$$

where $\mathbf{n}_1$, $\mathbf{n}_2$ and da_1, da_2 are the normal unit vectors and the area elements of the interface at $\mathbf{r}_1$ and $\mathbf{r}_2 = \mathbf{r}_1 + \mathbf{r}$, respectively. The surface integration $\int{}'$ is restricted in the region of the interface between two concentric spheres centered at $\mathbf{r}_1$ with radii r and $r+dr$. Let ds_2 and

ds_2' be line elements on the interface in $(\mathbf{n}_2, \mathbf{r})$ plane and $\mathbf{n}_2 \times \mathbf{r}$ direction respectively, as is shown in Fig.4. Using $\mathbf{n}_2\, da_2 = ds_2 \times ds_2'$, $dr = ds_2 \sin\alpha_2$ and then $da_2 = ds_2' dr/\sin\alpha_2$, where α_2 is the angle between $\mathbf{n}_2$ and $\mathbf{r}$, one finds another expression for Eq.(2.5),

$$4\pi r^2 g_1(r)dr = (1/V) \int da_1 \oint_{C(r)} (\mathbf{n}_1 \cdot \mathbf{n}_2)\, ds_2'/\sin\alpha_2, \tag{2.6}$$

where C(r) is the curved line on the interface attained as the intersection by the sphere centered at $\mathbf{r}_1$ with radius r, and $\oint_{C(r)}$ denotes line integration along C(r). In the case of the smooth interface system we have

$$\mathbf{n}_1 \cdot \mathbf{n}_2 \to 1 \quad \text{and} \quad \alpha_2 \to \pi/2 \quad \text{when} \quad r \to 0 ,$$

then the line integration in Eq.(2.6) is approximated by

$$\oint_{C(r)} (\mathbf{n}_1 \cdot \mathbf{n}_2)\, ds_2'/\sin\alpha_2 \simeq 2\pi r . \tag{2.7}$$

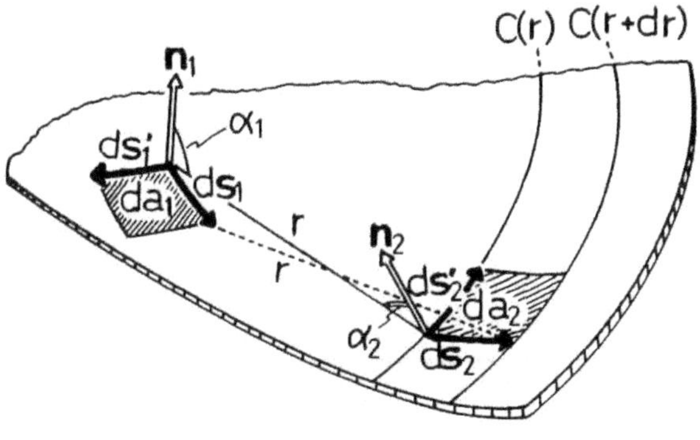

Fig.4 The geometry of interface area elements.

Let us show that the right-hand side of Eq.(2.6) is an odd function[16] of $\mathbf{r} = \mathbf{r}_2 - \mathbf{r}_1$, then of $r = |\mathbf{r}|$. This means that the expression is antisymmetric with respect to suffices 1 and 2. This can be shown as follows: Let $\mathbf{ds}_1$ be in ($\mathbf{r}$, $\mathbf{n}_1$) plane and $\mathbf{ds}_1'$ in $\mathbf{n}_1 \times (-\mathbf{r})$ direction so that $\mathbf{n}_1 da_1 = \mathbf{ds}_1' \times \mathbf{ds}_1$ and $|\mathbf{r} + \mathbf{ds}_2 - \mathbf{ds}_1| = r$. Then one finds following symmetric relation;

$$dr = ds_1 \sin\alpha_1 = ds_2 \sin\alpha_2 \ . \tag{2.8}$$

The anti-symmetric property becomes more apparent if one uses the notation of exterior products of differentials[17] for the triple integral element in Eq.(2.7) as follows;

$$\frac{da_1 ds_2'}{\sin\alpha_2} = \frac{ds_1' \wedge ds_1 \wedge ds_2'}{\sin\alpha_2} \underset{(1\leftrightarrow 2)}{\rightarrow} \frac{ds_2' \wedge ds_2 \wedge ds_1'}{\sin\alpha_1}$$

$$= \frac{ds_2' \wedge ds_1 \wedge ds_1'}{\sin\alpha_2} = -\frac{ds_1' \wedge ds_1 \wedge ds_2'}{\sin\alpha_2} \ , \tag{2.9}$$

where $\wedge$ denotes the exterior product and the symmetric relation Eq.(2.8) is used.

According to the above considerations the following general formulae for $g_1(r)$ and $g(r)$ are obtained;

$$g_1(r) = (A/2r) \times \{\text{even function of } r\} \ ,$$

and

$$\lim_{r\to 0} [\ \nabla^2 g(r) + A/2r\ ] = 0 \ , \tag{2.10}$$

where the definition of the area density Eq.(1.21) is used.

It is easily seen in the above derivation that the coefficient 1/2 (= $2\pi/4\pi$) of A/r should be replaced by $S(D-1)/S(D)$ for general dimension D, where $S(D)$ denotes the surface area of a unit sphere in D-dimensional Euclidean space. Then for D-dimensional smooth, random interface system we have

$$\lim_{r\to 0} [\ \nabla^2 g(r) + ((D-1)\gamma_D/4)\ A/r\ ] = 0 \ , \tag{2.11}$$

where the Gamma-function expression for S(D) are used, i.e.

$$\gamma_D = 2\,\Gamma(D/2)/[\Gamma(1/2)\Gamma((D+1)/2)] \tag{2.12}$$

$$= \begin{cases} 4/\pi & \text{for } D=2\ , \\ 1 & \text{for } D=3\ . \end{cases}$$

It is straightforward to obtain the following formula on the symmetry of g(r) by integrating Eq.(2.11) with use of

$$\nabla^2 = d^2/dr^2 + ((D-1)/r)(d/dr)$$

for isotropic situation and $g(0) = \phi(1-\phi)$;

$$g(r) = \phi(1-\phi) - (\gamma_D/4)\ A\ r \times G(r), \tag{2.13}$$

where G(r) is an even function of r and G(0)=1.

It should be noted that these symmetry and singularity can be observed by the small angle neutron or X-ray scattering experiments, if we re-formulate them with use of the structure function defined by

$$S(q) = \int g(r)\exp(i\mathbf{q}\cdot\mathbf{r})\ d\mathbf{r}\ , \tag{2.14}$$

which is normalized by the sum-rule

$$(2\pi)^{-D}\int S(q)\ d\mathbf{q} = g(0) = \phi(1-\phi)\ . \tag{2.15}$$

It can be easily shown with use of the formulae on the asymptotic expansions of Fourier transformation

$$\int r^N \exp(i\mathbf{q}\cdot\mathbf{r})\ d\mathbf{r}$$

$$\simeq -2^{D+N}\pi^{D/2-1}\,\Gamma((N+2)/2)\,\Gamma((N+D)/2)\,\sin(N\pi/2)\ q^{-(N+D)}\ , \tag{2.16}$$

that, corresponding to the singularity in $g(r)$ (or $g_1(r)$) due to the sharp profile of interface ($\xi \to 0$), there appears a long tail in $S(q)$ expressed as

$$\lim_{q \to \infty} q^{D+1} S(q) = \beta_D A \quad , \tag{2.17}$$

where

$$\begin{aligned} \beta_D &= (\gamma_D/4) \times (4\pi)^{D/2}\, \Gamma((D+1)/2)/\sqrt{\pi} \\ &= 2(4\pi)^{(D-2)/2}\, \Gamma(D/2) \qquad (2.18) \\ &= \begin{cases} 2 & \text{for } D=2 \,, \\ 2\pi & \text{for } D=3 \,. \end{cases} \end{aligned}$$

This corresponds exactly to the well-known Porod law.[18] Besides of it one finds another sum-rule formula

$$\int_0^{\infty} [\, q^{D+1} S(q) - \beta_D A \,]\, dq = 0 \quad , \tag{2.19}$$

which corresponds to Eq.(2.11). An exceptional case is the mono-dispersive spherical droplet system, where $q^{D+1}S(q)$ becomes sinusoidal around the level $\beta_D A$ at large q region.[16]

The formula Eq.(2.19) is illustrated in Fig.5 with use of a randomly scattered droplets model.[19] As is shown in Fig.5 the effect of this sum-rule is less apparent in the log-log plot which is usually used to analyze the experimental data and to test the Porod law. Unfortunately few report[20] on the scattering experiments in the present problem has used the Porod plot, i.e. $q^{D+1}S(q)$ vs. q plot. Note that the evidence of the sum-rule, if it is correct, appears in the range of smaller q than that in the Porod tail region. The Porod plot curve $q^{D+1}S(q)$ must not be monotonic increasing but should have an evident peak before approaching the asymptotic level.

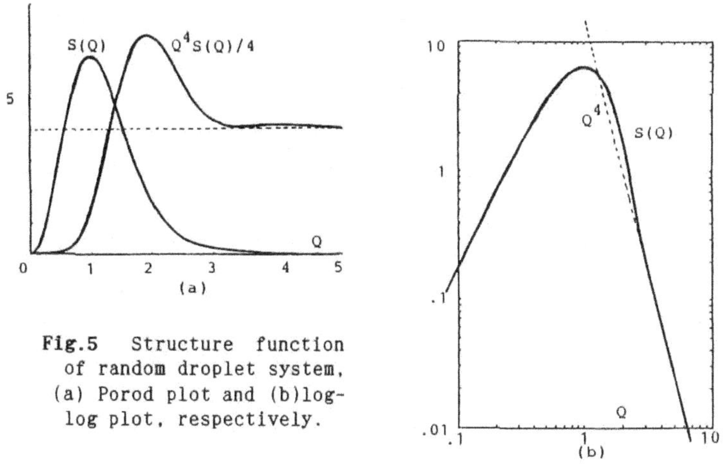

Fig.5 Structure function of random droplet system, (a) Porod plot and (b)log-log plot, respectively.

2-2 Effects of Curvature on the Structure Function

So far we have not taken into consideration the finite curvature of the interface. This affects the higher order terms in g(r) given by Eq.(2.13). Let us calculate it explicitly for D=3. Using the spherical coordinate (r, α, ψ) at $\mathbf{r}_1$, where $\alpha(=\alpha_1)$ is the polar angle from $\mathbf{n}_1$ direction and ψ is the angle around $\mathbf{n}_1$ measured from one of the principal axes of curvature at $\mathbf{r}_1$, the geometry shown in Fig.4 is re-formulated as follows: Let R_1^{-1} and R_2^{-1} be the principal curvatures at $\mathbf{r}_1$, and define Euler's curvature in ($\mathbf{n}_1$, ψ) sectional plane by[21]

$$\rho(\psi)^{-1} = R_1^{-1}\cos^2\psi + R_2^{-1}\sin^2\psi \quad . \tag{2.20}$$

Then the interface near $\mathbf{r}_1$ is approximated by

$$\zeta = r^2 \sin^2\alpha_1(\psi)/2\rho(\psi) \quad , \tag{2.21}$$

where ζ is the coordinate in $\mathbf{n}_1$ direction. On the other hand we have $\zeta = r\cos\alpha_1$ by definition, and we obtain following expressions for $\alpha_1(\psi)$:

$$\cos\alpha_1(\psi) = r/2\rho(\psi) + o(r^3/\rho^3) ,$$

and (2.22)

$$\sin\alpha_1(\psi) = 1 - r^2/8\rho(\psi)^2 + o(r^4/\rho^4) .$$

With use of Eq.(2.21) the normal vector at $\mathbf{r}_2$ is approximately given by

$$\mathbf{n}_2 = k\,(-(r/R_1)\sin\alpha_1\cos\psi,\ -(r/R_2)\sin\alpha_1\sin\psi,\ 1) , \qquad (2.23)$$

where normalization constant k is to be determined by $k = (\mathbf{n}_1\cdot\mathbf{n}_2)$ because $\mathbf{n}_1$ is expressed as (0,0,1) in this frame. Using

$$ds_2' = [(d\alpha_1/d\psi)^2 + \sin^2\alpha_1]^{1/2}\, r\, d\psi ,$$

and

$$\sin\alpha_2 = |\mathbf{r}\times\mathbf{n}_2|/r ,$$

the integrand in Eq.(2.6) becomes

$$(\mathbf{n}_1\cdot\mathbf{n}_2)\, ds_2'/\sin\alpha_2 = (1 - r^2/2\rho(\psi)^2 + \cdots)\, r\, d\psi . \qquad (2.24)$$

Then we obtain

$$g_1(r) = (A/2r)[1 - r^2/2\bar{R}_m^2 + \cdots] ,$$

and

$$g(r) = \phi(1-\phi) - (Ar/4)[1 - r^2/12\bar{R}_m^2 + \cdots] , \qquad (2.25)$$

where $\bar{R}_m$ is defined by

$$\bar{R}_m^{-2} = \int da_1\ (2\pi)^{-1} \oint \rho(\psi)^{-2} d\psi \ / \int da_1$$

$$= \int da_1 [(1/8)(3/R_1^2 + 2/R_1R_2 + 3/R_2^2)] \ / \int da_1 , \qquad (2.26)$$

for D=3. For general dimension D we find[22]

$$g(r) = \phi(1-\phi) - (\gamma_D/4)\ A\ r\ [1 - (D-1)r^2/24\bar{R}_m^2 + \cdots\ ] , \qquad (2.27)$$

where $\bar{R}_m$ is given by an invariant of principal curvatures $\{R_j: j=1,2,\cdots, D-1\}$ for D-dimension as

$$\bar{R}_m^{-2} = \int [2 \sum_{j=1}^{D-1} R_j^{-2} + (\sum_{j=1}^{D-1} R_j^{-1})^2]/(D^2-1) \, da_1 \; / \int da_1 \, . \qquad (2.28)$$

The Fourier transformation of Eq.(2.27) yields

$$S(q) = (\beta_D A/q^{D+1})[1 + (D^2-1)/8\bar{R}_m^2 q^2 + \cdots] \, . \qquad (2.29)$$

Thus the odd term of the curvature does not appear in the correlation function or the structure function of the smooth random interface systems. This is Babinet's reciprocity principle[18] on the complementary system: The correlation function defined by Eq.(2.2) is invariant by exchanging the A-rich and B-rich regions, i.e. by changing c to 1-c, but the direction of the radius of curvature should be inverted.

According to Eq.(2.29) one may obtain a criterion

$$q \gg \bar{R}_m^{-1} \quad , \qquad (2.30)$$

for the range of validity of the Porod law. It should be noted that $\bar{R}_m$ is another characteristic length independent of A^{-1} in general. The latter is related to the dominant wave number q_c by[16]

$$A \simeq [\pi\phi(1-\phi)/2] \, q_c \quad ,$$

where q_c is defined by

$$(2\pi)^{-D} \int_{q<q_c} S(q) \, d\mathbf{q} = (1/2) \times \phi(1-\phi) \quad .$$
$$(\text{or } q>q_c)$$

Though both should obey the same scaling if the dynamical scaling law mentioned in § 1 is expected, the case $\bar{R}_m \ll q_c^{-1}$ is still possible.

Suppose a droplet whose mean radius is well defined, say $\tilde{R}$, but whose surface is very complicated like the coast of fjord. We have $q_c^{-1} \sim \tilde{R} \gg \overline{R}_m \sim$ *the root mean square of the radius of curvature.* The limit $\overline{R}_m \to 0$ corresponds to the fractal surface, whose curvature cannot be defined almost everywhere as well as its surface area.

2-3 The Case of Fractal Surface

So far it has been assumed that we have finite curvatures almost everywhere on the interface. In the case of fractal surface, the area density A itself is indefinite as well as curvatures.

The surface fractal dimension F is introduced as follows: If the least number of spheres (or cubes) of radius r one needs to cover the surface completely is asymptotically given by

$$N(r) \simeq N_0 \, r^{-F} \quad , \qquad (2.31)$$

when $r \to 0$, F is called the surface fractal dimension,[9] and the range of it is given by

$$D - 1 \leqq F < D \quad . \qquad (2.32)$$

D is the dimension of the physical space where the object is embedded, then the lower boundary D-1 corresponds to the smooth surface case. On the other hand the upper limit F = D is attained only if we have mass fractal besides surface fractal.

Thus the interface area density A becomes scale dependent. Let us express it as

$$A(r) \simeq A_0 \, r^{(D-1)-F} \quad . \qquad (2.33)$$

Note that the probabilistic meaning of the correlation function is that[18]

$$\overline{c(0)c(\mathbf{r})} = g(r) + \bar{c}^2$$

= Probability of { a pair of points of distance r are both included in c = 1, i.e. the A-rich region } .

Fig.6 Fractal interface and smoothing or scaling.

In calculating this probability for given r, the detailed surface structure smaller than r may be smoothed out by averaging over various positions and orientations of the test rod of length r as is shown in Fig.6. This means that we should use a length scale unit of order r. In this approximation Eq.(2.27) becomes

$$g(r) \simeq \phi(1-\phi) - (\gamma_D/4)\ A_\square\ r^{D-F} \quad , \tag{2.34}$$

by replacing A in Eq.(2.27) by the scale dependent area density A(r) and neglecting higher order terms. Corresponding to the Porod law we find that[23]

$$S(q) \propto A_\square \ /\ q^{2D-F} \quad . \tag{2.35}$$

The coefficient neglected in Eq.(2.35) may be given by using the asymptotic relation Eq.(2.16), but it makes no sense to write down the explicit result because of the indefinite factor in the smoothing approximation. Only the exponent 2D-F in Eq.(2.35) makes sense. Note that

$$D < 2D - F \leq D + 1 \quad , \tag{2.36}$$

according to Eq.(2.32); that is, the slope of the asymptotic tail in log S(q) becomes slower than that of the Porod tail.

2-4 The Global Statsitical Quantities

So far we have investigated the general aspects of the correlation function and the structure function of random interface systems, mainly based on the local geometry. Therefore all results are limited in large wavenumber range like the Porod law. In this sub-section let us discuss briefly the opposite limit, i.e. the small wavenumber range based on the global geometry of the system.

Define a new field variable by

$$\rho(\mathbf{r}) = c(\mathbf{r}) - \bar{c} \quad , \quad (\bar{c} = \phi) \tag{2.37}$$

then Eq.(2.3) is rewritten as

$$(1/V) \int \rho(\mathbf{r}) = 0 \quad , \tag{2.38}$$

that is, $\rho(\mathbf{r})$ may be regarded as an globally neutral 'charge' density. Rewriting the structure function defined by Eq.(2.14) as

$$S(q) = \int [(1/V) \int \rho(\mathbf{r}_1)\rho(\mathbf{r}_1+\mathbf{r})\, d\mathbf{r}_1] \exp(i\mathbf{q}\cdot\mathbf{r})\, d\mathbf{r} \quad , \tag{2.39}$$

one finds

$$S(0) = (1/V) \; [\int \rho(\mathbf{r})\, d\mathbf{r}]^2 = 0 \quad , \tag{2.40}$$

which is the well known condition for the order parameter conserved system. Then the lowest order term of S(q) is q^2 term, whose coefficient is given by expanding $\exp(i\mathbf{q}\cdot\mathbf{r})$ in Eq.(2.39), that is, given by

$$- (1/4V) \int \int (\mathbf{r}_1-\mathbf{r}_2)^2 \, \rho(\mathbf{r}_1) \, \rho(\mathbf{r}_2) \, d\mathbf{r}_1 \, d\mathbf{r}_2$$

$$= (1/2V) \, [\int \mathbf{r} \, \rho(\mathbf{r}) \, d\mathbf{r}]^2 \quad . \tag{2.41}$$

Note that

$$\mathbf{p} = \int \mathbf{r} \, \rho(\mathbf{r}) \, d\mathbf{r} \quad , \tag{2.42}$$

is the 'dipole moment' of the neutral charge system $\{\rho(\mathbf{r})\}$ and may be rewritten as

$$\mathbf{P} = \mathbf{p}/V$$

$$= 2\phi(1-\phi)[(\int \mathbf{r}\rho_+ d\mathbf{r}/ \int \rho_+ d\mathbf{r}) - (\int \mathbf{r}\rho_- d\mathbf{r}/ \int \rho_- d\mathbf{r})] \quad , \tag{2.43}$$

where

$$\rho_+ = 2 \, (1 - \phi) \ , \quad \rho_- = 0 \quad \text{in } c(\mathbf{r}) = 1 \text{ region} \ ,$$

and

$$\rho_- = 2 \, \phi \qquad , \quad \rho_+ = 0 \quad \text{in } c(\mathbf{r}) = 0 \text{ region} \ .$$

Namely, **P** is the displacement vector from the center of negative charge region to that of positive charge region, then it may be used as a parameter of the global anisotropy of the random pattern system. In our isotropic system **P** remains microscopic, and then q^2 term in S(q) becomes negligible in the scaling stage where the characteristic length scale becomes semi-macroscopic. In fact it is possible to show that the dipole moment **p** is one of the dynamically conserved quantities in the order parameter conserved system: For this system we have a couple of fundamental equations, i.e. the continuity equation

$$\partial\rho/\partial t = - \nabla \cdot \mathbf{j} \quad ,$$

and the phenomenological transport equation

$$\mathbf{j} = - L \nabla \mu \quad ,$$

where μ is the chemical potential and L is Onsager's transport coefficient. Then we find by partial integration that

$$dp/dt = - \int \mathbf{r}(\nabla \cdot \mathbf{j}) \, d\mathbf{r}$$

$$= \oint_B [(\mathbf{j}\cdot\mathbf{n})\mathbf{r} + L \mu \mathbf{n}] \, dS \quad , \qquad (2.44)$$

where $\oint_B \cdots dS$ is the surface integration along the physical boundary B and $\mathbf{n}$ is the normal vector on it. When the system becomes infinitely large, the second term of the surface integration becomes negligible as usual. The first term may be also neglected if one employs the natural boundary condition $\mathbf{j}\cdot\mathbf{n} = 0$ on the mass flow.

On the other hand the next term, i.e. the q^4 term becomes semi-macroscopic, even if the system has global isotropy. The coefficient of q^4 term is related to the 'quadrupole moment' tensor $\{Q_{JK}\}$, and is easily shown to be of order L^4, where L is the characteristic length of the pattern. This is easily seen by calculating it for an example of the checkers-like pattern. Note that the non-zero value of scalar quantity does not contradict isotropy of the system and it is possible to create a linear, scalar quantity with $\{Q_{JK}\}$, e.g. $\mathrm{Tr}\, Q$.

Thus a general conclusion[24),25)]

$$S(q) \propto q^4 \quad , \qquad (2.45)$$

in the small q region, is expected at the scaling stage for globally isotropic systems.

In Fig.7 an example of the evolution of **P** in D=2 system is shown by using a computer simulation on 128×128 kinetic Ising model of exchange type.[28)] A completely random configuration of 50%-50% $\pm$ spins is used as the initial state. As is seen in the figure, **P** behaves like a Brownian motion and stays in microscopic range $|\mathbf{P}| \lesssim 1$, in spite of systematic growing of the characteristic length $A(t)^{-1}$ up to about 10 times of the initial microscopic scale. An explicit pattern at t = 100,000 MCS with natural boudary condition is shown in Fig.1**b** in § 1.

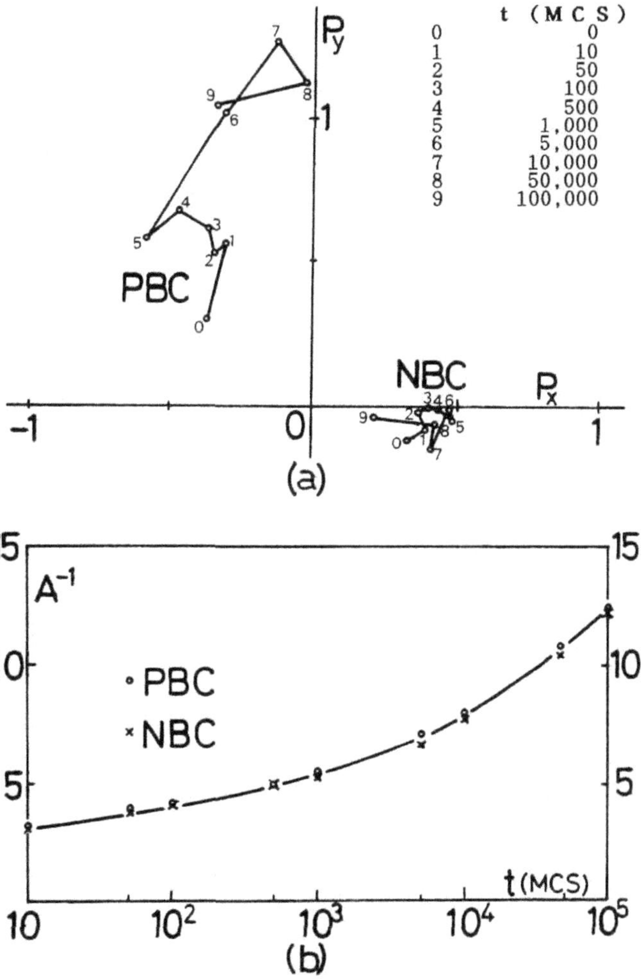

Fig.7 Conservation of isotropy, (**a**) the dipole moment **P** defined by Eq.(2.43) and (**b**) the inverse of interface line density. Natural boundary condition (**NBC**) and periodic one (**PBC**) for 128 × 128 exchange type Ising model is used.

§ 3. A STATISTICAL MODEL OF RANDOM INTERFACE

In this section a statistical model for random interface system is introduced to calculate explicitly the area density and the average of curvature invariants which appeared in the expressions of the correlation function or the structure function of the system in the previous section. The more general mathematical foundations of this section are found in Ref.10.

3-1 The Excursion Set of Random Stochastic Field

Let $\{u(\mathbf{r})\}$ be a stochastic scalar field in the Euclidean space $\mathbf{r} = (x_1, x_2, \cdots, x_D)$ of an arbitrary dimension D. Explicit properties of $\{u(\mathbf{r})\}$, such as statistics of stochastic variable u and spatial correlations of it, are to be given later when we wish to calculate geometrical quantities explicitly. Define the excursion set $\mathbf{U}$ and its boundary set $\partial\,\mathbf{U}$ as follows;

$$\mathbf{U} = \{\ \mathbf{r}\ |\ u(\mathbf{r}) < U\ \}\ ,$$

and (3.1)

$$\partial\,\mathbf{U} = \{\ \mathbf{r}\ |\ u(\mathbf{r}) = U\ \}\ .$$

This boundary set $\partial\,\mathbf{U}$ serves as a model of the random interface system in the present problem as is mentioned in § 1. The physical field $\{c(\mathbf{r})\}$ in § 2 may be defined by a nonlinear transformation from $u(\mathbf{r})$ as

$$c(\mathbf{r}) = \theta(U-u(\mathbf{r}))\ , \tag{3.2}$$

where $\theta(x)$ is the usual step function given by

$$\theta(x) = \begin{cases} 1 & \text{if } x > 0\ , \\ 0 & \text{otherwise}\ . \end{cases}$$

Note that the volume fraction of $\mathbf{U}$ and the area density of $\partial\,\mathbf{U}$

are given by

$$\phi(U) = (1/V) \int \theta(U-u(\mathbf{r}))\, d\mathbf{r} \quad , \tag{3.3}$$

and

$$A(U) = (1/V) \int |\nabla u|\, \delta(U - u(\mathbf{r}))\, d\mathbf{r} \ , \tag{3.4}$$

respectively, where the relations $d\mathbf{r} = da\, d\zeta$, $|\nabla u| = du/d\zeta$ and $\delta(u) = d\theta(u)/du$ are used.

Further the curvatures of ∂U are obtained as follows: Let $\{T_{ij}\}$ be the coefficient matrix of the orthogonal transformation to the normal coordinate system at $\mathbf{r}$, whose components $(x_1', x_2', \cdots, x_D')$ are chosen so that one of them, say x_D', is in the normal direction $\mathbf{n} = \nabla u/ |\nabla u|$, i.e.

$$\partial_i 'u = \sum_{j=1}^{D} T_{ij}\ \partial_j u = |\nabla u|\ \delta_{iD} \quad ,$$

where the abbreviation $\partial_i = \partial/\partial x_i$ is used. In this new coordinate system, the local form of the interface near $\mathbf{r}$ is expressed as

$$\zeta = (1/2) \sum_{i,j=1}^{D-1} (\partial_i \partial_j 'u/|\nabla u|)_r\ x_i' x_j' + \cdots \quad , \tag{3.5}$$

where $(\ )_r$ denotes the value at $\mathbf{r}$ and

$$\partial_i \partial_j 'u = \sum_{k,m=1}^{D} T_{ik}\ T_{jm}\ \partial_k \partial_m u \qquad (i,j=1,2,\cdots,D-1). \tag{3.6}$$

Then the principal curvatures R_1^{-1}, $R_2^{-1}, \cdots$, R_{D-1}^{-1} at $\mathbf{r}$ are given by the D-1 eigenvalues of the coefficient matrix

$$\Lambda = \{ (\partial_i \partial_j 'u/|\nabla u|)_r \} \qquad (i,j=1,2,\cdots,D-1). \tag{3.7}$$

The simplest examples of curvature invariants are the mean curvature and the Gauss total curvature defined by

$$H = (1/(D-1))\sum_{j=1}^{D-1} R_j^{-1} = (1/(D-1)) \ \mathrm{Tr}\ \Lambda \ , \qquad (3.8)$$

and

$$K = R_1^{-1}\ R_2^{-1}\ \cdots\ R_{D-1}^{-1} = \det\ \Lambda \ , \qquad (3.9)$$

respectively. Define n-th invariants related to the rank n principal minor of Λ by

$$K(n) = (1/{}_{D-1}C_n)\ \Sigma_S\ R_1^{-1}R_2^{-1}\cdots R_n^{-1} \ , \qquad (3.10)$$

where Σ_S denotes the symmetric summation over ${}_{D-1}C_n = (D-1)!/n!(D-1-n)!$ combinations of n different components of $\{R_j^{-1}; j=1,\cdots,D-1\}$.

Thus for the purpose of obtaining the area density and the spatially averaged curvature invariants we need the complete statistical properties of the composite fields $\{u(\mathbf{r}),\ \partial_i u(\mathbf{r}),\ \partial_i\partial_j u(\mathbf{r})\}$. However, an explicit calculation is intractable in general.

3-2 The Gaussian Random Field

A completely tractable model is the Gaussian field $\{u(\mathbf{r})\}$ whose probability is given by the (continuously) multi-variate normal distribution function. In the ordinary field theory the Gaussian field is a linear, trivial one. Its statistical geometry, however, is non-trivial as is shown in the followings.

Let $\{u(\mathbf{r})\}$ be a homogeneous, isotropic and non-erratic Gaussian field with expectations

$$\langle u(\mathbf{r})\rangle = 0 \ ,$$

and

$$\langle u(\mathbf{r}_1)u(\mathbf{r}_2)\rangle = \sigma(|\mathbf{r}_2-\mathbf{r}_1|) \ ,$$

where $\langle\cdots\rangle$ denotes the ensemble average, and the covariance or the correlation function $\sigma(r)$ is assumed to be regular at $\mathbf{r}=0$, i.e.

$$\sigma(r) = \sigma(0) - |\sigma''(0)| r^2/2! + \sigma^{(4)}(0) r^4/4! + \cdots , \tag{3.11}$$

is assumed. For simplicity let the amplitude of the field be normalized so that

$$\sigma(0) = \langle u^2 \rangle = 1 . \tag{3.12}$$

The non-erratic property Eq.(3.11) ensures that the interface $\partial \mathbf{U}$ is smooth almost everywhere as is mentioned soon later. The assumption of homogeneity makes the calculation of spatial averages very easy. For example, the volume fraction defined by Eq.(3.3) may be replaced by single point ensemble average as

$$\phi(U) = \langle \theta(U-u(\mathbf{r})) \rangle = (2\pi)^{-1/2} \int_{-\infty}^{U} \exp(-u^2/2)\, du , \tag{3.13}$$

which is the well known normal distribution function, or the error function defined as $\phi(-\infty)=0$ and $\phi(+\infty)=1$. The spatial correlation function of $c(\mathbf{r}) = \theta(U-u(\mathbf{r}))$ is calculable[4] with use of the integral representation for the step function

$$\theta(u) = (1/2\pi i) \int_{-\infty}^{+\infty} \exp(-i\lambda u)/(\lambda - i\varepsilon)\, d\lambda , \tag{3.14}$$

and the generating function

$$\langle \exp[i \int \lambda(\mathbf{r})\, u(\mathbf{r})\, d\mathbf{r}] \rangle$$

$$= \exp[-(1/2) \int \int \sigma(|\mathbf{r}_2-\mathbf{r}_1|)\, \lambda(\mathbf{r}_1)\lambda(\mathbf{r}_2)\, d\mathbf{r}_1 d\mathbf{r}_2] .$$

One of the simplest expression of the result is given by[10]

$$g(r) = \langle c(0)c(\mathbf{r})\rangle - \langle c\rangle^2$$

$$= (1/2\pi)\int_0^{\sigma(r)} \exp[-U^2/(1+x)]/\sqrt{1-x^2}\,dx$$

$$= (1/2\pi)\int_0^{\mathrm{Arcsin}\,\sigma(r)} \exp[-U^2/(1+\sin\alpha)]\,d\alpha\ . \tag{3.15}$$

From Eq.(3.15) one finds the following formula for U=0 which is well known in the probability theory[27];

$$\langle\theta(u(\mathbf{r}_1))\theta(u(\mathbf{r}_2))\rangle$$

$$= 1/4 + (1/2\pi)\ \mathrm{Arcsin}\ [\langle u(\mathbf{r}_1)u(\mathbf{r}_2)\rangle/\langle u^2\rangle]\ , \tag{3.16}$$

where $\phi = \langle c\rangle = 1/2$ is used. This form has been already attained in the present problem.[4),12)] By using the expansion

$$\mathrm{Arcsin}\ \sigma(r) = \pi/2 - \sqrt{|\sigma''(0)|}\ r$$

$$\times \{1 - [(\sigma^{(4)}(0) - \sigma''(0)^2)/24|\sigma''(0)|]\ r^2 + \cdots\}\ ,$$

Eq.(3.15) is expanded in the same form as Eq.(2.27) for smooth interface systems. Comparing the coefficients we find[18]

$$A = [\Gamma((D+1)/2)/\Gamma(1/2)\Gamma(D/2)]\sqrt{|\sigma''(0)|}\ \exp(-U^2/2)\ , \tag{3.17}$$

and

$$\bar{R}_m^{-2} = [\sigma^{(4)}(0) - (1 - U^2)\ \sigma''(0)^2]/(D-1)|\sigma''(0)|\ . \tag{3.18}$$

Note that these results for smooth interface system is based on the assumption of non-erratic field, i.e. is obtained with use of the expansion Eq.(3.11).

3-3 Expectations of Curvature Invariants

Though the complete form of the correlation function g(r) and its expansion coefficients A and $\bar{R}_m^{-2}$ have been obtained directly, it is still interesting to calculate the explicit forms of expectations of other curvature invariants, because they are useful in investigating the statistical geometry of random interface system as is shown in § 4.

Note that the area element of interface and the principal curvatures are local quantities represented by $u(\mathbf{r})$, $\mathbf{v}(\mathbf{r}) = \nabla u(\mathbf{r})$ and $u_{JK}(\mathbf{r}) = \partial_J \partial_K u(\mathbf{r})$ as is shown in 3-1. Then the normal probability function we need to calculate the ensemble averages is an extremely contracted one, i.e. that of [u, $\mathbf{v}$, $\{u_{JK}\}$] at a single point. The number of variables is reduced to

$$1(\text{scalar}) + D(\text{vector}) + [D + D(D-1)/2](\text{symmetric tensor})$$

$$= (D^2 + 3D + 2)/2 \,, \tag{3.19}$$

from continuous degrees of freedom! Further it can be easily shown that if $\{u(\mathbf{r})\}$ obeys the multi-variate Gaussian probability law, then the probability functions of

$$v(\mathbf{r}) = \lim_{d\mathbf{r} \to 0} [u(\mathbf{r}+d\mathbf{r}) - u(\mathbf{r})]/d\mathbf{r}$$

and

$$u_{JK}(\mathbf{r}) = \lim_{dx_K \to 0} [v_J(x_K+dx_K) - v_J(x_K)]/dx_K$$

are also Gaussian. With use of Eq.(3.11) of the covariance of $\{u(\mathbf{r})\}$ and assuming isotropy, one obtains the following expectations;

$$\left.\begin{aligned}
&\langle u\rangle = \langle v_J\rangle = \langle u_{JK}\rangle = 0 \,, \\
&\langle u^2\rangle = 1 \,, \quad \langle u\, v_J\rangle = 0, \quad \langle v_J\, v_K\rangle = |\sigma''(0)|\, \delta_{JK} \,, \\
&\langle u\, u_{JK}\rangle = -\, |\sigma''(0)|\, \delta_{JK} \,, \\
&\langle u_{JK}\, u_{mn}\rangle = (\sigma^{(4)}(0)/3)[\, \delta_{JK}\delta_{mn} + \delta_{Jm}\delta_{Kn} + \delta_{Jn}\delta_{Km} \,] \,.
\end{aligned}\right| \tag{3.20}$$

It is apparent from Eq.(3.20) that $\mathbf{v}$ is independent of u and $\{u_{JK}\}$. Then the probability density for $\mathbf{v}$ is explicitly given by

$$P(\mathbf{v})d\mathbf{v} = (2\pi|\sigma''(0)|)^{-D/2} \exp\{ -(1/2|\sigma''(0)|) \sum_{j=1}^{D} v_J^2 \} d\mathbf{v} ,$$

which immediately yields the following formulae

$$< \delta(\mathbf{v}) > = (2\pi|\sigma''(0)|)^{-D/2} , \qquad (3.21)$$

and

$$< v^N > = (2|\sigma''(0)|)^{N/2} \Gamma((D+N)/2)/ \Gamma(D/2) , \qquad (3.22)$$

where N > -D. These are used in the following calculations and in § 4. On the other hand the diagonal components of $\{u_{JK}\}$ are coupled with the scalar variable u. Using new notations defined by

$$y_{JK} = u_{JK} /|\sigma''(0)| , \qquad (3.23)$$

and

$$s = \sigma^{(4)}(0)/3|\sigma''(0)| , \qquad (3.24)$$

the non-zero variances and covariances with respect to u and $\{y_{JK}\}$ are listed as

$$<u^2> = 1 , \quad <u\ y_{JJ}> = -1 ,$$

and

$$<y_{JJ}\ y_{KK}> = <y_{JK}^2> = s\ (1 + 2\delta_{JK}) .$$

Since

$$<u\ (u + y_{JJ})> = <u^2> + <u\ y_{JJ}> = 0 ,$$

the probability density for u and $\{y_{JK}\}$ should have the following form;

$$P(u,\{y_{JK}\})\ du\ d\{y_{JK}\} = \mathcal{N}^{-1} \exp(- H)\ du\ d\{y_{JK}\} , \qquad (3.25)$$

with

$$H = (1/2)u^2 + (1/2)\sum_{j,k} G_{jk}(u + y_{jj})(u + y_{kk}) + (1/2s)\sum_{(jk)} y_{jk}^2 , \tag{3.26}$$

and

$$\mathcal{N} = (2\pi)^{(D+1)/2} (\det G)^{-1/2} (2\pi s)^{D(D-1)/2} . \tag{3.27}$$

The coefficient matrix $G = \{G_{jk}\}$ in the second term in Eq.(3.26) is determined by

$$(G^{-1})_{jk} = \langle (u + y_{jj})(u + y_{kk}) \rangle = s\ (1 + 2\delta_{jk}) - 1 , \tag{3.28}$$

and (jk) in the third term denotes the pair of $j \neq k$.

The quantities concerned with here are the surface averages of the quantities defined on the interface, which are denoted by $\overline{\overline{\ \ }}$ and replaced by the conditional ensemble averages as

$$\begin{aligned}\overline{\overline{Q}} &= (1/V) \int Q(\mathbf{r})\ \delta(\zeta(\mathbf{r}))\ d\mathbf{r} \\ &= (1/V) \int_{\partial U} Q\ da \\ &= \langle\ Q\ |\nabla u|\ \delta(U-u)\ \rangle ,\end{aligned} \tag{3.29}$$

where representations

$$d\mathbf{r} = da\ d\zeta ,$$

and

$$\delta(\zeta(\mathbf{r})) = \delta(U-u(\mathbf{r}))\ (du/d\zeta) ,$$

are used. As is mentioned above, $\mathbf{v} = |\nabla u|$ is independent of u and $\{y_{jk}\}$, then the average of the quantity with following form

$$Q = |\nabla u|^N F(u, \{y_{jk}\}) \quad ,$$

are shown to be given by

$$\bar{\bar{Q}} = (2\pi)^{-1/2} \exp(-U^2/2) < v^{N+1} >$$

$$\times < F(U, \{y_{jk} - U\, \delta_{jk}\} >_G \quad , \tag{3.30}$$

with use of Eq.(3.29) and Eqs.(3.25)~(3.27). Here a new expectation

$$< \cdots >_G = [(2\pi)^{1/2}/\mathcal{N}] \int \cdots \exp(-H')\, d\{y_{jk}\} \ ,$$

with (3.31)

$$H' = (1/2) \sum_{j,k} G_{jk}\, y_{jj} y_{kk} + (1/2s) \sum_{(j,k)} y_{jk}^2 \quad ,$$

is introduced. For example we have

$$<y_{jj} y_{kk}>_G = (\, G\,)^{-1}{}_{jk} = s(1+2\delta_{jk}) - 1 \ ,$$

and

$$<y_{jk}{}^2>_G = <y_{jk}{}^2> = s \quad \text{for } j \neq k \quad ,$$

then

$$< (y_{11} y_{22} - y_{12}{}^2)>_G = -1 \quad . \tag{3.32}$$

Another example is

$$< \det\{y_{jk} - U\, \delta_{jk}\} >_G = H_D(-U) \ , \tag{3.33}$$

where $H_D(x)$ is the Hermite polynomial of D-th degree. This formula is the only thing we need in calculating our expectations in the remainder of this section and the next section. The derivation of it is shown at the last of this section with use of the Gaussian properties and Eq. (3.32).

Now let us calculate the expectations for curvature invariants defined by Eq.(3.10) in 3-1. Note that the orthogonal transformation to the normal coordinate system does not change the Gaussian natures. The formulae Eq.(3.20) for the variances and covariances are also unchanged except for the fact that the dimension is reduced by 1. For example one may find

$$\begin{aligned} \langle u\, u_{ij}' \rangle &= \sum_{k,m} \langle T_{ik} T_{jm}\, u\, u_{km} \rangle \\ &= \sum_{k,m} \langle T_{ik} T_{jm} \rangle \langle u\, u_{km} \rangle \\ &= - \sum_{k,m} \langle T_{ik} T_{jm} \rangle\, \delta_{km}\, |\sigma''(0)| \\ &= - \delta_{ij}\, |\sigma''(0)| \quad , \end{aligned}$$

if one notices that $\{T_{jk}\}$ is determined by only $\nabla u = \{\partial_j u\}$ and does not include u and $\{\partial_j \partial_k u\}$. Then one may use Eq.(3.33) with $D = n$ to calculate the average of the n-th invariant related to the rank n principal minor of Λ defined by Eq.(3.10). Using Eq.(3.22) with $N = -(n-1)$ the result is written down as

$$\overline{\overline{K(n)}} = \lambda^{-(n+1)} [2\pi^{(n+1)/2}\, \Gamma((D-n+1)/2)/\Gamma(D/2)]\, \phi^{(n)}(-U), \tag{3.34}$$

where

$$\begin{aligned} \phi^{(n)}(x) &= (-d/dx)^n \exp(-x^2/2)/\sqrt{2\pi} \\ &= H_n(x) \exp(-x^2/2)/\sqrt{2\pi} \quad , \end{aligned} \tag{3.35}$$

is the usual definition of the Hermite polynomial, and a length scale

$$\lambda = \sqrt{2\pi\, \sigma(0)/|\sigma''(0)|}\ , \quad (\sigma(0)=1) \tag{3.36}$$

is introduced, which is, however, taken as $\lambda = 1$ in the remainder of this section for simplicity.

The average area density A is given[4] by $\overline{\overline{K(0)}}$ and it is exactly in agreement with Eq.(3.17), which has been obtained directly. Further we obtain explicit expressions for the averages of the mean curvature H and the total curvature K; that is

$$\overline{\overline{H}} = \overline{\overline{K(1)}} = 2\pi\, \phi^{(1)}(-U) = -\, 2\pi U \exp(-U^2/2)/\sqrt{2\pi}\ , \qquad (3.37)$$

and

$$\overline{\overline{K}} = \overline{\overline{K(D-1)}} = S(D)\ \phi^{(D-1)}(-U)\ , \qquad (3.38)$$

where $S(D) = 2\pi^{D/2}/\Gamma(D/2)$ is the surface area of a unit sphere in the D dimensional Euclidean space. These quantities have useful geometrical meaning as is discussed in the next section.

It is possible to obtain averages of another series of invariants related to

$$\mathrm{Tr}\,\Lambda = (1/|\nabla u|) \sum_j \partial_j{}^2 u = (|\sigma''(0)|/|\nabla u|) \sum_j y_{jj}\ ,$$

with the ordinary partition function method; that is

$$\overline{\overline{H^n}} = [2\pi^{(n+1)/2}\, \Gamma((D-n+1)/2)/\Gamma(D/2)]$$

$$\times\ (i/\beta)^n\, H_n(i\beta U)\ \exp(-U^2/2)/\sqrt{2\pi}\ , \qquad (3.39)$$

where

$$\beta^{-2} = ((D+1)/(D-1))\ s - 1\ .$$

The result $\overline{\overline{H^2}}$ coupled with $(\overline{\overline{H}})^2$ yields Eq.(3.18) for $\overline{R}_m{}^{-2}$ again.[16]

(Derivation of Eq.(3.33))

Note that $\det\{y_{JK} - U\,\delta_{JK}\}$ is an polynomial of U and may be expanded with use of the principal minors of $\det\{y_{JK}\}$. All ${}_DC_r = D!/r!(D-r)!$ principal minors $\det\{y_{JK}\}^{(r)}$ of the same rank r, have the equal expectations

$$a(r) = \langle \det\{y_{JK}\}^{(r)} \rangle_G\ ,$$

because of the isotropy. Then one may write down the averaged polynomial form as

$$< \det\{y_{JK} - U\,\delta_{JK}\} >_G = \sum_{r=0}^{D} {}_DC_r\; a(r)\; (-U)^{D-r} \quad . \tag{3.40}$$

Since $\{y_{JK}\}$ obeys the zero-mean Gaussian probability function, we find

$$a(2r-1) = 0 \quad ,$$

and

$$a(2r) = \Sigma_{\text{pairing}} < \begin{vmatrix} y_{11} & y_{12} \\ y_{21} & y_{22} \end{vmatrix} >_G < \begin{vmatrix} y_{33} & y_{34} \\ y_{43} & y_{44} \end{vmatrix} >_G \cdots\cdots$$

$$= (-1)^r\ (2r-1)!! \quad ,$$

where the summation Σ_{pairing} is taken over all $(2r-1)!! = 1\cdot3\cdot5\cdots\cdots(2r-1)$ different combinations of pairing $\{(12),(34),\cdots\cdots\}$, and the result Eq.(3.32) is used. Finally we obtain the following result;

$$< \det\{y_{JK} - U\,\delta_{JK}\} >_G = \sum_{r=0}^{[D/2]} (-1)^r\ {}_DC_{2r}\ (2r-1)!!\ (-U)^{D-2r}$$

$$= H_D(-U) \quad ,$$

where [D/2] is the usual Gauss notation, i.e. the integer part of D/2.

§ 4. STATISTICAL GEOMETRY AND PERCOLATION PROBLEM

Though the characteristics calculated in the previous section are purely local quantities, it is possible to use them to investigate some global natures of the interface system. For example, note that all of the expectations Eq.(3.34) of curvature invariant of any rank are positive when $U < -X_0$, where X_0 is the largest zero of $H_{D-1}(x)$. Then we can expect that the curvature is positive almost everywhere on the interface. In fact it is possible to show that the system of sufficiently small volume fraction is composed of simple droplets with ovaloid surfaces.[10] Among the related quantities the Euler characteristic[21] plays an important role, since it is a topological invariant related to the connectivity of the manifold.

4-1 The Euler Characteristic and the Morse Theorem

By the Gauss-Bonnet theorem, the Euler characteristic (EC) of a manifold **M** embedded in D-dimensional Euclidean space with smooth closed surface ∂ **M** is given by[21),28)]

$$\chi[\mathbf{M}] = S(D)^{-1} \oint_{\partial M} K(a)\, da \quad , \tag{4.1}$$

where $S(D) = 2\pi^{D/2}/\Gamma(D/2)$, and K(a) is the Gauss total curvature defined by Eq.(3.9) and da the area element of ∂ **M**. We have $\chi = 1$ for a simply connected solid body with closed surface, and χ is decreased by the number of attached handles and increased by the number of cavities for D=3. For example, we have χ[a nut] = +1, χ[a doughnut] = 0, χ[a contacted pair of doughnuts] = -1 and χ[an uncrashed shell of coconut] = +2, etc.

As is seen from the definition Eq.(4.1), EC of the excursion set **U** $= \{ \mathbf{r} \mid u(\mathbf{r}) < U \}$ of Gaussian random field has been calculated already in the previous section. Because our system is infinitely extended and has the translational symmetry, it is better to use EC density per unit volume defined by

$$\chi[\mathbf{U}]/V = (1/V)\, S(D)^{-1} \oint_{\partial U} K(a)\, da \quad . \tag{4.2}$$

With use of Eq.(3.38) the Gaussian expectation of it is given by

$$\chi_D(U) = \langle \chi[\mathbf{U}]/V \rangle = \lambda^{-D}\, \phi^{(D-1)}(-U) \quad , \tag{4.3}$$

where the unit of length $\lambda = \sqrt{2\pi\sigma(0)/|\sigma''(0)|}$ is recalled to elucidate the geometrical situation.

It makes the physical meaning of EC much clearer for us to use an algebraic definition of it in the present case. This is given by the Morse theorem; that is,

$$\chi[\,\mathbf{U}\,] = \sum_{k=0}^{D} (-1)^{k}\, C_{k}(\,\mathbf{U}\,) \quad . \tag{4.4}$$

Here $C_k(\mathbf{U})$ denotes the number of critical points $\nabla u = 0$ of index k included in $\mathbf{U}$, where the index k denotes the number of negative eigenvalues of the matrix $\{\partial_i \partial_j u\}$ at each critical point. The proof of this theorem[28),29)] is not straightforward, but the geometrical meaning is apparent as is illustrated for D=2 in Fig.8.

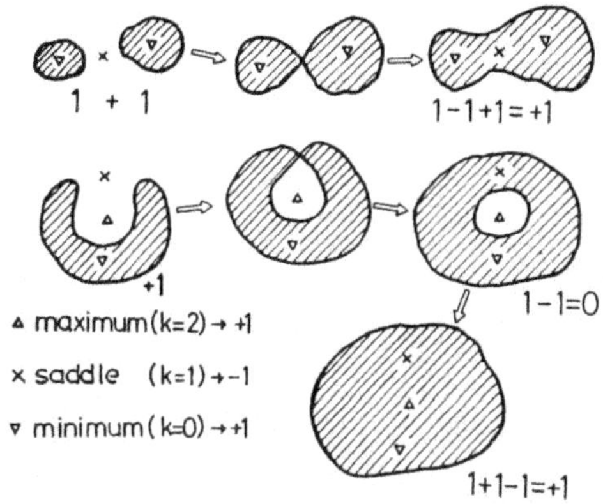

Fig.8 The Morse theorem for D=2. Note that a lake has at least one minimum (▽) in it.

The expectation of Eq.(4.4) is easily calculated as follows: Let $\mathbf{r}^*$ be a critical point of $\{u(\mathbf{r})\}$, i.e. $\nabla u(\mathbf{r}^*) = 0$. In counting the number of critical points we need the delta function $\delta(\mathbf{r}-\mathbf{r}^*)$, which is related to the field variable $\{u(\mathbf{r})\}$ by

$$\delta(\mathbf{r}-\mathbf{r}^*) = \delta(\mathbf{v})\ |\det\{\partial_i\partial_j u\}| \ , \tag{4.5}$$

with the definition of the Jacobian

$$\partial(v_1, v_2, \cdots, v_D)/\partial(x_1, x_2, \cdots, x_D) = \det\{\partial_i\partial_j u\} \quad .$$

Note that $\det\{\partial_i\partial_j u\} = (-1)^K|\det\{\partial_i\partial_j u\}|$ by the definition of the index. Then the expectation of Eq.(4.4) divided by V is given by

$$\chi_D(U) = (1/V)\int \theta(U-u(\mathbf{r}))\ \delta(\mathbf{v})\ \det\{\partial_i\partial_j u\}\ d\mathbf{r}$$

$$= \langle\ \theta(U-u(\mathbf{r}))\ \delta(\mathbf{v})\ \det\{\partial_i\partial_j u\}\ \rangle \quad , \tag{4.6}$$

or

$$d\chi_D(U)/dU = \langle\ \delta(U-u)\ \delta(\mathbf{v})\ \det\{\partial_i\partial_j u\}\ \rangle \quad . \tag{4.7}$$

In the case of Gaussian random field, Eq.(4.7) becomes

$$d\chi_D(U)/dU = \langle\delta(\mathbf{v})\rangle\langle\ \det\{\partial_i\partial_j u - U\ \delta_{ij}\}\rangle_G\ \exp(-U^2/2)/\sqrt{2\pi}$$

$$= \lambda^{-D}\ \phi^{(D)}(-U) \quad , \tag{4.8}$$

according to the formulae in the previous section. Integrating Eq.(4.8) with the boundary condition $\chi_D(-\infty)=0$, we obtain Eq.(4.3) again.

4-2 Topological Transition and Percolation of Random Pattern

Let us use the volume fraction $\phi(U) = \langle\ \theta(U-u)\ \rangle$ given by Eq.(3.13) instead of U itself. For D=2, $\phi(U)$ is the area fraction of the 'lake' or the 'sea' as is used in the followings. The explicit expressions for the EC density are

$$\chi_2(U) = \lambda^{-2}\ (-U)\ \exp(-U^2/2)/\sqrt{2\pi} \quad , \tag{4.9}$$

and

$$\chi_3(U) = \lambda^{-3}\ (1 - U^2)\ \exp(-U^2/2)/\sqrt{2\pi} \quad , \tag{4.10}$$

for D=2 and D=3, respectively, and are shown in Fig.9 for D=1～4.

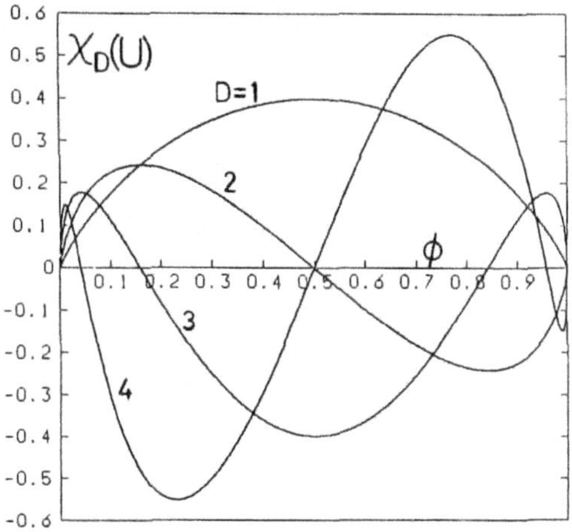

Fig.9 EC density for D=1～4. Length unit λ = 1 is used.

Suppose water is poured little by little into a random topography. For sufficiently small ϕ, only the isolated small ponds or lakes of χ = 1 would be found, where $\chi_D(U)$ (> 0) is essentially equal to the number density of ponds or lakes. With increasing ϕ, that is, raising the water level U, $\chi_D(U)$ increases till it attains maximum at $\phi = \phi_M$ (= 0.1586$\cdots$ for D=2) and turns to decreasing due to the coalescence of the lakes. If a lake coalesces by itself, it changes into an annulus (or solid torus for D=3) of χ = 0. Besides this causes the discrepancy between the number density of lakes and $\chi_D(U)$. At a certain value $\phi = \phi_C$ (= 0.5 for D=2) $\chi_D(U)$ changes its sign from plus to minus and the seas join into a network (χ < 0) percolating all over the world. For D=3 we would find one more change at a higher value of ϕ from a percolating network pattern to a foam-like pattern (χ > 0), which is just the reversed situation of the first change if the roles of the 'sea' and the 'land' are exchanged. Thus the changes of the sign of EC density correspond to phase transitions of mean topology of the random pattern.[30]

Apparently there are D-1 zeros of $\chi_D(U)$ or $H_{D-1}(-U)$. The value U_C =0 (or ϕ_C= 0.5) for D=2 just agrees with the exact percolation limit in continuous media.[31] The lower zero U_C=-1 (or ϕ_C=0.15865544$\cdots\cdots$) for D=3 is found to be very close to ϕ_C=0.16 obtained by computer simulation[32] on the Gaussian random field with use of various types of $\sigma(r)$, and to ϕ_C=0.157 predicted by a speculation from the lattice percolation problem.[33]

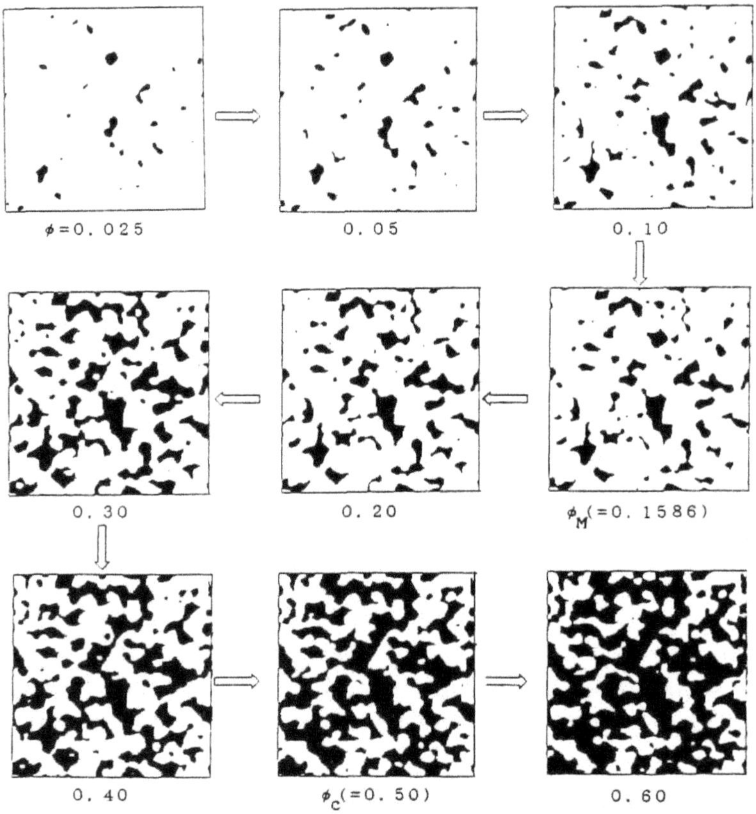

Fig.10 The development of the lake or the sea with increasing water level U (or ϕ). $\sigma(r) = \exp(-r^2/50)$ in 200×200 square is used.

This seems a strange coincidence if one notices that any local structure such as a lake or an island, however small, affects the value of EC. It is possible, however, to show that the positions of zeros are not changed by coarse graining or smoothing transformations: It should be noted here that the expression Eq.(4.3) (or explicitly Eqs. (4.9) and (4.10)) has a universal form, i.e. it does not depend on the detailed natures of original Gaussian field except that the amplitude is determined by the length scale $\lambda = \sqrt{2\pi\sigma(0)/|\sigma''(0)|}$. Let us introduce a smoothing transformation by

$$\tilde{u}(\mathbf{r}) = \int K(\mathbf{r}-\mathbf{r}_1)\, u(\mathbf{r}_1)\, d\mathbf{r}_1 \quad , \tag{4.11}$$

with the averaging kernel K properly chosen so that the normalization $\langle\tilde{u}^2\rangle = 1$ is conserved; that is,

$$\langle\tilde{u}^2\rangle = \int\int K(\mathbf{r}_1) K(\mathbf{r}_2) \langle u(\mathbf{r}_1) u(\mathbf{r}_2)\rangle\, d\mathbf{r}_1 d\mathbf{r}_2$$

$$= \int\int K(\mathbf{r}_1) K(\mathbf{r}-\mathbf{r}_1)\, \sigma(\mathbf{r})\, d\mathbf{r}_1 d\mathbf{r} = 1 \ .$$

This new field $\{\tilde{u}(\mathbf{r})\}$ is also Gaussian, and the functional form of EC density $\chi_D(U)$ is not changed except for the decreasing of the amplitude, because of the universality mentioned above. Then the positions of the zeros are not changed. As is well known in the lattice percolation problem,[34] the system is expected to have a self similar structure just at the percolation limit. Then it can be said that at least mean topology must be invariant under scale transformation or coarse graining at that limit. We cannot find any invariant point other than the zeros of EC density $\chi_D(U)$, because we were able to define a characteristic length with $\chi_D(U)$ if $\chi_D(U) \neq 0$.

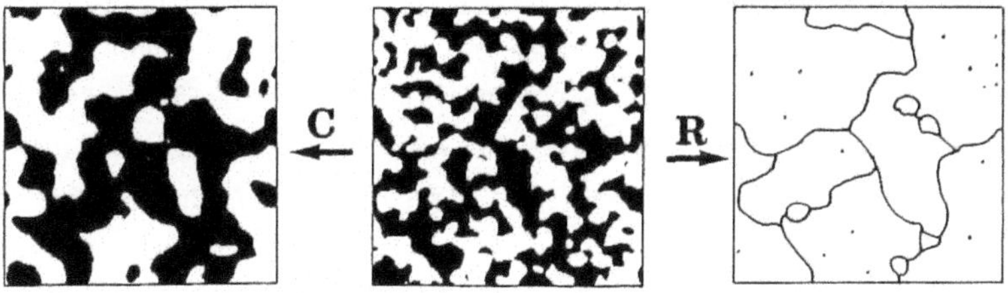

Fig.11 The coarse graining (**C**) and retraction (**R**).

We have one more reason authorizing this conclusion; that is, the zeros $\phi_C=1$ for D=1, $\phi_C=0.5$ for D=2 and $\phi_C=0.158655\cdots$ for D=3, and so on, are connected one another by analytic continuation as follows: With use of Eqs.(3.35) and (3.13), Eq.(4.3) is rewritten as

$$\chi_D(U) = (d/dU)^D \phi(U) \quad , \tag{4.12}$$

where length scale λ is neglected again. One may extend the dimension D in this expression to a continuous one by using the Riemann-Liouville fractional integral form defined by[38]

$$\chi_{D-F}(U) = \Gamma(F)^{-1} \int_{-\infty}^{U} \chi_D(u)\ (U - u)^{-(1-F)}\ du \ . \tag{4.13}$$

One may restrict F to the range $0 < F \leqq 1$ to avoid the ambiguity of singular integration because we know explicit form of $\chi_D(u)$ with Hermite polynomial for all non-negative integer D. This is essentially a unique analytic continuation into the continuous dimension D, except for the another choice of integration range $(U, +\infty)$, which merely corresponds to exchanging the roles of the 'sea' and the 'land'.

The branches of numerically computed zeros are shown in Fig.12. The sign of $\chi_D(U)$ is also shown. This is a phase diagram for the mean topology of the excursion set **U** in continuous dimension. In the region $\mathbf{D}_0$, any kind of percolating manifold does not exist. The lowest branch **a** is the ordinary percolation limit. Next branch **b**, starting from ϕ=1 at D=2, passes ϕ=0.5 at D=4, just as the branch **a** passes ϕ=0.5 at D=2. This is a reflection of the symmetry of the field. Then the branch **b** should be interpreted as a new transition line where the effective dimension of percolating manifolds changes again. To make this situation clearer let us use the topological dimension D_R of the deformation retract of percolating manifold. Note that EC is homotopy invariant as well as topologically invariant. After retraction, the percolating manifold becomes a network backbone with negative EC and D_R=1 in the region $\mathbf{D}_1$, a foam-like object with positive EC and D_R=2 in the region $\mathbf{D}_2$, and so on. And fractal retracts are expected on the boundary lines separating these regions.

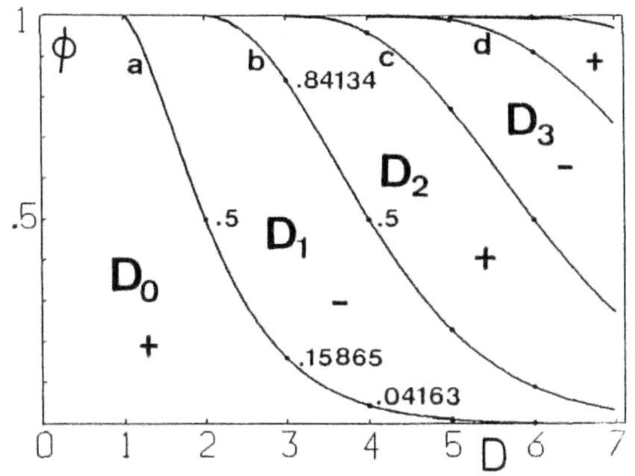

Fig.12 Phase diagram of the mean topology of excursion set obtained by analytic continuation of spatial dimension D.

As a concluding remark of this section, it should be noted that the Gaussian random field serves as the most fundamental percolation problem in continuous systems like the Bernoulli problem in lattice systems: Suppose one starts with spatially uncorrelated field $\{f(\mathbf{r})\}$, which need not be Gaussian. It is impossible to define any kind of manifold of the excursion set with this field, because manifold is a notion of continuity. Thus a coarse graining or smoothing procedure like Eq.(4.11) is unavoidable to define a percolation problem in continuum. Then a correlated Gaussian field $\{u(\mathbf{r})\}$ would be obtained due to the central limit theorem. In addition a critical value of percolation limit independent of this coarse graining procedure should be found, if it exists. As is mentioned previously, the zeros of EC density do not depend on the details of the coarse graining process defined by Eq.(4.11), which seems to have very general form. This fact suggests a possibility of constructing a kind of renormalization group method by which one would be able to compute the critical indeces. However, such a renormalization procedure should be applied not to the original Gaussian field $\{f(\mathbf{r})\}$ or $\{u(\mathbf{r})\}$, but to the physical field $\{c(\mathbf{r})\}$ defined by $\{\theta(U-u(\mathbf{r})\}$.

References

1) Gunton, J.D. and Droz, M., "Introduction to the Theory of Metastable and Unstable States", Lecture Notes in Physics 183. (Springer-Verlag, 1983)
2) Furukawa, H., Adv. in Phys. **34**, 703 (1985).
3) Kawasaki, K. and Ohta, T., Physica **118a**, 175 (1983).
4) Ohta, T., Ann. of Phys. **158**, 31 (1984).
5) Lifshitz, I. M. and Slyozov, V. V., Jour. Phys. Chem. Solid **19**, 35 (1961).
6) Cahn, J. W. and Hilliard, J. E., Jour. Chem. Phys., **28**, 258 (1958).
7) Kadanoff, L. P., Goetze, W., Hamblen, D., Hecht, R., Lewis, E. A. S., Palciauskas, V. V., Rayl, M., Swift, J., Aspnes, D. and Kane, J., Rev. Mod. Phys. **39**, 395 (1967).
8) Allen, S. M. and Cahn, J. W., Acta Metal. **27**, 1085 (1979).
9) Mandelbrot, B. B., "The Fractal Geometry of Nature", (Freeman, 1982).
10) Adler, R. J., "The Geometry of Random Field", (John-Wiley & Sons, 1980).
11) Ziman, J. M., "Models of Disorder", (Cambridge Univ. Press, 1979).
12) Kawasaki, K., Yalabik, M. C. and Gunton, J. D., Phys. Rev. **B17**, 455 (1978).
13) Ohta, T., Jasnow, D. and Kawasaki, K., Phys. Rev. Letters **49**, 1223 (1982).
14) Langer, J. S., Bar-ron, M. and Miller, H. D., Phys. Rev. **A11**, 1417 (1975).
15) Tomita, H., Prog. Theor. Phys. **59**, 1116 (1978).
16) Tomita, H., Prog. Theor. Phys. **75**, 482 (1986).
17) Monroe, M. E., "Modern Multidimensional Calculus", (Addison-Wesley, 1963).
18) Porod, G., in "Small Angle X-ray Scattering", ed. Glatter, O. and Kratky, O., (Academic Press, 1982).
19) Tomita, H., Prog. Theor. Phys. **71**, 1405 (1984).
20) Spooner, S., in "Kinetics of Aggregation and Gelation", ed. Family, F. and Landau, D. P., (North-Holland, 1984).

21) Weatherburn, C. E., "Differential Geometry of Three Dimensions", (Canbridge Univ. Press, 1927).
22) Tomita, H., Prog. Theor. Phys. **72**, 656 (1984).
23) Bale, H. D. and Schmidt, P. W., Phys. Rev. Letters **53**, 596 (1984).
24) Yeung, C., Phys. Rev. Letters **61**, 1135 (1988).
25) Furukawa, H., Phys. Rev. **B40**, 2341 (1989).
26) Kawasaki, K., Phys. Rev. **145**, 224 (1966); in "Phase Transition and Critical Phenomena", Vol.2. ed. Domb, C. and Green, M. S., (Academic Press, 1972).
27) Kendal, M. G. and Stuart, A., "The Advanced Theory of Statistics", (Charles Griffin, 1958).
28) Nash, C. and Sen, S., "Topology and Geometry for Physicists", (Academic Press, 1983).
29) Milnor, J., "Morse Theory", (Princeton Univ. Press, 1963).
30) Tomita, H. and Murakami, C., in "Dynamics of Ordering Processes in Condensed Matter", ed. Komura, S. and Furukawa, H., (Plenum Press, 1988).
31) Zallen, R. and Scher, H., Phys. Rev. **B4**, 4471 (1971).
32) Skal, A. S., Shklovskii, B. I. and Efros, A. L., JETP Letters **17**, 377 (1973).
33) Zallen, R., "The Physics of Amorphous Solids", (John Wiley & Sons, 1983).
34) Stauffer, D., "Introduction to Percolation Theory", (Taylor & Francis, 1985).
35) Erdelyi, A., "Tables of Integral Transformations", Chap.13. (MacGraw Hill, 1954).

DIFFUSION-LIMITED AGGREGATION AND RELATED PROBLEMS

Mitsugu MATSUSHITA

Department of Physics
Chuo University
Kasuga, Bunkyo-ku, Tokyo 112, Japan

ABSTRACT: Starting from the explanation of diffusion-limited aggregation (DLA) model, we discuss various aspects of DLA; properties of the fractal structure, a variety of related phenomena, theoretical attempts to derive the fractal dimension expression, inherent stochasticity, extension and generalization of the model, lattice anisotropy effect, growth probability distribution and effective growing surface. These problems are investigated mainly from the unified viewpoint based on our own mean-field theoretic approach.

1. INTRODUCTION

There exist various types of pattern formation in nature. In particular, the formation of large aggregates from small diffusing particles is a topic of considerable recent interest,[1-12] since it can be seen in a wide variety of physical, chemical and even biological processes. To understand the underlying mechanism is, therefore, very important in both pure science and technology. Among many nonequilibrium growth models proposed so far, particularly investigated has been the diffusion-limited aggregation (hereafter abbreviated as DLA) model introduced first by Witten and Sander.[13,14]

A DLA cluster grows from a seed particle through sticking of Brownian particles, one at a time, which are launched far away from the cluster. As discussed in detail later, it is now wellknown from both computer simulations and some theoretical approaches that DLA clusters have randomly ramified self-similar structure with universal fractal dimensionality D in that it seems to depend only on the Euclidean dimension d of the space in which the aggregation processes take place.

There are many phenomena related to DLA. The DLA model (or variously modified versions) has extremely wide applicability. This may be partly due to the model simplicity and partly due to the abundance of pattern-forming phenomena in the Laplace field. In spite of the model simplicity, however, the theoretical study has been confined to a phenomenological stage and the model itself is not well understood on a detailed level. The collaboration among experiments, numerical simulations and theoretical approaches are clearly needed.

DLA is one of the prototype models in the statistical physics of pattern formation. It should be stressed, therefore, that the development of the study of DLA contributes to the resolution of theoretically unsolved problems in many other far-from-equilibrium growth models proposed so far. In this review we would like to discuss various striking properties of DLA such as the inherent stochasticity, lattice anisotropy effect and multi-fractality. After describing briefly the model and related phenomena observed experimentally, we will try to present our own numerical results and theoretical (though still phenomenological) approaches to DLA.

2. MODEL

For simplicity let us take a square lattice as the pattern-forming stage and explain original Witten-Sander model[13)] for DLA. For more details see, e.g., Refs. 10, 15-17. The simulation is started by occupying a single site at the origin to represent a stationary "seed" particle. Mobile particles are then launched, one at a time, from random positions on a far-distant "launching" circle whose center is at the origin. The particles follow random-walk trajectory on the lattice. If the particle reaches a surface site of a cluster having already grown from the seed, i.e., an empty site with one or more occupied nearest-neighbor sites, it stops, stays at the site and becomes a member of the cluster. In other words, the particle sticks permanently to the cluster. The extension to higher dimensions or off-lattice simulations is straightforward.

The probability that a particle arrives for the first time from far away on a circle whose center is at the origin is equal at every site on the circle, so far as it encloses the cluster. Hence the launching circle need not have very large radius. Use is usually made of a radius of the launching circle of $R_L=(R_{max}+5)$ lattice units, where R_{max} is the maximum cluster distance from the origin. On the other hand, if a particle moves a long distance away from the cluster, the particle is discarded to save computer time. A new particle is then launched at a new random position on the launching circle. The radius R_K of the "killing circle" at which particle trajectories are terminated should be large enough to ensure that a particle reaching the circle has negligible probability of returning the cluster. This is particularly important for the two-dimensional (2-d) simulations, and recent, large-scale simulations[16,17]) typically use $R_K=100\ R_{max}$. In practice, however, $R_K \approx 3\ R_{max}$ seems to be adequate for small clusters. For higher dimensions R_K can be made smaller.

Figure 1 shows a typical 10^5-particle DLA cluster obtained by 2-d off-lattice simulations. As seen from the figure, DLA clusters have open, ramified random structures with no loops, consisting of branches of various sizes. They seem to have no characteristic lengths except the particle (branch thickness) and overall cluster sizes. In fact they are statistically self-similar. Since random-walkers have spreading random trajectories, they are easily captured by outwardly growing branches of the cluster (screening effect) and cannot enter deeply into "fjords" (outwardly open regions). DLA clusters seem to maintain

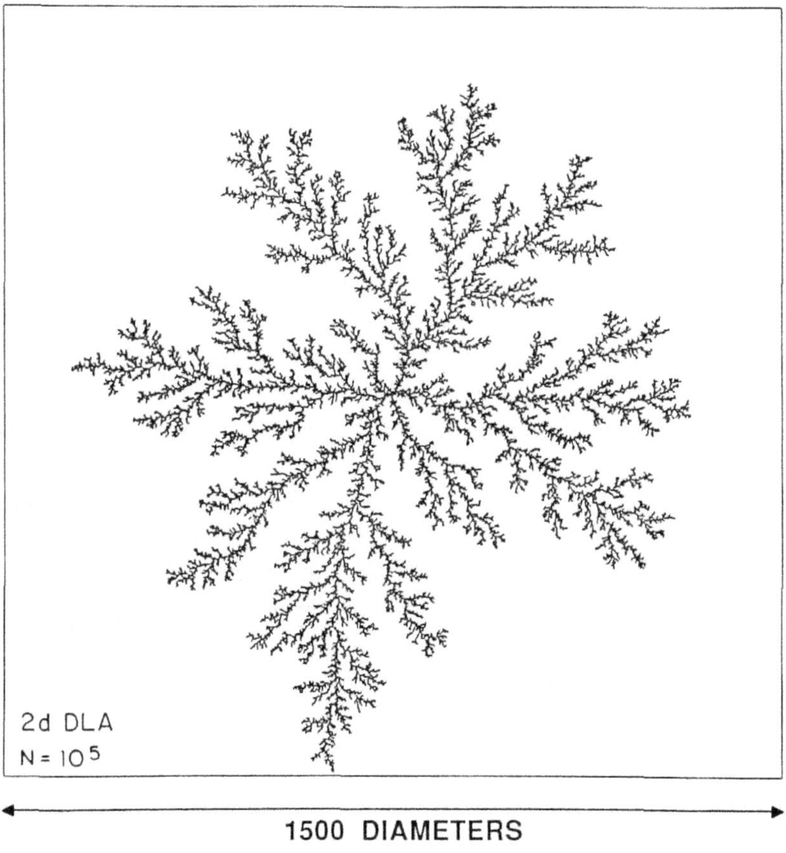

Fig. 1

their self-similar structure through the growth of branches and their screening effect.

Systematic attempts were first done by Meakin[15] to simulate the DLA model on hypercubic lattices for dimensions in the range 2-6. He confirmed the self-similarity of DLA clusters and obtained the dependence of the fractal dimension D on the spatial dimension d. This stimulated many theoretical approaches to predict the dependence of D on d. Based on their mean-field theoretic arguments, Muthukumar[18] and Tokuyama and Kawasaki[19] derived the expression

$$D = \frac{d^2+1}{d+1}. \tag{1}$$

Hereafter we call this MTK formula. This not only gives pretty good agreement with the simulation results, but also satisfies rigorous causality bound[20]; $d-1 \leq D \leq d$. In the following sections we will generalize eq.(1).

A number of modified versions of DLA were also proposed, e.g., by changing sticking probability of random walkers on the cluster surface to a value less than one[14] or by depositing random-walkers on a line or plane seed or substrate (diffusion-limited deposition; DLD). The global fractal structure in the former case does not alter except for the fact that branches become thick. In the latter case an entire pattern consists of individual clusters (trees) grown on the substrate, unlike usual DLA clusters. These individual clusters, however, have the same fractal structures as the original DLA clusters.

3. RELATED PHENOMENA

The growth mechanism of a DLA cluster is essentially composed of two processes; Brownian motion of particles and their local sticking to the cluster surface. It is well-known that Brownian motion can be described by a diffusion equation in the continuum limit. DLA is, therefore, a model idealizing the aggregation of diffusing particles onto an immobile cluster. Moreover, in DLA the cluster growth is so slow compared with the Brownian motion that the time-differential term in the diffusion equation can be neglected (quasi-static approximation). Hence DLA has the potentiality to explain many pattern growth phenomena occurring in the field which satisfies a Laplace equation. [For more extensive review on the subject in this section, see Refs. 10 and 21.]

In fact, it has been shown experimentally that various phenomena of random pattern formation are related to DLA. For instance, metal-leaves[22)] and electrodeposits[23)] grown under the diffusion-limited condition are aggregates of metallic ions, and their structures can be well described by two- and three-dimensional DLA. In particular, the screening effect inherent in DLA were clearly observed in the former experiment.

Static electric field is one of the most typical Laplace fields. Electric discharge such as lightening progresses slowly, however fast it looks, compared with characteristic electromagnetic variation rate, i.e., the light speed. Dielectric breakdown can, therefore, be explained by DLA. In fact, some Lichtenberg figure which is a sort of surface discharge was reported to be an example of DLA patterns.[24)] Strictly speaking, however, the dielectric breakdown should be investigated in terms of the generalized DLA, as will be discussed later.

Under some conditions DLA-like patterns are found in the sputter-deposited thin films of $NbGe_2$[25)] and Au.[26)] As in other thin-film systems, the growth processes observed here may involve such complex physical and chemical ones that the fundamental mechanism is not clear.

The Hele-Shaw cell is made of two parallel plates, between which a high-viscosity fluid such as glycerine is sandwiched. If a low-viscosity fluid such as air is injected through a small hole at the center of one of the plates, the interface between the two fluids becomes unstable. This phenomenon called Saffman-Taylor instability or viscous fingering[27] is also an example of DLA in two dimensions[28-30]: The pressure field in the high-viscosity fluid satisfies a Laplace equation and the local growth velocity of finger surface is proportional to the local pressure gradient at the surface.[27] The viscous fingering in porous media can also be explained by DLA.[31,32]

Perhaps the best-known diffusion-limited aggregation is dendritic crystal growth, exemplified by beautiful six-fold symmetric snowflakes. In fact, under some appropriate conditions one can observe dendritic crystals similar to DLA clusters.[33-37] The interrelation between more or less regular dendrites usually observed and DLA-like random fractal dendrites is very interesting, and will be discussed in more detail later.

When applying two-dimensional pressure on a phospholipid monolayer on the water/air interface, the crystallization of the constituent molecules takes place above some critical pressure. This process can actually be observed under a microscope if some fluorescent substance is contained in the phospholipid as an impurity.[38, 39] Under the condition of an appropriate amount of the impurity the dendritic crystals exhibit two-dimensional DLA-like patterns. The reason is that the diffusion on the free water/air interface of the impurity molecules which are excluded out of the phospholipid dendritic crystals controls the crystallization process.

Plaster is slightly soluble in water. A route in the plaster dissolved by water which is injected at appropriate flow rates was found to be describable in terms of

DLA.[40,41] In this case also, the pressure field of water permeated into the whole plaster satisfies a Laplace equation.

As seen from several examples cited above, DLA can explain random pattern formation of surprisingly many and seemingly very different phenomena. It is interesting and important not only to find other concrete examples of DLA but also to elucidate experimentally the crossover between DLA and other pattern formation mechanisms. To do so we might know essential properties of pattern formation more deeply.

4. HTM THEORY

4.1 Various Theoretical Attempts

Unfortunately there still do not exist fully convincing theories for DLA at present. Many theoretical attempts have, however, been made to derive analytical expressions of the fractal dimension D. We review here a few of them.

Tokuyama and Kawasaki[19] applied to this problem a mean-field theory analogous to Flory's theory of single polymer chains,[42] and derived that elegant MTK formula, eq.(1). Muthukumar[18] obtained it a little earlier by using the coherent potential approximation. Although several assumptions included in their theories seem to be unjustifiable, the MTK formula (1) yields an excellent agreement with computer simulation results.

Instead of the theory of single polymer chains, Hentschel[43] resorted to a Flory-type theory[44] for branched polymers which look certainly more similar to DLA clusters, and derived the following fractal dimension expression

$$D = \frac{5d^2+(2D_w-4)d+4D_w}{5d+(5D_w-4)} . \qquad (2)$$

Here d is the spatial dimension and D_w the fractal dimension of the particle trajectory. The formula can now be applied to the extended DLA in which diffusing particles follow fractal trajectory such as Lévy flight. In the case of usual random walk (Brownian motion; D_w=2) which corresponds to ordinary DLA, however, this expression ($D=(5d^2+8)/(5d+6)$) violates the causality bound ($D \geq d-1$)[20] unless $d \leq 14$. Note that eq.(1) satisfies this bound for any d. In addition, this relation gives not only quite a large disagreement with computer simulation results[45,46] for $1 \leq D_w \leq 2$, but also wrong behaviors for the case of d=1 and D_w arbitrary (D must be one), and D_w=1 and d arbitrary (D must be equal to d).

A DLA cluster grown on a square lattice becomes diamond-shaped when the cluster size N (the number of constituent particles) is around 10^5. Interesting attempt by Turkevich and Scher[47] and by Ball et al.[48] is based on the idea that the anomalous Laplace field in the vicinity of corners of a large-scale diamond-shaped cluster is responsible for the fractal growth. Assuming more generally that the cluster shape is a regular polygon, they obtained the fractal dimension expression as

$$D = \frac{3\pi-\theta}{2\pi-\theta} , \qquad (3)$$

where θ is an interior angle of the polygon. In fact, the case of diamond shape ($\theta=\pi/2$) yields D=5/3, in agreement with eq.(1) for d=2. Although eq.(3) gives a good agreement with computer simulation results, their theory does not seem to be decisive: Within their theory θ cannot be determined. In addition, more recent large-scale simulations reveal that a DLA cluster changes its shape from diamon-like to cross-like for $N>10^6$ on a square lattice.[49] Their theory is also meaningless for experimentally very important case in which diffusing and aggregated particles take off-lattice positions (off-lattice DLA) and, hence, an isotropic pattern is formed. Nevertheless, the idea of Turkevich and Scher is noteworthy in that they paid attention to the growth probability at the growing interface; it led to the study of multifractality in the growth probability distribution of DLA clusters.

4.2 HTM Theory

Based on the assumption that a DLA cluster is self-similar and resorting to a mean-field like argument, Honda, Toyoki and Matsushita (abbreviated as HTM)[50] have generalized the MTK formula (4.1). Let us discuss their theory in detail.

Let us first derive a few relations which hold under the sole condition that an outwardly growing pattern such as a DLA cluster is self-similar with the fractal dimension D. First of all, we have

$$N \sim R^D, \tag{4}$$

where N is the number of particles constituting the pattern and R is its radius of gyration. Let us define a radial "density" $\rho_r(r)$ of aggregated particles on a shell at r from the center of the pattern;

$$N \sim \int^R \rho_r(r)dr. \tag{5}$$

Then it must satisfy from (4) the following relation;

$$\rho_r(r) \sim r^{D-1}. \tag{6}$$

Let the pattern grow further by additional ΔN ($1 \ll \Delta N \ll N$) particles. The self-similar growth requires

$$N+\Delta N \sim (R+\Delta R)^{D}, \tag{7}$$

where ΔR is the increment of the radius of gyration. Then, as schematically shown in Fig.2, the radial "density" also increases as

$$\rho_r(r) \rightarrow \rho_r(r)+\Delta\rho_r(r), \tag{8}$$

because some particles can intrude within some penetration depth into various sizes of "fjords" surrounded by protrud-

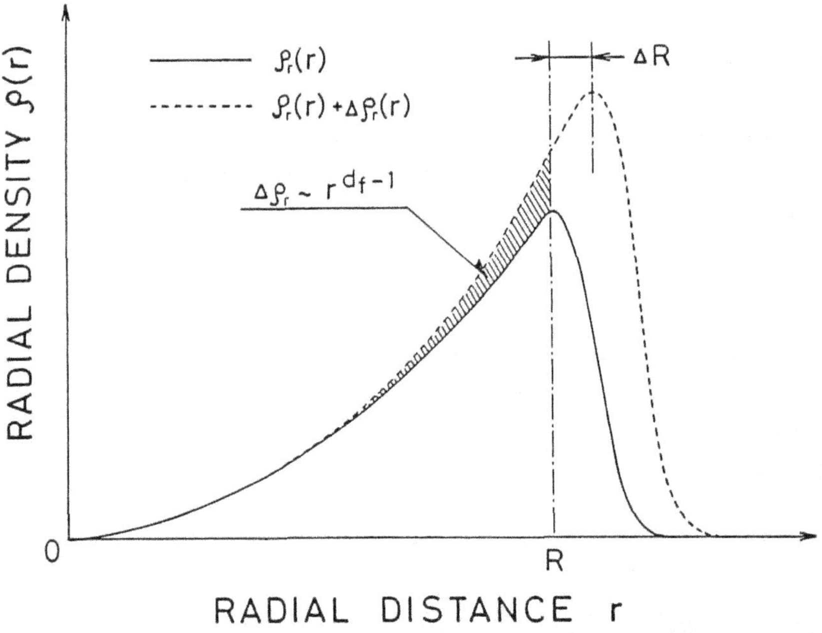

Fig. 2

ing outer branches.[51] From (4)-(8) the "density" increment $\Delta\rho_r$ must take the functional form of

$$\Delta\rho_r(r) \sim r^{D-1}, \tag{9}$$

which is the same one as that of ρ_r itself. This is the first important relation for the increment of the radial "density" of aggregated particles, obtained from the requirement that an outwardly growing pattern in question be self-similar.

The surface density $\sigma(r)$ of aggregated particles at the shell r can also be obtained easily;

$$\sigma(r) = \frac{\rho_r(r)dr}{S\ (r)dr} \sim r^{D-d}, \tag{10}$$

because $\rho_r(r)dr$ is the number of aggregated particles between two surfaces $S(r) \sim r^{d-1}$ and $S(r+dr)$ in the d-dimensional space.

One of the prominent features of a DLA cluster is that one can always recognize the existence of several main branches accompanied by characteristic void regions (main "fjords"), as shown by broken-line circles in Fig.3. In other words, a DLA cluster seems to maintain its self-similar structure through the growth of main branches with main fjords by their side[47,48]: Random-walkers coming from far away stick almost exclusively to the main branches. However, the structure of the front part of a growing pattern consisting of particles which have aggregated most newly (indicated by unhatched area of $\Delta\rho_r(r)$ in Fig.2), or tip part of the main branches which has grown through sticking of the majority of random walkers approaching from far away, is always incomplete from the viewpoint of the self-similar structure. The complete pattern formation may be accomplished a little inside the growing pattern or main fjords through sticking of the minority of random-walkers (indicated by hatched area in Fig.2) who happen to stray into there[51]: The majority of walkers are consumed around the front part of a cluster to make up the backbone of the cluster, whereas the minority finish to construct the complete self-similar structure a little inside the fjords.

An alternative way of disclosing the secret of DLA may, therefore, be to elucidate the mechanism of maintaining the self-similar structure of a growing pattern through the investigation on the growth (or diminution) of main fjords by the minority of stray walkers, instead of the growth of main branches, however paradoxical it sounds. In fact, (6),

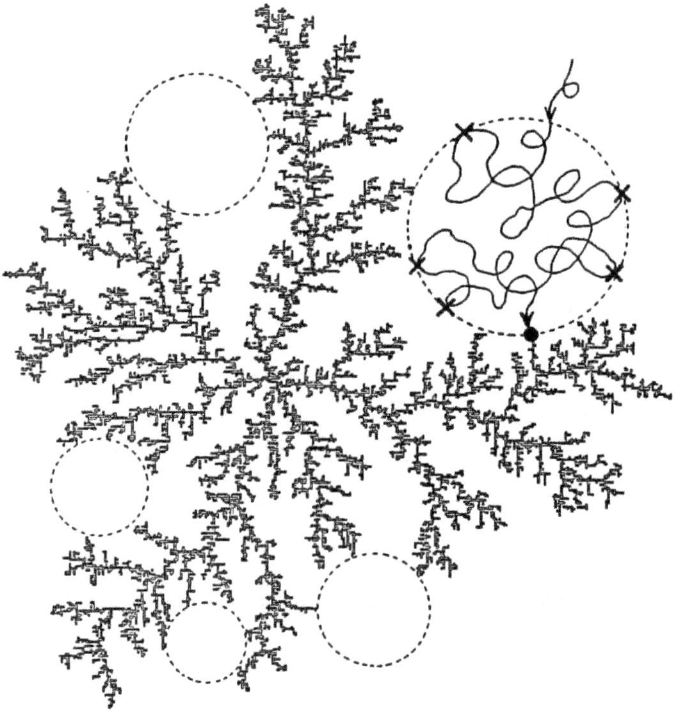

Fig. 3

(9) and (10) are only valid for $r<R$, as seen qualitatively in Fig.2. Let us call voids in the main fjords effective capture regions, shown by broken-line circles in Fig.3. The typical value of the diameter of the regions may be given by the penetration depth ℓ of random-walkers.

Random-walkers coming into these effective capture regions are almost certainly captured by any of the surface sites of a main branch, as schematically shown in Fig.3. This means that the increment of the radial density $\Delta\rho_r$ is proportional to the volume of the effective capture region ℓ^d;

$$\Delta\rho_r \sim \ell^d. \tag{11}$$

Many computer simulations tell us that the number of main branches and main fjords in DLA, i.e., the number of main effective capture regions, is almost constant at any growth stages (e.g., 4~5 for 2-d DLA). This fact may be considered to be one consequence of the self-similar growth of any growing patterns.

The relation (11) gives the number of random-walkers which are to stick to the surface sites of main branches and ultimately to be incorporated into the DLA cluster. In that sense this relation corresponds to that for the number of molecules which are to collide with a wall of a container and ultimately to contribute to give rise to the pressure in the kinetic theory of gases.

Let us next discuss the relation concerned with one particular random-walker. Following previous analogy of the gas pressure, one should next investigate on how one random-walker sticks to the cluster. A random-walker entering by chance into one of the effective capture regions continues to random-walk inside the region until it happens to come to one of nearest-neighbor sites of the cluster, as schematically shown in Fig.3. Let N_w be the number of steps of the walker inside the region. Then $N_w \sim \ell^{D_w}$, where D_w is the fractal dimension of the random-walker trajectory. The surface N_{ws} of a cloud made by the trajectory (indicated by x's in Fig.3) can then be estimated as

$$N_{ws} \sim \ell^{D_w - 1}. \tag{12}$$

Since the random-walker inside the region almost certainly sticks to the cluster, the aggregation probability which is given by the product of N_{ws} and the surface density σ of the cluster particles satisfies

$$N_{ws} \cdot \sigma \sim 1. \tag{13}$$

This leads to the balance between the walker trajectory and the cluster structure: If D_w is large, the cloud of the random-walker trajectory is dense. Then the surface density σ and ultimately D should become small. This tendency sounds quite reasonable. The relation (13) can also be interpreted as "chemical reaction" of one random-walker with one perimeter site which is to be occupied as the reaction product.

We can now estimate the penetration depth ℓ. The substitution of (10) and (12) into (13) gives

$$\ell \sim r^{\zeta}; \quad \zeta = \frac{d-D}{D_w - 1}. \tag{14}$$

In the case of $D_w=2$ (ordinary DLA) the exponent ζ reduces to that of Muthukumar's "screening" length ($\zeta = d-D$)[18] rather than that of Tokuyama and Kawasaki's screening length ($\zeta = (d-D)/2$)[19], although the present derivation is quite different from both of them.

We have now finished preparing everything to derive the expression of the fractal dimension D. Subsituting (14) into (11) and then comparing the resultant expression with (9), we obtain

$$D = \frac{d^2+(D_w-1)}{d+(D_w-1)} \qquad (15)$$

for the fractal dimensionality of DLA. Let us call it HTM formula. It should be noted that so far we have not imposed any restrictions on the values of both the embedding spatial dimension d and the dimension of the random-walker trajectory D_w. In fact, as is clearly seen in TABLE I, it

TABLE I. A comparison of theoretical values of fractal dimensionality due to HTM theory and those obtained by computer simulations.

d	D_w	D: HTM theory	D: Simulations	Ref.
2	2	5/3 = 1.67	1.71 ± .07	10)
3	2	5/2 = 2.50	2.50 ± .08	10)
4	2	17/5 = 3.40	3.31 ± .10	10)
5	2	13/3 = 4.33	4.20 ± .16	10)
6	2	37/7 = 5.29	~5.35	10)
2	4/3	13/7 = 1.86	1.85 ~ 1.88	45.46)
2	5/3	7/4 = 1.75	1.80 ~ 1.85	(Levy flight)
~1.89	~2.8	~1.46	1.40 ± .05	45,52)
~2.5	~3.5	~1.75	1.75 ± .10	(percolation clusters)
~1.89	?	1.60 (if D_w=d)	1.51 ~ 1.60	43)
~1.89	1	~1.89	~1.89	43)
d(=2,3)	1	d	~d	53)

turns out that eq.(15) gives an excellent agreement with a variety of computer simulations; not only ordinary DLA (d: positive integer, D_w=2) but also DLA of diffusing particles with Lévy flight trajectory[45,46] (d: positive integer, $1 \leq D_w \leq 2$) and DLA on a fractal pattern such as Sierpinski carpet[45] or percolation cluster[45,52] (d, D_w: any positive number ≥ 1). In the case of ordinary DLA (D_w=2), eq.(15) reduces to MTK formula, eq.(1). Equation (15) also satisfies the causality bound[20] described before.

5. STOCHASTICITY IN DLA

As shown so far, in spite of its model simplicity DLA has extreme fertility in explaining various experiments. Where does the fertility which fascinates us so much come from? Let us here discuss the stochasticity inherent in DLA and try to disclose the fundamental characteristics.[54] This will lead to a clue of the generalization of DLA and the deeper understanding of physical phenomena such as dendritic crystal growth.

First let us consider Monte Carlo method of numerically solving a Laplace equation

$$\nabla^2 \Phi = 0 \tag{16a}$$

with the boundary condition

$$\Phi(Q) = f(Q), \ Q \in \Gamma \text{ (boundary)}. \tag{16b}$$

Here Γ is a given boundary, Q any point on it, and f(Q) a function specifying the boundary condition. As schematically shown in Fig.4, let a random-walker start from a point P inside the region surrounded by the absorption-type boundary Γ. We define random-walker propagator p(P;Q) as the probability that the random-walker reaches a point Q on the boundary Γ. Then it satisfies

$$p(P;Q) = \frac{1}{z} \sum_{i=1}^{z} p(P_i;Q), \qquad (17)$$

where P_i is one of z nearest-neighbor points of P. The reason is that a random-walker starting from P and reaching Q inevitably passes one of z nearest-neighbor points of P and the probability of passing any one is all the same. Equation (17) is just a discrete version of a Laplace equation. This is also easily verified by performing the Taylor-series expansion of (17) around the point P. Since we assumed that the boundary Γ is of absorption-type, we have

$$p(Q;Q') = \delta_{QQ'} \quad (Q,Q' \in \Gamma) \qquad (18)$$

for the boundary condition of p(P;Q).

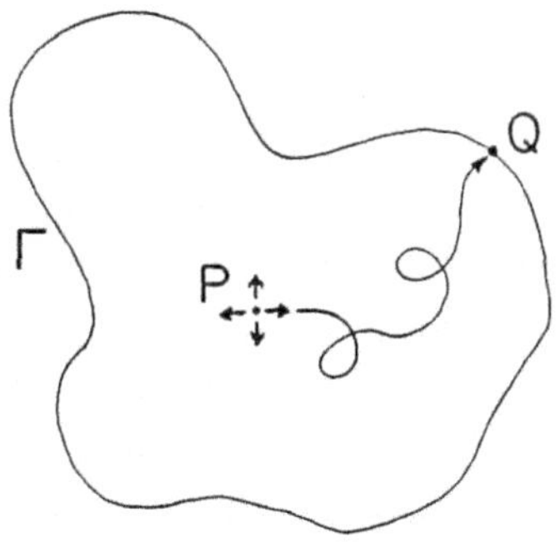

Fig. 4

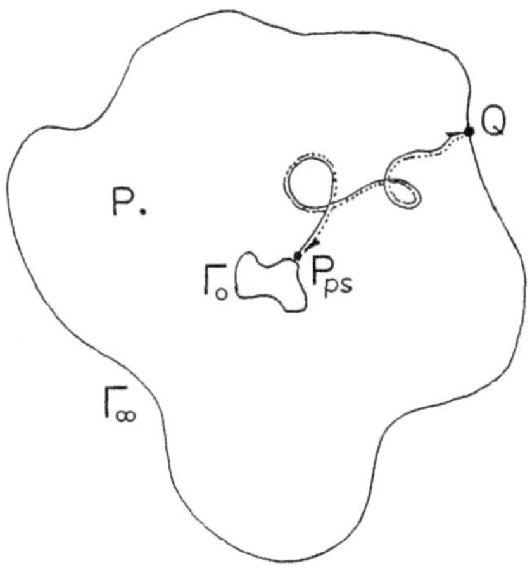

Fig. 5

The solution of the Laplace equation with the boundary condition, (16a,b), at a point P is then given by

$$\Phi(P) = \sum_{Q\in\Gamma} f(Q)p(P;Q). \tag{19}$$

As is easily shown, p(P;Q) ensures from (17) that this Φ(P) satisfies the Laplace equation, and f(Q) ensures the boundary condition from (18). Hence, if the probability p(P;Q) can be estimated by any means, we can obtain the solution Φ(P) through (19). In the Monte Carlo method we obtain p(P;Q) by launching a sufficient number N_P of random-walkers from a point $P(\notin\Gamma)$ and counting the number n_Q of walkers who arrive at any point Q on the boundary Γ. Then p(P;Q) is estimated as $\lim_{N_P\to\infty}(n_Q/N_P)$. This method has the advantage of enabling us to solve a Laplace equation, no matter how complicated a given boundary Γ is.

In order to investigate the relation between the Monte Carlo method described above and DLA and the fundamental characteristics of DLA, let us here consider the properties of the Laplace field Φ(P) under the boundary conditions compatible with the DLA problem. In DLA the surface of the cluster itself gives the absorption-type boundary, because random-walkers reaching perimeter sites P_{ps} of the cluster immediately become part of the cluster, i.e., $\Phi(P_{ps})=0$ ($\Phi_0=0$). We call the boundary Γ_0. There is another boundary Γ_∞ surrounding and far from Γ_0 which corresponds to the launching circle in DLA, as shown in Fig.5. [We have already mentioned that the launching circle may be close to the cluster to save the CPU time of simulations. It should be noted that we are here discussing the original Witten-Sander DLA model. Or Γ_∞ may be regarded as the killing circle which must be very far from the cluster.]

Let us now take the boundary condition that Φ(P) is constant on both Γ_0 and Γ_∞, and put $f(Q)=\Phi_0$ $(Q\in\Gamma_0)$ and Φ_∞

$(Q \in \Gamma_\infty)$, compatible with the DLA problem. One can easily obtain the relation

$$\Phi(P)-\Phi_0 = (\Phi_\infty-\Phi_0) \sum_{Q\in\Gamma_\infty} p(P;Q) \tag{20}$$

by using the equality

$$\sum_{Q\in\Gamma_0,\Gamma_\infty} p(P;Q) = 1, \tag{21}$$

which means that a random-walker having started from P eventually reaches any point Q on the boundary surrounding P.

Let us take as P a perimeter site P_{ps} on the cluster surface Γ_0. Then the left-hand side of eq.(20) is proportional to the gradient of Φ, and eq.(20) becomes

$$|\nabla\Phi(P_{ps})| \sim \sum_{Q\in\Gamma_\infty} p(P_{ps};Q). \tag{22}$$

Since the probability for a random-walker to come from P to P' via any possible path t, $p_t(P;P')$, is exactly equal to $p_t(P';P)$, and $p(P;P')=\Sigma_t p_t(P;P')$, we have in general $p(P;P')=p(P';P)$. This is a consequence of the time reversal nature of random walks. Applying this relation in the right-hand side of (22), we obtain

$$|\nabla\Phi(P_{ps})| \sim \sum_{Q\in\Gamma_\infty} p(Q;P_{ps}). \tag{23}$$

The right-hand side represents the probability that a random-walker having started from the boundary Γ_∞ reaches a perimeter site P_{ps} on the boundary Γ_0. The relation (23) means, therefore, that the flow of random-walkers at a point P_{ps} is proportional to the gradient of concentration field at the point.

Based on the results obtained above, let us rethink of DLA. In the case of the Monte Carlo method to solve a Laplace equation with the boundary condition for boundaries at rest, one must perform sufficient amount of trials of releasing random-walkers. In DLA, however, as soon as a random-walker having started from a point on the far distant boundary Γ_∞ reaches a perimeter point P_{ps} on the cluster surface Γ_0, it sticks to the cluster and stays there as a member of cluster constituents: In DLA, only one trial is performed to grow one perimeter site. Hence, DLA is summed up to have two important characteristics. First, from eq.(23), the growth probability of a DLA cluster at a perimeter site P_{ps}, $p_g(P_{ps})$, is proportional to the gradient of the Laplace field at the site;

$$p_g(P_{ps}) = \sum_{Q\in\Gamma_\infty} p(Q;P_{ps}) \sim |\nabla\Phi(P_{ps})|. \quad (24)$$

Secondly, since a DLA cluster grows through sticking, at most one at a site, of random-walkers, the selection of growth sites is stochastic with no averaging procedures. DLA is, therefore, regarded as the pattern formation in which a growing pattern is under the influence of strong random perturbations in a Laplace field and its growth probability is given by (24).

From these discussions, conversely, the following very interesting problems arise immediately: (1) How does the pattern structure change if one takes as the growth probability some function of the concentration gradient, instead of (24)? This is an important problem leading to the generalization of DLA. (2) Are there not any ways to weaken the random fluctuations? Or, how does the pattern change if applying some regular perturbations competing with the random fluctuations? This inspires us to introduce some sort of averaging methods to determine growth sites such as

Monte Carlo averaging[10,12] or to investigate the effect of lattice anisotropy on the pattern structure. In particular, the last effect clearly suggests the relevance to dendritic crystal growth. We will discuss these problems in more detail in the follwing sections.

One important remark: In order to investigate proper characteristics of DLA clusters the exterior boundary Γ_∞ must be very far from the cluster surface Γ_0. Only then, as seen from (24), the growth probability p_g does not depend on the details of Γ_∞. Otherwise, a cluster grows under the strong influence of the shape of Γ_∞. Taking two-dimensional simulations as an example, while the launching circle from which random-walkers are released may be close to a cluster as far as it does not cross the cluster, the radius of the killing circle where random-walkers moving too far away from the cluster are discarded must be large enough.

6. GENERALIZATION OF DLA

6.1 Generalized DLA (η-model)

From the viewpoint of general pattern formation problem there is no reason to restrict the local growth probability to (24) even within the pattern formation in a Laplace field. In fact, Niemeyer et al. (NPW)[24] proposed the dielectric breakdown model in which the local growth probability of a breakdown pattern at a perimeter site P_{ps} is given by

$$p_g(P_{ps}) \sim |\nabla\Phi(P_{ps})|^\eta, \tag{25}$$

where η is any positive number, and the electrostatic field $\Phi(P)$ outside the pattern satisfies a Laplace equation. Hence, the field Φ corresponds exactly to the random-walker concentration discussed in the previous section. Comparing (25) with (24), the case of $\eta=1$ corresponds to ordinary DLA model. Moreover, the model asymptotically includes Eden model in that when $\eta \to 0$ every perimeter site of a growing pattern has the same growth probability. In this sense the NPW dielectric breakdown model can be called generalized DLA. [This is also called η-model.]

As seen from (25), the generalized DLA can be applied to cases where a cluster does not necessarily grow through real material flow into the cluster. For instance, in the dielectric breakdown neither electrons nor ions are flowing in the undischarged region. In this region only the quasi-static electric field which satisfies a Laplace equation is changing according to the progress of discharge. Since the probability of progress of discharge at the surface of already discharged pattern is not necessarily proportional to electric field $|\nabla\Phi|$ there, it is more general to assume (25) and regard η as being dependent on the property of dielectrics. In the viscous fingering also, the growth is not governed by real material flow into fingers. In this case, however, the growth velocity is proportional to the gradient of pressure which satisfies a Laplace equation (Darcy's law). Hence, the viscous fingering has $\eta=1$, i.e., it is regarded as an example of ordinary DLA.

6.2 Monte Carlo Simulation of the Generalized DLA

NPW grew generalized DLA clusters as follows: First solve Laplace equation $\nabla\Phi=0$ numerically, then determine a growth site stochastically according to (25), and repeat the procedure. In the following, however, we show how to simulate the generalized DLA by means of pure Monte Carlo method,[54-56] as in ordinary DLA. This method can reduce the CPU time of simulations more than NPW method[24,57] for simple values of η such as $\eta=1/2$ or 2. Besides, it will turn out to give an important clue to derive the expression for the fractal dimension of the generalized DLA.[54)]

Let us take, e.g., $\eta=2$. Launch a random-walker far away from a cluster, following usual procedure of DLA simulations. If the random-walker reaches a perimeter site, P_{ps}, of a growing cluster, it is released again from there. When and only when that walker goes far away from the cluster, the site P_{ps} is occupied and becomes part of the growing cluster, as schematically shown in Fig.6. If the walker released from P_{ps} touches any perimeter site of the cluster, kill the walker, leave the site P_{ps} unoccupied and launch another walker from far away. In this case the local growth probability at P_{ps} is given by $p_g(P_{ps})\sim|\nabla\Phi(P_{ps})|^2$, because this event is equivalent to the simultaneous arrival of two walkers at the same site. Here again use is made of the symmetry relation of the random-walker propagator; $p(P;P')=p(P';P)$. This simulation method can easily be

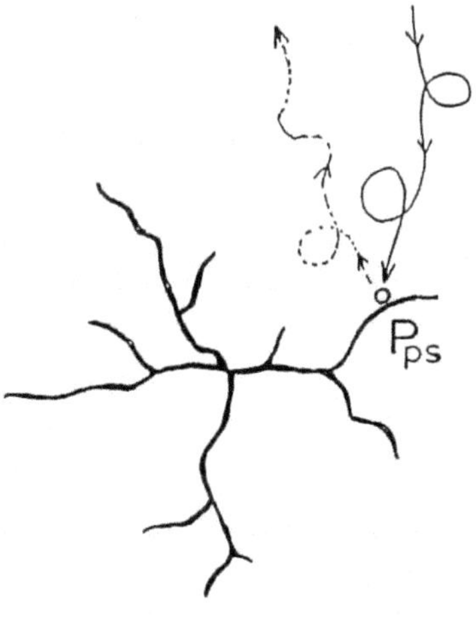

Fig. 6

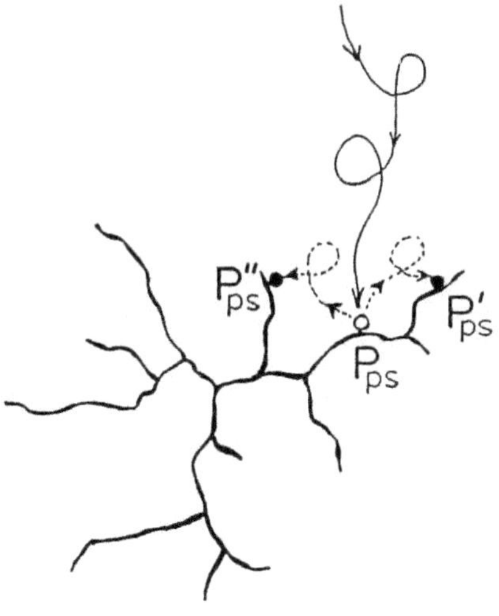

Fig. 7

extended to the case

$$p_g(P_{ps}) \sim |\nabla\Phi(P_{ps})|^n, \tag{26}$$

where n is any positive integer. It can be easily imagined that the larger the value of n is, the more stringy the cluster becomes, since the sreening effect is enhanced more for larger n.

On the other hand, the case of $\eta<1$ is not as clear as above. Nevertheless we propose the following simulation procedure: Let us take $\eta=1/2$. Again launch a random-walker from far away. If the walker reaches a perimeter site P_{ps}, then release two random-walkers from there. When and only when both of them touch any perimeter sites P_{ps}' and P_{ps}'' as schematically shown in Fig.7, the two sites P_{ps}' and P_{ps}'' (not P_{ps}) are occupied and become part of the growing pattern. Growth events at the two points P_{ps}' and P_{ps}'' are independent from each other, and the total event is determined by the event that the first walker reaches the point P_{ps}. Hence we have the relation $p_g(P_{ps}')\cdot p_g(P_{ps}'')\sim |\nabla\Phi(P_{ps})|$. Since the two points P_{ps}' and P_{ps}'' are expected statistically to be close to the point P_{ps}, and DLA itself contains some self-averaging nature from random growth processes, one may not lose any essential or global features of the pattern by assuming $P_{ps}'\approx P_{ps}''\approx P_{ps}$. Hence we obtain $p_g(P_{ps})\sim|\nabla\Phi(P_{ps})|^{1/2}$ for this case. This procedure can also be easily extended to the case

$$p_g(P_{ps}) \sim |\nabla\Phi(P_{ps})|^{1/m}, \tag{27}$$

where m is any positive integer. And a pattern is expected to become more compact as m is increased more.

Combining two simulation procedures (26) and (27), we can simulate the case (25) with $\eta=n/m$, where n and m are any positive integers. Of course this method is useful only when n and m are small. If η is large or has complicated fractional part, it is wiser to use NPW method.[24)]

The most important advantage of the present method is that it enables us to carry out off-lattice as well as on-lattice simulations for the generalized DLA. Independently of this, Meakin proposed the following simulation method[58)]: Place a counter at every perimeter site of a cluster, and record the number of random-walkers who have arrived the perimeter. When the number reaches some specified value k (any positive integer) at any perimeter site, the site is occupied and becomes part of the cluster. At the same time, reset the number of counters to zero at every perimeter site including newly emerged perimeter sites. This corresponds to the simulation of the generalized DLA with η=k. However, this method can be used only for on-lattice simulations. In particular, since the effect of background lattice anisotropy appears seriously when η>1,[55,58)] our method described above is indispensable to the investigation of inherent properties of the generalized DLA clusters.

In Fig.8 we show typical simulation results obtained by using our method for several values of η.[55)] In the figure the values of η are, respectively, (a) 1/3, (b) 1/2, (c) 1 (ordinary DLA), (d) 2, (e) 3. As clearly seen from the figure, patterns become gradually stringy as η is increased.

6.3 Fractal Dimension Expression for the Generalized DLA

The argument in the previous subsection, especially the simulation method of the generalized DLA, suggests the possibility of extending HTM theory developed in Sec.4 to the generalized DLA. In this subsection we will derive the fractal dimension expression for the generalized DLA.

The local growth probability of a perimeter site P_{ps} of a growing cluster given by (26) is the probability of simultaneous arrival of n random-walkers at the site P_{ps}.

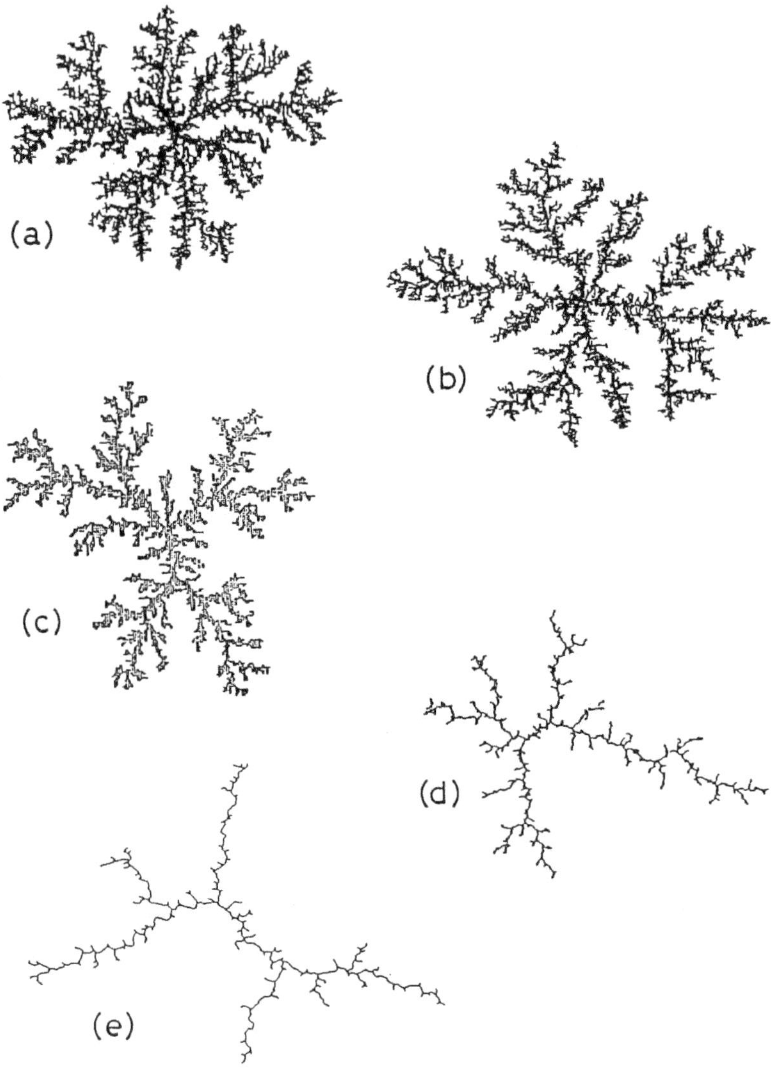

Fig. 8

From the viewpoint of chemical reaction this event can be regarded as the reaction of n random-walkers with one perimeter site which is to become an occupied site as the reaction product. Then, for this case one must replace (13) by $N_{ws}^{n} \cdot \sigma \sim 1$. On the other hand, (27) can be interpreted as the chemical reaction of one random-walker with m perimeter sites which are to be occupied as the reaction products. In this case (13) must be replaced by $N_{ws} \cdot \sigma^{m} \sim 1$. Hence the general case given by (25) with $\eta = n/m$ (n,m: positive integers) can be regarded as the chemical reaction of n random-walkers with m perimeter sites to be occupied, which leads to

$$N_{ws}^{n} \cdot \sigma^{m} \sim 1. \tag{28}$$

It should be noted that the relations (4)-(11) are very general in that they are valid for every growing fractal. The relation (12) is just for random-walkers only. They all do hold even after generalizing DLA. Hence the substitution of (10) and (12) into (28) yields the penetration depth

$$\ell \sim r^{\zeta'}; \; \zeta' = \frac{d-D}{\eta(D_w-1)}, \tag{29}$$

for the generalized DLA, where $\eta = n/m$. Substituting this into (11) and then comparing the resultant with (9), we finally obtain

$$D = \frac{d^2 + \eta(D_w-1)}{d + \eta(D_w-1)}, \tag{30}$$

as the fractal dimension expression for the generalized DLA model.[54)]

As expected, the formula (30) is very simple and natural extension of the HTM formula (15). Namely, putting

$\eta=1$ in (30) yields (15), which reduces to the MTK formula (1) by putting $D_w=2$ (ordinary DLA). In addition, since the formula gives $D=d$ for $\eta=0$, it explains the fact that Eden clusters are compact. Asymptotically, $D\to1$ as $\eta\to\infty$, which is also quite a reasonable result. In Fig.9 we show η-dependence of the fractal dimension D given in (30) for d=2 and $D_w=2$ (indicated by a broken line) as well as some simulation results (indicated by vertical bars).[55] The agreement is good.

Recently Satpathy[59] carried out the generalized DLA simulations in three dimensions and obtained the results shown in TABLE II. To compare with them, we also show theoretical results obtained by putting d=3 and $D_w=2$ in (30) and by Turkevich and Scher (TS),[60] who extended their theory[47] to higher dimensions. Our theoretical results are in much better agreement with the simulation results.

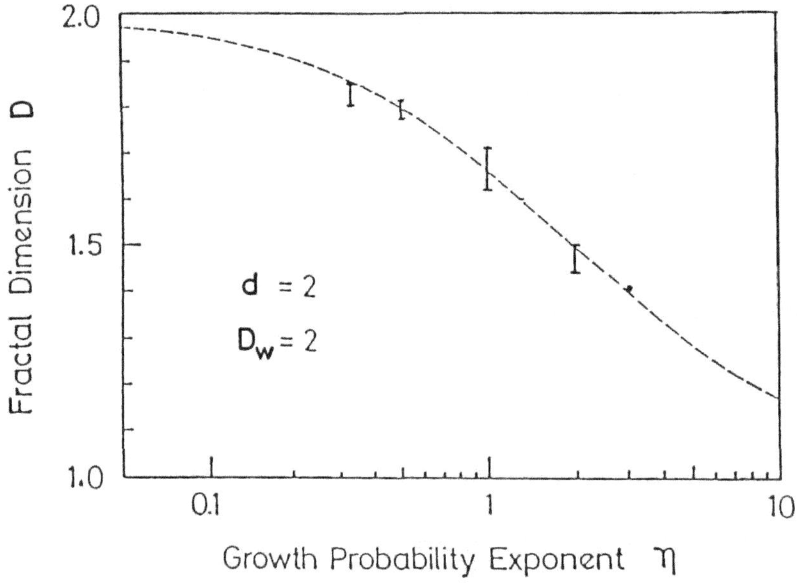

Fig. 9

More important fact derived from (30) may be that putting η=1 gives the same result D=d as putting D_w=1. This confirms very naturally the fact suggested by recent computer simulations that Eden model (η=0) and ballistic aggregation model (D_w=1) belong to the same universality class.

TABLE II. Comparison of simulation and theoretical results for the generalized DLA in three dimensions.

η	Simulation[a]	Present theory[b]	TS theory[c]
1	2.48± .06	2.50	2.50
2	2.11± .06	2.20	2.00
3	1.96± .08	2.00	1.50
4	1.75± .06	1.86	1.00

a: Ref.59, b: Ref.54, c: Ref.60.

7. DLA AND DENDRITIC CRYSTAL GROWTH

As described before, DLA is regarded as cluster growth in the diffusion field, because particles sticking to a cluster approach there from far away through pure random walk (Brownian movement). Crystal growth which is one of the most familiar phenomena not only in science and technology but in everyday life is also typical pattern formation in the diffusion field. There must be some close relationship between the two, which will be discussed in this section.

The mode of crystal growth is classified roughly into two categories. One is interface-controlled crystal growth, which is found for small degrees of supercooling in melt growth or weak supersaturation for solution growth. In this case the crystal morphology is determined mainly by the kinetics of atoms or molecules on crystal surfaces such as surface diffusion or migration. A growing crystal has a simple non-fractal structure with flat facets such as a cube. Hence one may denote it as "reaction-limited" crystal growth (RLCG). The other is diffusion-limited crystal growth (DLCG). In this case supercooling or supersaturation is so strong that flat facets cannot be maintained when a crystal is growing (e.g., owing to fast aggregation of diffusing solute on a crystal surface for solution growth). This brings about interfacial instability[61,62] and eventually leads to dendritic crystal growth. Here we consider only the latter, diffusion-limited crystal growth or dendritic crystal growth.

In contrast to irregularly branched and chaotic DLA patterns as shown in Fig.1, dendritic crystals have more or less regularly branched structures exemplified by snow crystals.[63] One can easily imagine that the difference comes from the existence of crystalline anisotropy. It is, therefore, very interesting to understand under what conditions DLA-like or dendritic crystal growth occurs and to experimentally realize the two contrasting types of crystal growth in one material.

Recent extensive studies of dendritic crystal growth models,[64] called geometrical model[65,66] and boundary-layer model,[67,68] revealed numerically a crossover from stable tip growth to unstable tip-splitting as crystalline anisotropy is decreased. Although these are extremely simplified growth models, tip-stable, tip-oscillating and tip-splitting types of dendritic crystal growth predicted by them were observed in supersaturated NH_4Cl solution.[69] Varying the degree of anisotropy for the Hele-Shaw cell in their viscous fingering experiment which is considered as the hydrodynamic analog of two-dimensional dendritic crystal growth, Ben-Jacob et al.[70] demonstrated the existence of two regimes of growth clearly reminiscent of irregular DLA and regular dendritic crystal growth. Sawada et al.[71] and Grier et al.[72] observed in the experiment of electrodeposition between two parallel plates that morphology of zinc-leaf which usually has DLA structure[22] exhibits transition-like behavior from DLA to dendritic growth when increasing applied voltage or electrolyte solution concentration. It is now wellknown that incorporating the effect of surface tension one way or another into the on-lattice DLA simulations gives rise to more or less regular dendrite-like patterns.[10,12]

All these results imply that the diffusion-limited crystal growth can still be classified into two groups; (i) DLA-like crystal growth when random fluctuations are very strong or lattice anisotropy is very weak,[54] and (ii) dendritic crystal growth when otherwise. Crystalline anisotropy is usually so strong that DLA-like crystal growth is difficult to be observed. This may be why this type of crystal growth is not familiar experimentally. [In nature, however, one often finds natural dendrites of manganese dioxide grown on the cleaved surface of shale or limestone which are clearly reminiscent of two-dimensional DLA patterns. One example is seen on the slab surface of the holotype fossil of Archaeopteryx lithographica found in the lithographic Solnhofen Limestone near Pappenheim (Bavaria) in 1861.[73]]

From this viewpoint Honjo et al.[21,33] investigated the growth of ammonium chloride dendrites from supersaturated aqueous solutions. The growth cell consists of two parallel glass plates (similar to Hele-Shaw cell in the viscous fingering) with uniform spacing of about 5 μm. If both glass plates have smooth surfaces, a regular dendrite such as that shown in Fig.10(a) is formed.[69] On the other hand, if the upper surface of the lower glass plate is roughened (scratched by abrasives), an irregular pattern similar to that associated with two-dimensional DLA is obtained. A typical example is shown in Fig.10(b). Honjo et al.[33] investigated the growth by analyzing digitized micrographs. They found that the dependence of the radius of gyration on the mass (pattern area) of NH_4Cl during the dendrite growth process (such as shown in Fig.10(b)) gave a fractal

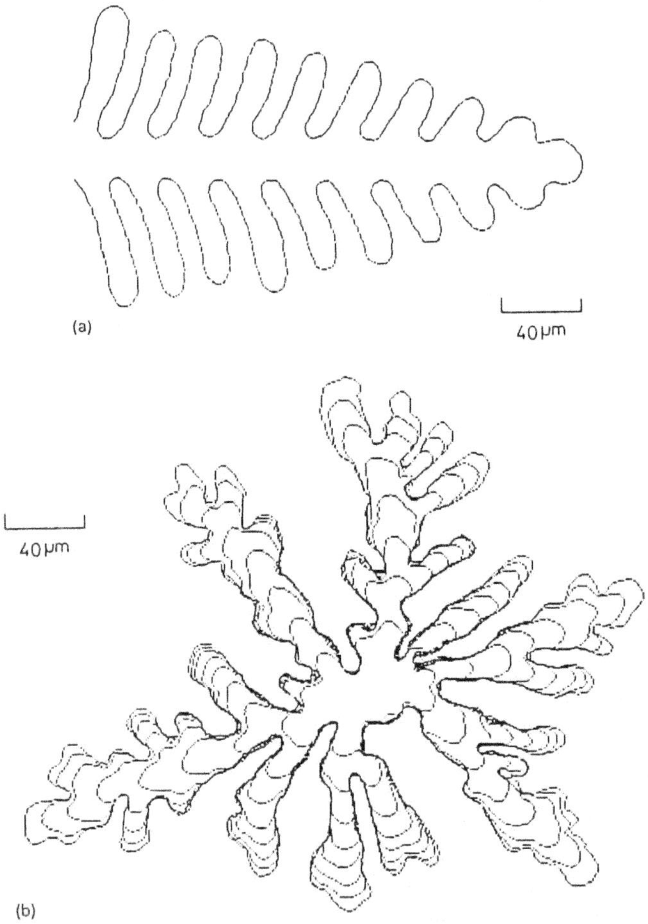

Fig. 10

dimension of about 1.67, which is in good agreement with the value expected for two-dimensional DLA. This clearly suggests that the irregularity of the cell surface effectively weakens the crystalline anisotropy to give rise to DLA-like random fractal growth. Honjo et al.[34] extended their experiment to the case of melt growth of succinonitrile by using the same experimental setup as above. By increasing both the degrees of roughness of the glass plate and supercooling they found characteristic morphological change in growth patterns from asymmetric tip-splitting (reminiscent of DLA patterns) through symmetric tip-splitting to tip-oscillating modes.

These experiments clearly confirm the statement described above that the diffusion-limited crystal growth can still be classified into at least two groups due to relative strength of random fluctuations and lattice anisotropy; DLA-like growth with randomly branched fractal structures and dendritic growth with more or less regularly branched structures.

8. EFFECTIVE GROWING SURFACE

As discussed in a first few sections, growing random fractal patterns are commonly seen in many physical, chemical and biological processes.[1-12] The fractal dimension is usually used to characterize a fractal pattern. It is, however, obvious that it cannot fully characterize such a growing pattern as a DLA cluster. The surface of a growing random pattern must play an important role in the formation of the fractal structure. In fact, surface properties have recently been studied extensively for a variety of models such as Eden,[74-77] DLA,[74,78-84] diffusion-limited cluster-cluster agregation,[79] screened-growth,[79,80,83] and ballistic deposition[76,80,85] models. Indeed, of practical importance are the surface sites which can actually grow, instead of the entire surface. In this section we will discuss the generalized dimension D_q which can characterize the effective surface of a growing fractal pattern, and derive the expression of the information dimension D_1 of the unscreened surface of the pattern on the basis of HTM theory and its generalization discussed in Secs. 4 and 6.[86)]

A small region (box) of size ε around the i-th site on the perimeter $\partial\Omega$ of a growing random pattern Ω has its own measure, the growth probability $p_i(\varepsilon)$. Then the investigation of the scaling properties of the growth probability distribution (GPD) $\{p_i(\varepsilon)\}$ may give us a clue to fully characterize growing random patterns.[87)] Or, more generally, there exists an intriguing problem of how to define the effective surface of randomly ramified fractal objects where random-walkers launched far away from them as probe particles can really reach.[88)] This effective "unscreened" surface may be much smaller than the entire surface. In this case also each perimeter site has its own probability (harmonic measure) for the random-walkers to reach there, and the probability distribution (PD) $\{p_i(\varepsilon)\}$ can be defined. Hereafter we will pay attention to the scaling properties of this PD of a fractal object, whether it is a growing pattern or not.

From the set $\{p_i(\varepsilon)\}$ one can construct an infinite hierarchy of generalized dimension D_q as follows[89)]:

$$D_q = \lim_{\varepsilon\to 0}\{(q-1)^{-1}\ln[\sum_{i=1}^{N(\varepsilon)}[p_i(\varepsilon)]^q](\ln\varepsilon)^{-1}\}, \qquad (31)$$

where $N(\varepsilon)$ is the number of boxes of size ε necessary to cover the perimeter (or the support of the probability measure) $\partial\Omega$. The fractal dimension of $\partial\Omega$ is given by $D_0 = \lim_{q\to 0}D_q$. As a consequence of the definition (31), the generalized dimension D_q with $q>0$ is relevant to the surface sites with larger probability, and important to the growing-pattern problems. In any case, one can extract detailed information on the probability distribution from the D_q spectrum by tuning q. If the PD on $\partial\Omega$ is uniform, i.e., $p_i(\varepsilon)=N(\varepsilon)^{-1}$, then $D_q=D_0$ for any q. In general, $D_q \geq D_{q'}$ for $q<q'$.[89] The so-called information dimension D_1 ($=\lim_{q\to1} D_q$) is rewritten in a more convenient form as

$$D_1 = \lim_{\varepsilon\to0}\{[\sum_{i=1}^{N(\varepsilon)} p_i(\varepsilon)\ell n p_i(\varepsilon)](\ell n\varepsilon)^{-1}\}. \qquad (32)$$

D_2 and higher D_q's are related to the scaling of two-point and higher correlation functions of the PD on $\partial\Omega$.

An alternative way to define the information dimension D_1 is the following.[90] Suppose that probe particles are launched far away from a fractal pattern of N particles and undergo random walks until they happen to come in contact with the surface, which absorbs them without changing the pattern. Let us denote the fractal dimensions of the pattern and of the trajectory of random-walkers as D and D_w, respectively. Because of screening, a relatively small number N_u of "unscreened" surface sites can absorb the bulk of these probe particles. This number is scaled as

$$N_u \sim R^{D_1}, \qquad (33)$$

where R is the size of the fractal pattern, say, the radius of gyration. For two dimensions it has been verified both phenomenologically[90] and mathematically[91] that $D_1=1$, independent of D, when Brownian particles ($D_w=2$) are used as probe particles. On the basis of the mean-field argument, Coniglio and Stanley[88] derived the expression

$$D_1 = D-1+\frac{d-D}{D_w}, \qquad (34)$$

where d is the spatial dimension. Unfortunately this expression does not give a correct value for d=2 and $D_w=2$.

We will now derive a new formula for D_1 by resorting to our mean-field theoretic arguments[50,54] discussed in Secs. 4 and 6.

Let us again use the radial density $\rho_r(r)=\rho(r)r^{d-1}$. From the relation $N\sim\int^R\rho_r(r)dr\sim R^D$ [eqs.(4) and (5)], one has $\rho_r(r)\sim r^{D-1}$ [eq.(6)] (or $\rho(r)\sim r^{D-d}$). These relations are still valid in the present case. Let us also use a mean penetration depth ℓ of probe particles invading into a fractal pattern from outside. The arguments given in Sec.4 leads to exactly the same expression of ℓ as eq.(14). The number of unscreened sites of the pattern is then given by

$$N_u \sim \rho_r(R)\ell(R) \sim R^{D-1}\ell(R). \tag{35}$$

Substituting eq.(14) into this and comparing resultant expression with (33), we have

$$D_1 = D-1+\frac{d-D}{D_w-1}. \tag{36}$$

The difference from (34) seems to be trivial, but turns out to be very important.

When Brownian particles are used as probe particles ($D_w=2$), then from (36) one has the simple formula

$$D_1 = d-1, \tag{37}$$

independent of the fractal dimension of the pattern. This gives the correct value $D_1=1$ for d=2.[90,91] Meakin and Witten[78] found numerically that $D_1=1$ and 2 for DLA patterns in d=2 and 3, respectively, which agrees well with the present expression. [It should be noted that in DLA the harmonic measure is equivalent to the GPD.] For the extended DLA of launching random-walkers following trajectories with a fractal dimension D_w, we have derived in Sec.4 the fractal dimension expression $D=(d^2+D_w-1)/(d+D_w-1)$ [eq.(15)]. Substituting this into (36), one has the information dimension for the GPD of extended DLA patterns

$$D_1 = \frac{(d+1)(d-1)}{d+D_w-1}. \tag{38}$$

In particular, for ballistic aggregates ($D_w=1$) we obtain $D_1=d-1/d$. This is in good agreement with recent simulations by Meakin et al.[85] for d=2, but not for d=3.

In the generalized DLA or η-model discussed in Sec.6, the growth probability at any perimeter site P_{ps} of the pattern is given by $p_g(P_{ps}) \sim |\nabla\Phi(P_{ps})|^{\eta}$, where Φ is the Laplace field outside the pattern. In particular, cases with the exponent η=1 and 0 correspond to ordinary DLA and the Eden model, respectively. In this case the penetration depth ℓ is given by eq.(29). Substituting this into (35) and comparing resultant expression with (33), we have

$$D_{1\eta} = D-1+\frac{d-D}{\eta(D_w-1)} \tag{39}$$

for the information dimension of the PD due to the growth rule of the η-model. If we use as D in (39) the fractal dimension formula, eq.(30), for this generalized DLA or η-model, we have

$$D_{1\eta} = \frac{(d+1)(d-1)}{d+\eta(D_w-1)} \tag{40}$$

as the information dimension of the GPD for the generalized DLA. This implies that the Eden model (η=0) and the ballistic aggregation model ($D_w=1$) belong to the same universality class in the sense that not only the fractal dimensions of the bulk structure [eq.(30)] but also the information dimension of the surface [eq.(40)] are equal for both models. This is now firmly confirmed by computer simulations. In Fig.11 is shown the η-dependence of $D_{1\eta}$ for the pattern of the generalized DLA in two dimensions, $D_{1\eta}=3/(2+\eta)$, for d=2 and $D_w=2$, as well as a few preliminary simulation results. The numerical results indicated by open circles in the figure were obtained as follows. First, patterns of any specific η value (η=1,2,3) consisting of N particles are obtained by off-lattice Monte Carlo procedures discussed in Sec.6.[55] Second, the Laplace field Φ outside the patterns is solved numerically by relaxation method, and the GPD is calculated at any perimeter site according to $p_g(P_{ps})\sim|\nabla\Phi(P_{ps})|^{\eta}$ [eq.(25)]. In Fig.11 we used the patterns of $N=7\times10^4$, 3×10^3 and 5×10^2 for η=1,2 and 3, respectively.

It should be noted, however, that the applicability of eqs. (38) and (40) to the Eden (η=0) and ballistic aggrega-

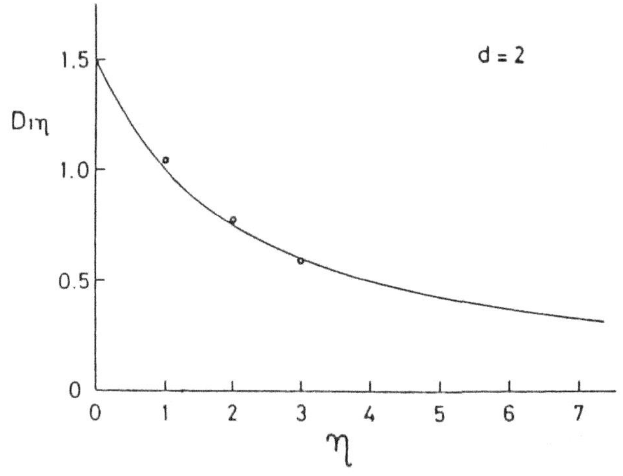

Fig. 11

tion (D_W=1) models is dubious, because cluster surfaces of these models are now wellknown to be self-affine, and not self-similar.[76,77,85,92]

9. SUMMARY

Starting from the explanation of DLA model and a variety of related phenomena, we have discussed various aspects of DLA; the derivation of the fractal dimension expression on the basis of mean-field theoretic arguments, elucidation of the characteristic stochasticity, extention and generalization of the theory, profound relationship to dendritic crystal growth, and surface properties. We have not fully discussed the effect of background lattice anisotropy on the cluster structures.[10,12,16,17,49,93,94] It is still an open question whether the lattice anisotropy eventually leads to self-affine structure or new self-similar one when the cluster size becomes very large.[95] We have also not fully discussed the multifractal nature of DLA clusters,[10,12,96] except the information dimension.

There remains one more interesting problem which was not taken here. The structures generated by various growth models including DLA can often be subdivided, in a natural way, into individual trees or clusters of connected particles. The structures of individual clusters, their size distribution, and the way in which they fit together to form an entire pattern are interesting questions.[97-99] Many studies related to this problem have been done for various models, which are waiting to be summarized in a systematic way.

In retrospect, the present high level of interest in growing fractal patterns is a consequence of the introduction of the concepts of fractal geometry by Mandelbrot[100] and the DLA model by Witten and Sander.[13] Indeed, DLA has provided us with the basis for understanding many other growth models as well. Despite the model simplicity and importance of DLA, we still do not have a nice analytic theory for it. Theoretical attempts presented so far are all more or less phenomenological. To establish the analytic theory may be one of the biggest challenges to unsolved statistical physics problems. It is certainly rewarding in that it gives invaluable breakthrough to many other, very difficult problems in fractal growth models.

ACKNOWLEDGEMENTS: The author would like to thank K. Honda, Y. Hayakawa, H. Honjo, H. Kondoh, S. Sato, H. Toyoki and S. Ohta for many valuable discussions.

REFERENCES

1) Family, F. and Landau, D. P.(eds.), "Kinetics of Aggregation and Gelation" (North-Holland, Amsterdam, 1984).
2) Stanley, H. E. and Ostrowsky, N.(eds.), "On Growth and Form: Fractal and Nonfractal Patterns in Physics" (Martinus Nijhoff, Dordrecht, 1986).
3) Pietronero, L. and Tosatti, E.(eds.), "Fractals in Physics", Proceedings of the Sixth Trieste International Symposium on Fractals in Physics, ICTP, Trieste, Italy (North-Holland, Amsterdam, 1986).
4) Jullien, R. and Botet, R., "Aggregation and Fractal Aggregates" (World Scientific, Singapore, 1986).
5) Herrmann, H. J., Phys. Rep. 136, 154 (1986).
6) Sander, L. M., Nature 322, 789 (1986).
7) Witten, T. A. and Cates, M. E., Science 232, 1067 (1986).
8) Meakin, P., Ann. Rev. Phys. Chem. 39, 237 (1988).
9) Meakin, P., Adv. Colloid Interface Sci. 28, 249 (1988).

10) Meakin, P., "Phase Transitions and Critical Phenomena", Domb, C. and Lebowitz, J. L.(eds.), Vol. 12 (Academic Press, New York, 1988), p.335.
11) Feder, J., "Fractals" (Plenum Press, New York, 1988).
12) Vicsek, T., "Fractal Growth Phenomena" (World Scientific, Singapore, 1989).
13) Witten, T. A. and Sander, L. M., Phys. Rev. Lett. 47, 1400 (1981).
14) Witten, T. A. and Sander, L. M., Phys. Rev. B 27, 5686 (1983).
15) Meakin, P., Phys. Rev. A 27, 604, 1495 (1983).
16) Meakin, P., J. Phys. A 18, L661 (1985).
17) Ball, R. C. and Brady, R. M., J. Phys. A 18, L809 (1985).
18) Muthukumar, M., Phys. Rev. Lett. 50, 839 (1983).
19) Tokuyama, M. and Kawasaki, K., Phys. Lett. 100A, 337 (1984).
20) Ball, R. C. and Witten, T. A., Phys. Rev. A 29, 2966 (1984).
21) Matsushita, M., "Fractal Approach to Heterogeneous Chemistry: Surfaces, Colloids, Polymers", Avnir, D.(ed.) (Wiley, Chichester, 1989), Chap. 3.1.3.
22) Matsushita, M., Sano, M., Hayakawa, Y., Honjo, H. and Sawada, Y., Phys. Rev. Lett. 53, 286 (1984).
23) Brady, R. M. and Ball, R. C., Nature 309, 225 (1984).
24) Niemeyer, L., Pietronero, L. and Wiesmann, H. J., Phys. Rev. Lett. 52, 1033 (1984).
25) Elam, W. T., Wolf, S. A., Sprague, J., Gupser, D. U., van Vechten, D., Barz, Jr., G. L. and Meakin, P., Phys. Rev. Lett. 54, 701 (1985).
26) Smith, D. T., Valles, Jr., J. M. and Hallock, R. B., Phys. Rev. Lett. 54, 2646 (1985).
27) Saffman, P. G. and Taylor, G. I., Proc. Roy. Soc. A245, 312 (1958).

28) Nittmann, J., Daccord, G. and Stanley, H. E., Nature 314, 141 (1985).
29) Daccord, G., Nittmann, J. and Stanley, H. E., Phys. Rev. Lett. 56, 336 (1986).
30) Van Damme, H., Obrecht, F., Levitz, P., Gatineau, L. and Laroche, C., Nature 320, 731 (1986).
31) Paterson, L., Phys. Rev. Lett. 52, 1621 (1984).
32) Måløy, K. J., Feder, J. and Jøssang, T., Phys. Rev. Lett. 55, 2688 (1985).
33) Honjo, H., Ohta, S. and Matsushita, M., J. Phys. Soc. Jpn. 55, 2487 (1986).
34) Honjo, H., Ohta, S. and Matsushita, M., Phys. Rev. A 36, 4555 (1987).
35) Radnoczi, Gy., Vicsek, T., Sander, L. M. and Grier, D., Phys. Rev. A 35, 4012 (1987).
36) Ben-Jacob, E., Deutscher, G., Garik, P., Goldenfeld, N. D. and Lareah, Y., Phys. Rev. Lett. 57, 1903 (1986).
37) Liu, B. X., Huang, L. J., Tao, K., Shang, C. H. and Li, H.-D., Phys. Rev. Lett. 59, 745 (1987).
38) Miller, A., Knoll, W. and Möhwald, H., Phys. Rev. Lett. 56, 2633 (1986).
39) Miller, A. and Möhwald, H., J. Chem. Phys. 86, 4258 (1987).
40) Daccord, G. and Lenormand, R., Nature 325, 41 (1987).
41) Daccord, G., Phys. Rev. Lett. 58, 479 (1987).
42) Flory, P. J., "Principles of Polymer Chemistry" Cornell Univ. Press, Ithaca, 1971), Chap. 12; de Gennes, P. G., "Scaling Concepts in Polymer Physics" (Cornell Univ. Press).
43) Hentschel, H. G. E., Phys. Rev. Lett. 52, 212 (1984).
44) Isaacson, J. and Lubensky, T. C., J. Phys. Lett. (France) 41, L469 (1980).
45) Meakin, P., in Ref.1, p.91.
46) Meakin, P., Phys. Rev. B 29, 3722 (1984).

47) Turkevich, L. A. and Scher, H., Phys. Rev. Lett. 55, 1026 (1985); Phys. Rev. A 33, 786 (1986).
48) Ball, R. C., Brady, R. M., Rossi, G. and Thompson, B. R., Phys. Rev. Lett. 55, 1406 (1985).
49) Meakin, P., Ball, R. C., Ramanlal, P. and Sander, L. M., Phys. Rev. A 35, 5233 (1987).
50) Honda, K., Toyoki, H. and Matsushita, M., J. Phys. Soc. Jpn. 55, 707 (1986).
51) One can see by a beautiful color picture of a DLA cluster in Ref.45 how some particles can penetrate slightly into the interior of the cluster.
52) Meakin, P., Phys. Rev. B 29, 4327 (1984).
53) Meakin, P., J. Colloid Interface Sci. 96, 415 (1983).
54) Matsushita, M., Honda, K., Toyoki, H., Hayakawa, Y. and Kondo, H., J. Phys. Soc. Jpn. 55, 2618 (1986).
55) Hayakawa, Y., Kondo, H. and Matsushita, M., J. Phys. Soc. Jpn. 55, 2479 (1986).
56) Tolman, S. and Meakin, P., Physica A 158, 801 (1989).
57) Meakin, P., J. Theor. Biology 118, 101 (1986).
58) Meakin, P., Phys. Rev. A 33, 4199 (1986).
59) Satpathy, S., Phys. Rev. B 33, 5093 (1986).
60) Turkevich, L. A. and Scher, H., p.223 in Ref.3.
61) Mullins, W. W. and Sekerka, R. F., J. Appl. Phys. 34, 323 (1963); 35, 444 (1964).
62) Langer, J. S., Rev. Mod. Phys. 52, 1 (1980).
63) Bentley, W. A. and Humphreys, W. J., "Snow Crystals" (Dover, New York, 1962).
64) Langer, J. S., in "Chance and Matter; Les Houches Ecole d'Ete de Physique Theorique", Souletie, J., Vannimenus, J., and Stora, R. (eds.) (North-Holland, Amsterdam, 1987), pp.629-711.
65) Brower, R., Kessler, D., Koplik, J. and Levine, H., Phys. Rev. Lett. 51, 1111 (1983); Phys. Rev. A 29, 1335 (1984).

66) Kessler, D., Koplik, J. and Levine, H., Phys. Rev. A 30, 2820 (1984); ibid. 30, 3161 (1984); ibid. 32, 1930 (1985).
67) Ben-Jacob, E., Goldenfeld, N. D., Langer, J. S. and Schon, G., Phys. Rev. Lett. 51, 1930 (1983); Phys. Rev. A 29, 330 (1984).
68) Ben-Jacob, E., Goldenfeld, N. D., Kotliar, B. G. and Langer, J. S., Phys. Rev. Lett. 53, 2110 (1984).
69) Honjo, H., Ohta, S. and Sawada, Y., Phys. Rev. Lett. 55, 841 (1985).
70) Ben-Jacob, E., Godbey, R., Goldenfeld, N. D., Koplik, J., Levine, H., Mueller, T. and Sander, L. M., Phys. Rev. Lett. 55, 1315 (1985).
71) Sawada, Y., Dougherty, A. and Gollub, J. P., Phys. Rev. Lett. 56, 1260 (1986).
72) Grier, D., Ben-Jacob, E., Clarke, R. and Sander, L. M., Phys. Rev. Lett. 56, 1264 (1986).
73) Charig, A. J., Greenaway, F., Milner, A. C., Walker, C. A. and Whybrow, P. J., Science 232, 622 (1986).
74) Plischke, M. and Racz, Z., Phys. Rev. Lett. 53, 415 (1984); Racz, Z. and Plischke, M., Phys. Rev. A 31, 985 (1985).
75) Plischke, M. and Racz, Z., Phys. Rev. A 32, 3825 (1985).
76) Family, F. and Vicsek, T., J. Phys. A 18, L75 (1985).
77) Jullien, R. and Botet, R., J. Phys. A 18, 2279 (1985).
78) Meakin, P. and Witten, T. A., Phys. Rev. A 28, 2985 (1983).
79) Meakin, P., Phys. Rev. A 32, 453 (1985).
80) Meakin, P., Stanley, H. E., Coniglio, A. and Witten, T. A., Phys. Rev. A 32, 2364 (1985).
81) Halsey, T. C., Meakin, P. and Proccacia, I., Phys. Rev. Lett. 56, 854 (1986).
82) Amitrano, C., Coniglio, A. and di Liberto, F., Phys. Rev. Lett. 57, 1016 (1986).

83) Meakin, P., Coniglio, A., Stanley, H. E. and Witten, T. A., Phys. Rev. A 34, 3325 (1986).
84) Hayakawa, Y., Sato, S. and Matsushita, M., Phys. Rev. A 36, 1963 (1987).
85) Meakin, P., Ramanlal, P., Sander, L. M. and Ball, R. C., Phys. Rev. A 34, 5091 (1986).
86) Matsushita, M., Hayakawa, Y., Sato, S. and Honda, K., Phys. Rev. Lett. 59, 86 (1987).
87) Halsey, T. C., Jensen, M. H., Kadanoff, L. P. Procaccia, I. and Shraiman, B. I., Phys. Rev. A 33, 1141 (1986).
88) Coniglio, A. and Stanley, H. E., Phys. Rev. Lett. 52, 1068 (1984).
89) Hentschel, H. G. E. and Procaccia, I., Physica 8D, 435 (1983); Procaccia, I., Phys. Scr. T9, 40 (1985).
90) Grassberger, P., unpublished.
91) Makarov, N. G., Proc. London Math. Soc. 51, 369 (1985).
92) Wolf, D. E. and Kertesz, J., Europhys. Lett. 4, 561 (1987); Kertesz, J. and Wolf, D. E., J. Phys. A 21, 747 (1988).
93) Meakin, P. and Family, F., Phys. Rev. A 34, 2558 (1986).
94) Matsushita, M., Family, F. and Honda, K., Phys. Rev. A 36, 3518 (1987).
95) Meakin, P., Phys. Rev. A 36, 332 (1987).
96) Stanley, H. E. and Meakin, P., Nature 335, 405 (1988).
97) Racz, Z. and Vicsek, T., Phys. Rev. Lett. 51, 2382 (1983).
98) Matsushita, M., Hayakawa, Y. and Sawada, Y., Phys. Rev. A 32, 3814 (1985).
99) Matsushita, M. and Meakin, P., Phys. Rev. A 37, 3645 (1988).
100) Mandelbrot, B. B., "Fractals: Form, Chance and Dimension" (Freeman, San Francisco, 1977); "The Fractal Geometry of Nature" (Freeman, New York, 1982).

Formation and Transition of Dynamical Patterns in an Evaporating Drop

Ryuji Takaki

Tokyo University of Agriculture and technology,
Fuchu, Tokyo 183, Japan

ABSTRACT. A review is given on observations and analyses of formation and transition of dynamical patterns in an evaporating drop. A drop of liquefied gas, placed on a horizontal floor with room temperature, deforms to a disk shape owing to the gravity and the surface tension, and begins to vibrate self-inducedly in the horizontal direction with a large amplitude. In the plane view the vibrating drop forms a polygonal shape instantaneously. The present author and his co-workers observed this motion in liquid-nitrogen, liquid-oxygen and liquid-argon. The drop was observed to select mode of vibration uniquely according to its size and to make a transition to other modes as it reduced size through evaporation. Mechanism of this pattern transition was investigated precisely by the use of image processing technique applied on 16 mm movie film. By changing temperature of the surrounding air, it was confirmed that the temperature gave no large effect on the phenomenon within a certain temperature range. Prediction of vibrational frequency is made based on the inviscid flow theory and a small amplitude approximation, result of which agrees with the measurement by the author's group. A computer simulation of the pattern transition process is made based on a model equation including effects of mass reduction, self-induced oscillation and nonlinearity. The result simulates well the observed behavior of a drop. These researches provide a new aspect of pattern dynamics of a highly inequilibrium system.

1. INTRODUCTION

Let a drop of water be placed on a horizontal hot plate, then it will deform to a thin disk owing to the gravity. If the temperature of the hot plate is sufficiently high, the water will be in an overheated state and the vapor coming from the lower surface of the drop will have enough vigor to levitate the drop from the plate. At the same time, the drop will acquire a form of circular disk, in which the hydrostatic pressure and the surface tension balance with each other.

However, according to the observation by Holter etal.,[1)] such a drop never stayed at rest but began a characteristic vibration, where the thickness of the drop remained constant and its contour in the plane view oscillated around a circular shape with large amplitude. Number of waves along the contour changed according to the drop size. This vibration did not decayed until it disappeared through evaporation, because it could move without friction from the plate by the presence of a vapor sheet under it and also because the evaporation itself played a role to maintain the vibration. In spite of the fact that it is quite an attractive phenomenon and that it is commonly observed in laboratories, there seems to be no investigation following this work. Motivated by this situation the present author undertook a similar experiment in collaboration with Adachi.[2)]

Behavior of a spherical drop floating in the space has been an interesting problem in many fields, such as cosmology,[3)] nuclear physics, [4,5)] chemical engineering,[6)] etc., and several attempts have been made to levitate a drop in the air.[7-9)] On the other hand, the overheated drop floating from the floor deforms largely from a spherical shape and is not suited for application in these fields, but is considered a good topic for investigation of dynamical behavior of a system very far from thermal equilibrium.

In our experiment, instead of a water drop on a hot plate, drops of liquefied gas were placed on a horizontal plate with room temperature, because it was easier to control temperature condition. We observed self-organization of regular patterns in vibrational mode and measured vibrational frequency. We also found an interesting phenomenon of sudden transition of a vibrational mode to another, which occurred as the drop size was reduced through evaporation. This mode transition was recorded by the use of a high-speed cinecamera, but in order to have a precise description of the process we needed to extract quantitative data. Therefore, an image processing technique was applied to the images in the film and spectral analysis was made for instantaneous drop shapes, from which we could guess a mechanism of mode transition.[10-12] Results of these researches are explained in Sec. 2.

Analyses of small amplitude vibrations of a spherical drop and a circular column are made by Rayleigh,[13,14] and the following results for frequencies ω_d and ω_c are obtained:

$$\omega_d{}^2 = \frac{(n-1)n(n+2)\sigma}{r_0{}^3\rho}, \quad \text{for a spherical drop,} \tag{1}$$

$$\omega_c{}^2 = \frac{(n-1)n(n+1)\sigma}{r_0{}^3\rho}, \quad \text{for a circular column,} \tag{2}$$

where n is the degree of spherical harmonic function for the sphere case and the number of waves for the column case, σ is the surface tension coefficient, ro is the mean radius, and ρ is the liquid density. For the case of a flattened drop as in our experiment, there has been no theory to predict its frequency.

The author and Adachi[15] developed a linear theory for this frequency based on the invicsid hydrodynamics and the shallow water approximation, and obtained results agreeing well with our experiment,[2] as is explained in Sec. 3. Although this linear analysis is successful for the short time behavior, such as the vibrational frequency, it can not be applied to problems of nonlinear effect or those coming from the thermal inequilibrium, such as determination of vibrational amplitude or analysis of transition of vibrational mode.

Since solution of a complete set of governing equations is hopeless for this complicated phenomenon, a feasible way to understand its basic properties would be a model simulation which takes important factors into account, i.e. self-excitation, decrease of drop volume, nonlinear interaction, frictional force, etc.[10,12] In Sec. 4 the basic ideas and results of the simulation are explained, and it is shown that behavior

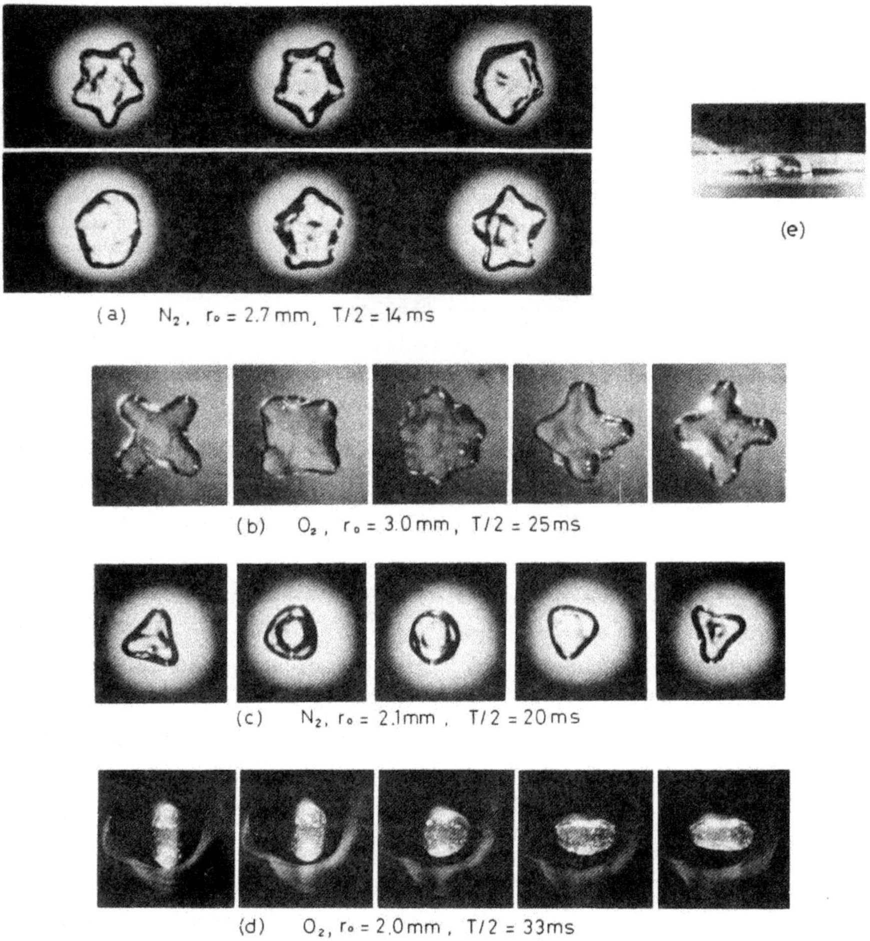

Fig. 1. Plane views of drops in series during half period of vibration, taken by a high-speed camera. ro is the mean radius of the drop, T is the period of vibration. (a), (c), (d) with 400 frames/s, (b) with 4000 frames/s, (e) a side view of 4-mode. (from Adachi & Takaki[2]))

of a drop observed in our experiment is understood to a certain degree by this analysis.

In the last section stress is laid on significance of this phenomenon both for a practical application and for a general understanding of pattern formation problems, which is a main driving force for the present author and co-workers to continue this investigation.

2. OBSERVATION

2.1 Mode Selection and Vibrational Frequency

In the experiment [2]) a drop of liquid-nitrogen, liquid-oxygen and liquid-argon with volume 0.1-0.2 cm^3 was put carefully at the center of a slightly concave lens with radius of curvature 10 cm, which was placed horizontally and kept at room temperature. Motion of the drop was recorded by the use of 16 mm cinecamera from above. The lens was used in order to keep the drop within the frame of the camera. Immediately after it was put on the lens, the drop began vibrating, where the thickness of the drop was nearly constant (1.5 mm) but the contour in the plane view showed a standing wave, as is shown in Fig. 1 for liquid-nitrogen and liquid-oxygen drops. This type of motion is called here a polygonal vibration. If an angular momentum in the vertical direction was given to the drop initially, the wave travelled along the periphery. However, we took care not to give an angular momentum so as to assure easy observation. The amplitude of vibration remained constant and had nearly the same value at any run of experiment.

The mode of vibration is indicated by the number of waves along the periphery, and the mode with n waves is called n-mode. The mode was found to be selected almost uniquely depending on the drop size, i.e. for smaller drop the mode number n was smaller. Then, what will happen if the drop size decreases through evaporation ? Actually, the mode of vibration was maintained for a certain time while the vibrational frequency increased, but soon the drop began to take a random motion and after short time convergd to another mode of vibration with smaller n. After 2-mode the drop became nearly spherical and disappeared. Variations of the mean radius and the vibrational frequency during this process are shown in Fig. 2 for liquid-oxygen. In particular, the variation of the frequency shows appearances of regular modes and random motions in turn for one run of experiment. The transition seemed to occur so that the period of vibration was maintained at values around 0.04 sec.

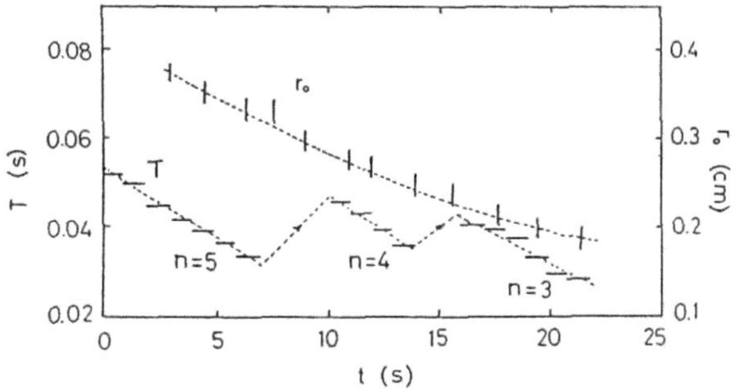

Fig. 2. Variations of mean radius r_0 and period of vibration T during one run of experiment for liquid-oxygen. (from Adachi & Takaki[2])

Strange to say, there was a clear distinction in the process of mode transition among three materials of drop. For drops of liquid-oxygen and liquid-argon the mode number decreased by unity, while for a drop of liquid-nitrogen the first mode was in most cases 5-mode and made a transition to 3-mode, then to 2-mode. This situation is expressed as follows in terms of asterisks indicating probability of mode appearance:

O_2 :	6*	→	5***	→	4**	→	3*	→	2***
Ar :			5**	→	4*	→	3*	→	2*
N_2 :			5***	→		→	3***	→	2**

where ***, ** and * mean "always", "often" and "sometimes", respectively. Modes with n>6 was not observed. Even if initial drop size was made larger, the drop showed random motion and no regular vibrational mode until it lost mass through evaporation. Quantitative measurement of mode transition was made by the use of a image processing technique and explained in the next subsection. *It is not clear* why the liquid-nitrogen did not show 4-mode. The model simulation in Sec. 4 seems to give a hint to this problem, but it is not yet solved.

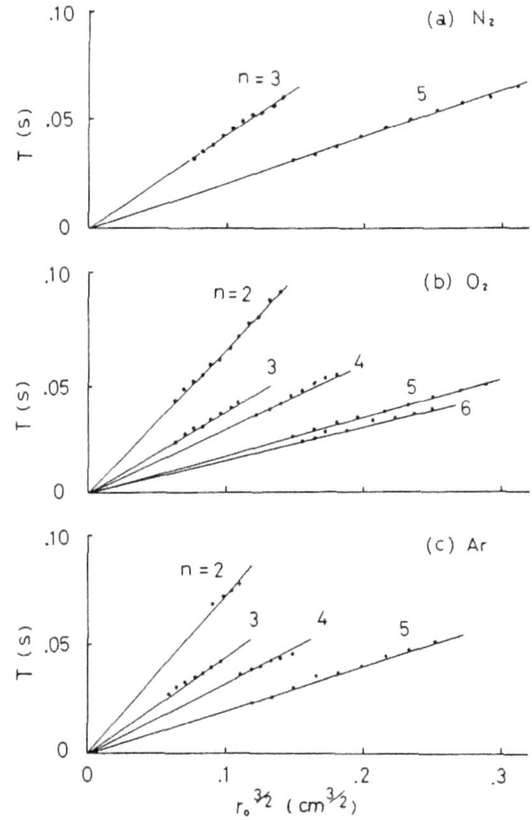

Fig. 3. Measured values of vibrational period. The solid lines indicate the $r_o^{3/2}$ dependence. (from Adachi & Takaki[2])

Periods of vibration of several modes were measured [2]) and shown in Fig. 3. The data show clear dependence on $r_o^{3/2}$ (r_o is the mean drop radius in the plane view). Although the mechanism of fluid motion is complicated, the dependence of the period on the drop radius is obtained easily by a dimensional analysis. Suppose the dynamics is influenced by the radius r_0, the surface tension coefficient σ and the liquid density ρ, then the period T should be expressed as

$$T = \text{const.}\ (r_0^3\rho/\sigma)^{1/2}, \qquad (3)$$

where the constant may depend on the mode number n.

The gravitational acceleration g must be playing an essential role in this problem, but it does not appear directly in the mechanism of vibration for the following reason. According to observation, the drop formed a thin disk and the thickness remained constant during vibration (about 1.5 mm). This fact suggests that the gravity balanced the effect of surface tension at the periphery to form a thin disk, and this balance was maintained throughout the drop motion, so that the liquid made little vertical motion during vibration. This idea is supported by the following estimation of drop thickness.

Let us denote the thickness by h_0, and assume that the radius of curvature at the periphery in the vertical cross section is $h_0/2$. Then the condition of balance of static pressure due to the gravity and pressure due to the surface tension leads to the following relation:

$$\frac{1}{2}\rho g h_0 \ \text{(averaged static pressure)} = \frac{\sigma}{h_0/2}\ . \qquad (4)$$

By substituting material values of liquefied gases, we have h_0=1.8 mm. Rough agreement of this estimation with measurement (1.5 mm) supports the understanding of mechanism of polygonal vibration, i.e. the gravity balanced the surface tension concerned with the curvature in the vertical cross section, and the surface tension concerned with the curvature in the plane view produced a restoring force when the drop shape in the plane view deviated from circle. Therefore, the vibrational frequency depended little on the gravity.

This situation is similar to the mechanism for vibration of a circular column analysed by Rayleigh, the frequency of which is given by eq.(2), as will be examined precisely in Sec. 4.

2.2 Mode Transition Mechanism

In order to investigate mechanism of the mode transition, images of a vibrating drop in 16 mm film taken with 400 frames/s were analysed by the use of image processing technique. The block diagram of the image processing system is shown in Fig. 4. Samples of image processing were 140 images of a liquid-nitrogen drop during an interval in transition from 5-mode to 3-mode, i.e. successive 20 frames at seven discrete times. Image analyses of other cases could not be made, because the task of image processing was very much time-consuming.

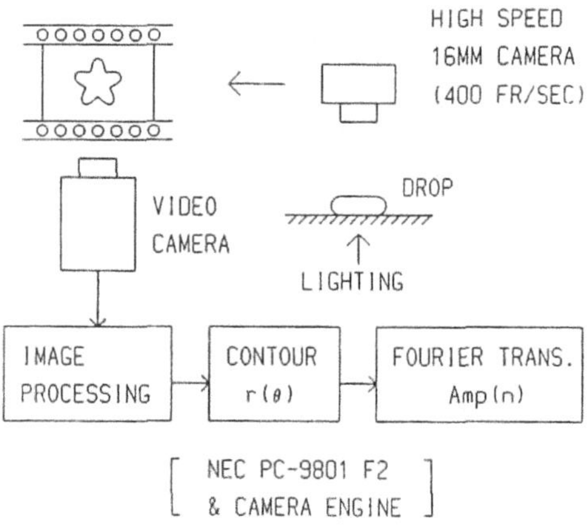

Fig. 4. Block diagram of image processing. (from Takaki, etal.[12])

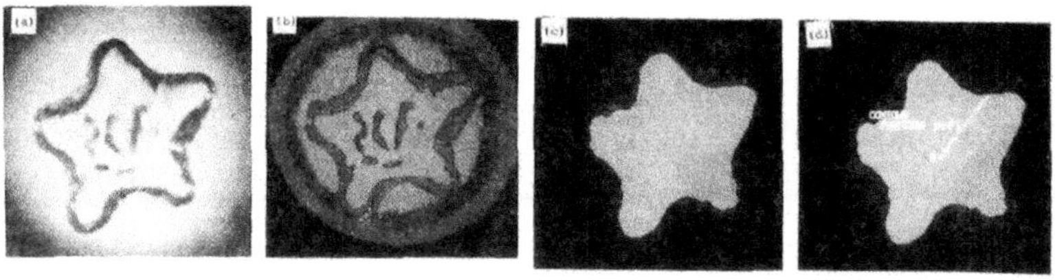

Fig. 5. Procedure of image processing. (a)original, (b)digitized, (c)reformed, (d)definition of contour function R(Θ). (from Takaki, etal.[12])

Figure 5 shows typical stages of image processing. An original image (a) was digitized to 4 gray levels, regions of certain gray levels (b) was extracted (this choice was established after some trials), cracks in the periphery were closed by a suitable way and the inner region was painted (c), and finally a contour function $R(\theta)$ was defined by drawing radii from the center of mass of the image (d). It is normalized by the mean radius.

The contour function $R(\theta)$ was analysed in the following two ways. One was the Fourier decomposition of $R(\theta)$. It should be noted here that the n-th Fourier component X_n does not immediately give an energy of the n-mode, because the contour function has an information of shape at a certain instant and not of fluid velocity. In order to obtain the energy, we make use of the fact that the mean kinetic energy is equal to the mean potential energy for a simple harmonic oscillation. For an image of a drop, X_n corresponds to the displacement of n-mode, and $X_n{}^2$ to the potential energy. Therefore, average of $X_n{}^2$ over 20 successive frames, defined as follows, will give the energy I_n of n-mode:

$$I_n = \{ \sum_{i=0}^{19} X_n(i)^2 \} / 20 , \tag{5}$$

where $X_n(i)$ is the n-th Fourier component at the i-th frame.

Another quantity extracted from $R(\theta)$ is a self-correlation function of drop shape defined as

$$F(\Phi) = \int_0^{2\pi} R(\theta+\Phi)R(\theta)d\theta \Big/ \int_0^{2\pi} R(\theta)^2 d\theta . \tag{6}$$

This function gives a measure of randomness of an instantaneous drop shape. If the drop is deformed to a regular wave similar to a regular polygon, $F(\Phi)$ will be a periodic wave. If the drop has a random shape, F will have smaller value except at $\Phi=0$.

Figure 6 shows variations of absolute values of $X_n(i)$ at typical three times during mode transition. Below each figure is shown a real time, and the abscissa covers a period of about 50 ms. At t=1.8 s the drop was in 5-mode, and through a transient state at t=3.5 s the drop attained 3-mode at t=4.74 s. The amplitude of this 3-mode seems larger than the initial 5-mode. But, it is only apparent, because the contour function R(θ) was normalized by the mean radius, which decreased with time through evaporation. Actually, the amplitudes of the 5- and 3-modes were nearly equal.

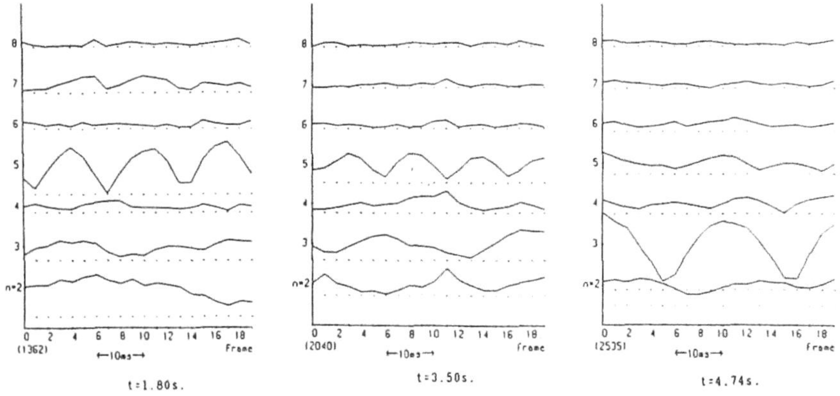

Fig. 6. Variations of absolute values of Fourier components obtained from processed images of a drop of liquid-nitrogen obtained at typical three times during the transition from 5-mode to 3-mode. Numbers along the ordinate indicate the mode, and base lines are shown by dotted lines. Vertical scales are arbitrary but proportional to each other. The three numbers near the origins of coordinates indicate serial numbers of frames in the 16 mm film. (from Takaki, etal.[12])

The spectra at seven times calculated by eq.(5) are shown in Fig. 7. From this figure we see clearly that the initial 5-mode disappeared and a 3-mode appeared. But, at the same time, it is found that the

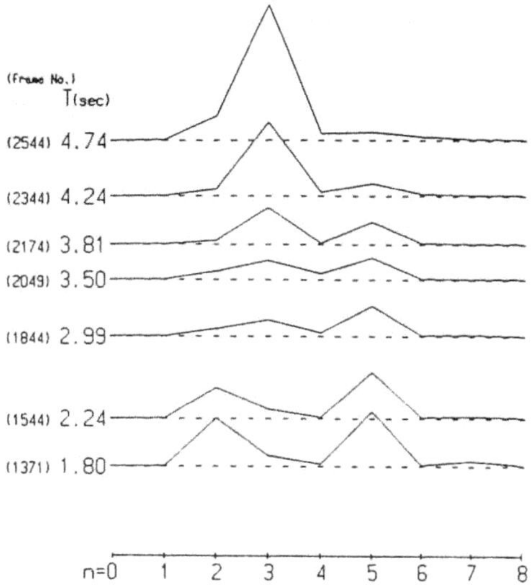

Fig. 7. Variation of the energy spectrum during mode transition of a liquid-nitrogen drop. (from Takaki, etal.[12])

initial drop had also a component of 2-mode with a considerable strength, which decayed as the 3-mode arose. Since a 2-mode corresponds to deformation of a circular drop to an elliptic shape, the mixture of 5- and 2-modes in the initial state means an elongation of the initial polygonal shape with n=5 to a certain direction. The elongation of initial drop can be recognized in Fig. 5(a), but it was not paid attention by us until the image analysis was made, because we are apt to watch the vertices and not an overall deformation when we are shown a figure with many vertices.

Although this variation of spectrum was obtained for one event of experiment, the result seems to suggest that a certain kind of nonlinear interaction was playing an important role in the mode transition mechanism and that the three components, i.e. 5-, 3- and 2-modes, were important actors. Furthermore, this interaction is considered to be expressed by second order terms of dynamical variables as in the Navier-Stokes equation, for the following reason. If an n-mode is expressed by a contour function $\exp(\pm ni\theta)$, then from 5- and 2-modes the second order nonlinear terms yield new terms with $\exp(\pm 7i\theta)$ and $\exp(\pm 3i\theta)$, thus allowing a 3-mode. This picture of mode transition is included in the model simulation in Sec. 4.

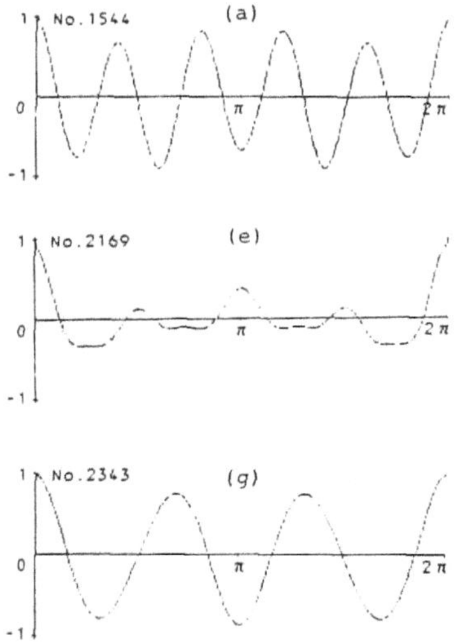

Fig. 8. Self-correlation functions $F(\Phi)$ at three times during mode transition of a liquid-nitrogen drop. The number at each figure is the frame number of 16 mm film. (from Takaki, etal.[12])

Examples of self-correlation functions $F(\Phi)$ from the 5-mode, the intermediate state and the 3-mode are shown in Fig. 8. As is expected, they varied nearly sinusoidally in the 5- and 3-modes. On the other hand, in the intermediate state $F(\Phi)$ had relatively lower values corresponding to rather irregular drop shape. Recovery of a regular vibration from this random state reminds us of a complicated mechanism in mode interaction.

2.3. Dependence on the Temperature of the Surrounding Air

The experiments explained so far were made in the situation where the horizontal floor and the surrounding air were at room temperature. The temperature can be changed easily by setting up the apparatus within a heater or a refrigerator. In our recent experiment (not published) the ambient temperature To was varied from -87°C to 200°C, and behavior of a drop of liquid-nitrogen was observed by a video-camera in the same way as we did by the use of 16 mm cinecamera. Selected mode was identified by the naked eyes from the video image and frequency of vibration were measured by image processing from video images. Results of these measurements are summarized below.

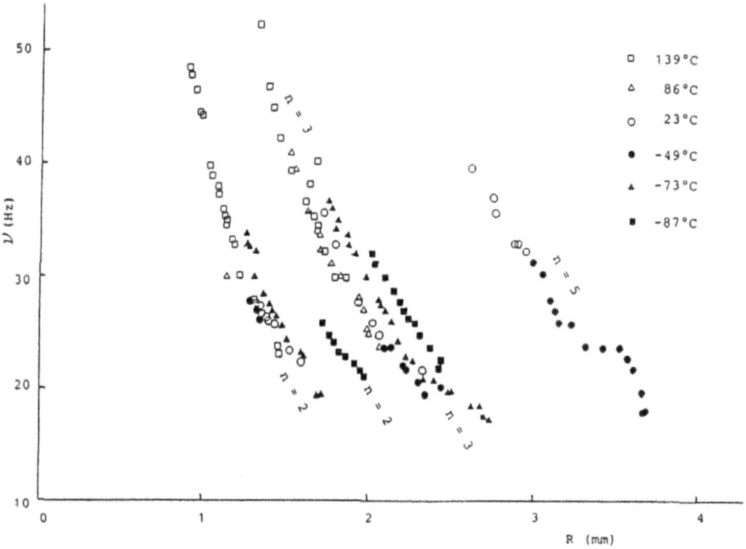

Fig. 9. Frequency of vibration plotted against the drop size.

Dependence of the frequency on the drop size is shown in Fig. 9, where data for several ambient temperatures are plotted by different symbols. It is remarkable that the data for all temperatures higher than -80°C lie on a single curve, hence the frequency did not depend on the ambient temperature but solely on the drop size. An only effect of different temperature was the long time-scale behavior, such as mode transition and evaporation, which had longer time scale for higher temperature.

Now, if the dynamics of vibration is governed in the first approximation by the drop size (the inertia effect) and the surface tension (the restoring force), then this result means that the surface tension of the drop was nearly constant independent on the ambient temperature To. Since the surface tension depends on the surface temperature, it is concluded that the surface was kept at a common temperature irrespective of the ambient condition. This conclusion may not be so surprising because the boiling temperature has a constant value for a fixed atmospheric pressure. However, what we were observing was the system very far from thermal equilibrium, and it is not obvious whether the drop surface was at the boiling temperature (for liquid-nitrogen -195.8°C). The present result at least means that the surface was kept at a constant temperature for any evaporation speed at the surface.

Figure 10 is a diagram showing processes of mode transition at various temperatures for drops of liquid-nitrogen. From this figure we can see that the process is the same in a wide temperature range $-80^{\circ}C < To < 150^{\circ}C$. It should be remarked that there is a region with $150^{\circ}C < To$ and $1.5mm < ro < 3mm$, where each mode had a regular pattern but was not selected uniquely so that the mode transition seemed to occur rather randomly. Probably, the process depended much on the initial condition. Moreover, 6-mode and 4-mode, which had never been observed in room temperature, appeared for conditions in this region. In that sense this region can be called chaotic.

There is at present no theory to explain why the transition processes in this region are so chaotic. It is not even clear by what quantities this chaotic nature should be described. On the other hand, for very low temperature drops did not show vibration. The critical temperature for occurrence of vibration is not yet obtained. Sometimes drops in the air with $To = -80^{\circ}C$ touched to the floor and did not move. From observations so far we may say that the critical temperature is located around $-100^{\circ}C$.

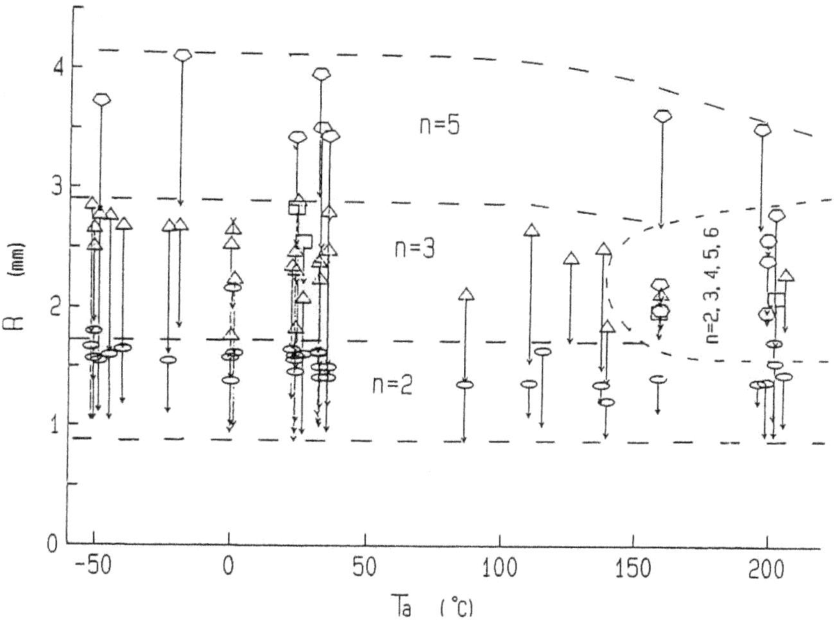

Fig. 10. Diagram of mode transition patterns.

Summary of results obtained by our experiments is given here. A drop placed on a horizontal floor with a highly overheated condition deformed to a thin disk, and its plane shape began a characteristic vibration. Mode of vibration, characterized by the number of waves n, was almost uniquely determined by the drop size. The mode number n decreased as the drop size decreased through evaporation. Process of this mode transition depended strangely on material of the liquid, but not on the degree of overheating, so long as it was within a certain temperature range. Dynamics of mode transition observed in a liquid-nitrogen drop seemed to be governed by a certain kind of nonlinear interaction between several modes. For a very high overheating the transition occurred in a chaotic manner.

3. PREDICTION OF VIBRATIONAL FREQUENCY

3.1. Governing Equations

Among many features of this phenomenon, the mechanism of periodic vibration, which has a short time scale, can be analysed by the hydrodynamics.[15] In particular, vibrational frequency for n-mode is predicted relatively easily based on several assumptions.

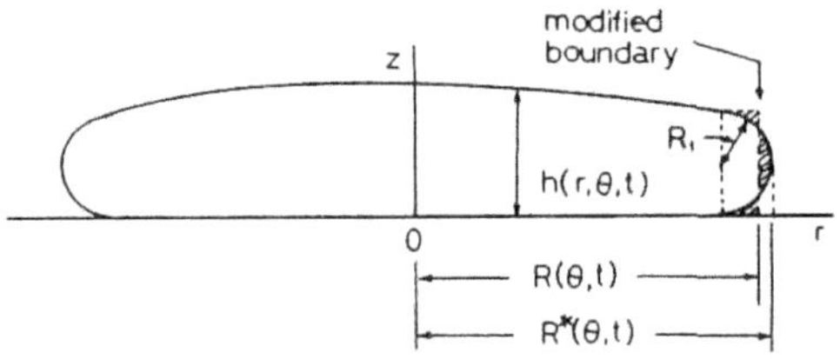

Fig. 11. Assumed geometry and definitions of variables. (from Takaki & Adachi[15])

Assumptions employed here are listed below.

(i) Aspect ratio h_o/R_o, where h_o is the mean thickness of the drop and R_o is the mean radius in the plane view, is sufficiently small, so that the drop is looked upon as a thin liquid layer. Fluid motion is looked upon as a surface wave on this layer and the shallow water theory is applied. Depth of the liquid is denoted by $h(r,\theta,t)$.

(ii) The liquid layer is prevented from spreading by the surface tension at the peripheral region. The contour of drop shape in the plane view is denoted by $R*(\theta,t)$. But, boundary conditions for shallow water motion of the liquid layer are applied not at $r=R*(\theta,t)$ but at a certain modified boundary $r=R(\theta,t)$, defined by (see Fig. 11)

$$R*(\theta,t) = R(\theta,t) + \beta\, h(R,\theta,t), \quad \text{where } \beta = \frac{1}{2} - \frac{\pi}{8} = 0.107. \tag{5}$$

The modified boundary is so defined that the semi-circle at the real periphery in the vertical cross section has the same area as the rectangular ending at the modified boundary as shown in Fig. 11. By introducing this boundary, inertia effect of fluid at periphery can be treated properly.

(iii) The lower boundary of the liquid layer remains horizontal and suffers no frictional force from the floor.

(iv) The liquid flow is incompressible and inviscid, hence the potential flow theory is applied.

(v) Effect of evaporation is neglected, hence all material parameters and the drop volume is constant during vibration.

Assumptions (i), (ii) and (iii) do not contradict our observations, while (iv) and (v) need some examination. Typical Reynolds number for vibrating drop is O(100), hence the dynamics can be analysed in terms of inviscid flow. Both effects of damping due to viscosity and driving force coming from evaporation belong to long time-scale phenomena, and can be neglected here. The volume change during one period of vibration is also negligible.

Basic equations in the shallow water theory are the continuity equation and the momentum equation, which are written in the cylindrical coordinates as follows:

$$\frac{\partial h}{\partial t} + \frac{1}{r}\frac{\partial (rhu_r)}{\partial r} + \frac{1}{r}\frac{\partial (hu_\theta)}{\partial \theta} = 0, \qquad (6)$$

$$\frac{\partial}{\partial t}(\rho hu) + \frac{\partial}{\partial r}(u_r \rho hu) + \frac{1}{r}\frac{\partial}{\partial \theta}(u_\theta \rho hu) = -\,\mathrm{grad}\,P, \qquad (7)$$

where $u=(u_r\ ,\ u_\theta)$ is the velocity vector in the horizontal direction, P is the pressure integrated along the vertical direction and ρ is the liquid density. The integrated pressure P comes from the gravity and the surface tension at the upper surface.

Pressure and kinematical conditions should be posed at the modified boundary. From assumption (ii) the integrated pressure P balances the surface tension at the periphery, i.e.

$$P = h\sigma\left(\frac{1}{R_1} + \frac{1}{R_2}\right), \qquad (8)$$

where $R_1=h(R(\theta,t),\theta,t)/2$ is the radius of semi-circle at the periphery in the vertical cross section, and R_2 is the radius of curvature of the contour expressed by $r=R(\theta,t)$. The kinematical condition is written as

$$u_r = u_\theta \frac{1}{r}\frac{\partial R}{\partial \theta} + \frac{\partial R}{\partial t}, \qquad \text{at } r = R(\theta,t). \tag{9}$$

3.2. Static State

For the static state we have $u=0$, $R_1=h_0/2$, $R_2=R_0$ (constant), and eq.(7) becomes

$$\frac{1}{2}\,GH^2 = 2 + \frac{1}{H^{-1}+\beta}, \qquad \text{where } G = \frac{\rho g R_0{}^2}{\sigma}, \quad H = \frac{h_0}{R_0}. \tag{10}$$

The quantities G and H are nondimensional parameters indicating the relative importance of the gravity at the upper surface and the aspect ratio, respectively.

From assumption (i) and eq.(10), we have

$$H \ll 1, \quad G \gg 1, \quad GH^2 \cong 4, \quad GH = O(H^{-1}) \gg 1. \tag{11}$$

These estimations are useful when limiting case is considered in the next subsection.

3.3. Small Amplitude Vibration

Let the radial coordinate and the time are normalized by R_0 and a time scale $T_0 = \sqrt{(\rho R_0{}^3/\sigma)}$, respectively. The same notations are used for normalized quantities. Then, dynamical variables are written as

$$h = R_0 H(1 + \tilde{h}(r,\theta,t)), \qquad R = R_0\{1 + \tilde{R}(\theta,t)\}$$

$$u_r = \frac{R_0}{T_0}\frac{\partial \tilde{\Phi}}{\partial r}, \qquad u_\theta = \frac{R_0}{T_0}\frac{\partial \tilde{\Phi}}{r\partial \theta}, \tag{12}$$

where $\tilde{\Phi}$ is the velocity potential, and quantities with tilde are nondimensional and assumed to be small perturbations. Substituting eq.(12) into the governing equations (6)-(9) and linearizing them with respect to the perturbations, we obtain

$$\frac{\partial \tilde{\Phi}}{\partial t} + GH\,\tilde{h} - H\,\Delta\tilde{h} = 0, \tag{13}$$

$$GH\,\tilde{h} - H\,\Delta\tilde{h} = \frac{\tilde{h}}{1+\beta H} - \frac{\tilde{R} + \beta H\tilde{h} + \partial^2(\tilde{R}+\beta H\tilde{h})/\partial\theta^2}{(1+\beta H)^2}, \quad \text{at } r=1, \tag{14}$$

$$\frac{\partial \tilde{\Phi}}{\partial r} = \frac{\partial \tilde{R}}{\partial t}, \qquad \text{at } r = 1. \tag{15}$$

Much of mathematical process to derive these equations is abbreviated, such as expressing R_2 in terms of $R(\theta,t)$. For precise the reader may refer Ref. 15.

For vibration of n-mode, the variables are written as

$$(\tilde{\Phi},\ \tilde{h},\ \tilde{R}) = (\hat{\Phi}(r),\ i\hat{h}(r),\ i\hat{R})\ \exp(in\theta + i\Omega t), \tag{16}$$

where quantities with hat are real. After substituting these into eqs.(13)-(15) and neglecting terms of $O(H^2)$ and eliminating h and R, we finally have the following equation and boundary condition for Φ:

$$(\Delta_n + w^2)\hat{\Phi} = 0, \qquad \text{where } w^2 = \Omega^2/GH, \tag{17}$$

$$(n^2-1)\hat{\Phi}' = w^2\{(1+\beta H)^2 GH - (1+\beta H) - (n^2-1)\beta H\}\hat{\Phi}, \quad \text{at } r = 1. \tag{18}$$

Solution of eq.(17) free from singularity at the origin is the Bessel function of the first kind. i.e.

$$\hat{\Phi} = A\,J_n(wr), \tag{19}$$

where A is an arbitrary constant. Substitution of this solution into eq.(18) leads to an eigenvalue equation for Ω, as follows:

$$\frac{J_n'(w)}{J_n(w)} = \frac{GHw\{(1+2\beta H) - (GH)^{-1} + O(H^2)\}}{n^2 - 1}, \tag{20}$$

where the prime denotes derivative. This equation gives an expression of the vibrational frequency as a function of n. This is the third formula following eqs. (1) and (2) valid for a flattened liquid drop.

In the limit $GH \to \infty$ (hence $H \to 0$ and $w \to 0$), eq.(20) reduces to

$$\Omega^2 = n(n^2 - 1). \tag{21}$$

This result coincides with the formula (2) for a circular column, as has been suggested in Sec. 2.1. For a finite value of GH, we have, after neglecting terms of $O(H)^2$,

$$\Omega^2 = n(n^2 - 1) \Big/ \{(1+2\beta H) + \frac{n-3}{2GH}\}. \tag{22}$$

Let us compare this result with experimental data.[2] For parameter H in eq.(22), experimental values are employed (for example, $0.38<H<0.52$ for a drop of liquid oxygen in 6-mode vibration), and GH is determined by the use of eq.(10). In order to normalize experimental values of frequency, we need the values of σ/ρ for three materials. However, it is not certain whether we can use values at the boiling points, since the liquid surface was not at thermal equilibrium. After some trials, theoretical values of frequency are found to be fitted to experimental ones if we choose $\sqrt{(\sigma/\rho)}$ = 2.9, 3.2, 2.9 ($cm^{3/2}/s$), for liquid-nitrogen, -oxygen and -argon, respectively. These values are smaller than those at the boiling points, 3.4, 3.5, 3.1, respectively, by several percent. Figure 12 shows comparison of theoretical values with experiment along with Rayleigh's formulae (1) and (2). The narrow width in theoretical values obtained here comes from those of experimental values of H used for their evaluations.

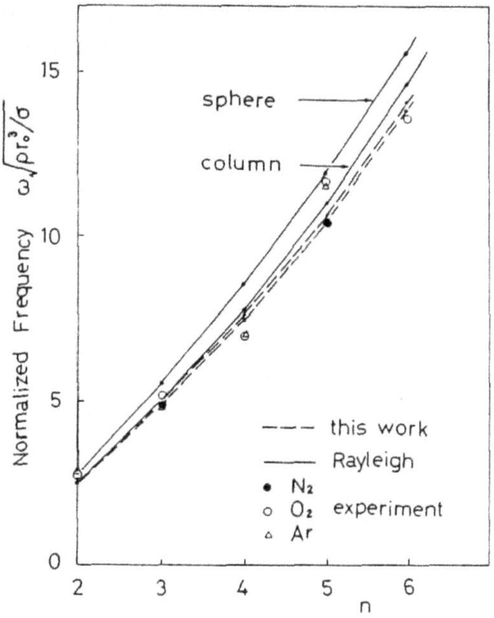

Fig. 12. Comparison of theoretical values of frequency with experiment. Rayleigh's formulae are also shown. Width in the theoretical result comes from that in experimental values of H. (from Takaki & Adachi[12])

As is seen from this figure, the present theory agrees better with experiment than Rayleigh's, especially for larger n. Although the agreement in values is owing to assumed surface tension coefficients, the agreement in the gradient between theoretical and experimental curves is an evidence of success of the present theory. Note that both formulae by Rayleigh yield asymptotic behavior of frequency $n^{3/2}$ for $n \gg 1$, while eq.(22) gives a behavior n^1 for $n \gg 1$ agreeing with experiment. On the other hand, the Rayleigh's formula for a sphere agrees best to experiment of 2-mode vibration. This result is convincing, because the flattenning of drop is not considered in Rayleigh's analysis and the drop in 2-mode vibration was small enough to take a spherical form on the average.

After all, the present hydrodynamical analysis gives a good prediction of the vibrational frequency for relatively higher mode number (larger drop), and suggests also that the surface tension coefficient of a highly overheated liquid is smaller appreciably than that in thermal equilibrium. Since the surface tension decreases with temperature, it means that the liquid surface had a higher temperature than the boiling point.

4. MODEL SIMULATION

4.1. Choice of Model Equation

Effects of nonlinearity and nonequilibrium, neglected in the linear analysis, are treated here by a suitable model simulation.[10,12] The model should include

(i) self-excitation of oscillation,
(ii) mode selection according to drop size,
(iii) decrease of drop size,
(iv) nonlinear interaction of modes,
(v) dissipative effect.

Among physical systems with sustained oscillation, the forced oscillation, the Van der Pol oscillation and the parametric oscillation are most familiar to us. The forced oscillation requires an external force, contradicting the constraint (i), and omitted. The Van der Pol equation allows a self-sustained oscillation. But it is not employed here, because preferable nonlinear term is of the second order, as is mentioned in the previous section, while the equation has the third order nonlinear term. We try to set up a model as a parametric oscillation.

The Mathieu equation given below is the simplest one for parametric oscillation.

$$\frac{d^2u}{dt^2} + (\omega_0{}^2 + \alpha \cos(2\omega t))\, u = 0 \; . \tag{23}$$

As is well known, this equation has an unstable solution for $\omega=\omega_o$, which satisfies constraint (i). However, it should be modified further, in order to satisfy other constraints.

The variable u in eq.(22) stands for displacement of drop contour and should be a function of θ and t. It is expressed by a Fourier series,

$$u = \Sigma\, u_n(t)\exp(in\theta), \tag{24}$$

where each component u_n is considered to be governed by eq.(23). Let us normalize length and time by the initial drop radius ro and the time scale To (see Sec. 3). Since the drop size is decreasing (constraint (iii)), the proper angular frequency ω_o decreases also with time. The theoretical result (22) might be used for evaluation of ω_o. However, since it has a rather complicated form for application to a simple model, an empirical formula from our measurement[2]) is used, which, in a rough approximation, is expressed under the present normalization as follows:

$$\omega_o = 2.5(n - 1)(1 + \beta t)^{3/2}, \quad \beta = 0.05. \tag{25}$$

On the other hand, the drop had a tendency to keep nearly the same frequency (~10) during successive transition process. Therefore, the unstable frequency ω is fixed at this value.

Constraint (iv) is satisfied by adding a nonlinear term of the second order. It is assumed that this nonlinear term is expressed as

$$NL_n(u) = \text{n-th Fourier component of } (\varepsilon\, u\dot{u}), \tag{26}$$

where ε is a constant and the dot over u denotes the time derivative. Although there is an ambiguity in determination of nonlinear term, the above special choice has a certain reasoning. The term $\varepsilon u\dot{u}$, if combined with $\omega_o^2 u$, yields $(\omega_o^2+\varepsilon\dot{u})u$, and gives a modification of the proper angular frequency due to drop motion. Now, in the period where the periphery of drop penetrates into the ambient air, i.e. when $\dot{u}>0$, a thin boundary layer of gas flow is formed on the surface of the periphery, which will exert a strong shear stress on the surface and induce more evaporation, hence the surface tension will be increased through surface cooling. On the contrary, in the period where the periphery moves back (for $\dot{u}<0$), it is covered with a wake flow with low shear stress, which will induce less evaporation and decrease the surface tension. Thus, this effect of surface motion gives the nonlinear term expressed by eq.(26) with $\varepsilon>0$.

Lastly, a friction term $-\gamma\dot{u}$ is added to eq.(23), where γ is a constant. This term is added in order to suppress divergence of unstable solution. It proves later, however, that solution in the case with $\gamma=0$ behaves properly if smaller value of α is given. After all, the equation for model simulation is

$$\ddot{u}_n + \gamma\dot{u}_n + NL_n(u) + \omega o^2\{1 + \alpha\cos(2\omega t)\}\, u_n = 0, \qquad (27)$$

where ωo and $NL_n(u)$ are given by eqs.(25) and (26), respectively, and ω is fixed at 10.

Some comments are given on physical meanings of parameters included in eq.(27). In the numerical simulation the initial drop is assumed to take 5-mode, as is memtioned in Sec. 4.2. On the other hand, in the experiment the controllable parameters were the ambient temperature and the drop size, and the mode number was selected according to the drop size. Therefore, the initial drop size is considered to be fixed to a value allowing the 5-mode vibration. Thus, the only controllable parameter corresponding to the simulation is the ambient temperature. The parameter α in the simulation is related to self-excitation and stands for the degree of thermal inequilibrium, hence it corresponds to the ambient temperature. The parameter ε, representing the degree of nonlinearity, should be uniquely determined if precise mechanism of drop motion is clarified. But, since such a mechanism is not included in the present model, ε must be treated also as a free parameter. The friction constant γ should be determined, if the flow field around the drop is obtained. But, γ is considered here also as a free parameter.

Equation (27) is integrated numerically by the use of the Runge-Kutta-Gill method. Components with $n \geq 7$ are neglected, and five modes with $2 \leq n \leq 6$ are treated. As initial conditions, a large value is assigned for u_5 and small values for other components; it reflects the observation of mode transition beginning from 5-mode. The numerical scheme proved to become unstable after a certain time interval. It is because the coefficient ω_0^2 increases with time. When any component exceeds unity, i.e. the amplitude becomes larger than the drop radius, the computation is stopped.

4.2. Results of Model Simulation

Figure 13 shows three examples of solution for $\gamma=0.02$, 0.1 and 0.2, where other parameters are fixed at $\varepsilon=13$, $\alpha=0.26$, and initial conditions are

$$u_5 = 0.5, \quad u_n = 0.05 \ (n \neq 5), \quad \dot{u}_n = 0. \tag{28}$$

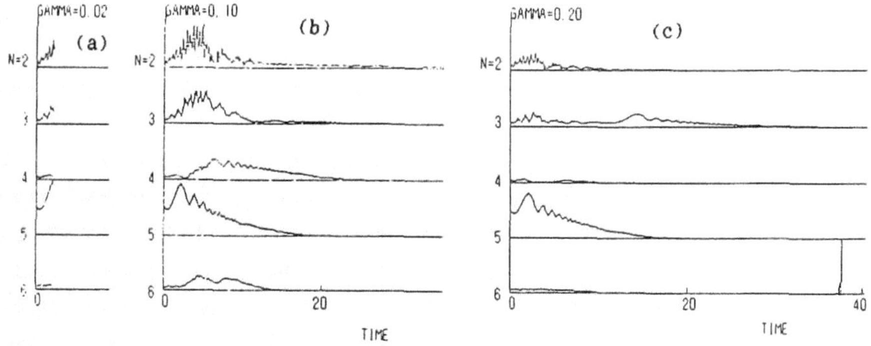

Fig. 13. Variations of amplitudes of several modes from model simulation with $\varepsilon = 13$ and $\alpha = 0.26$. Distances between base lines are unity. (a): $\gamma = 0.02$, (b) $\gamma = 0.1$, (c): $\gamma = 0.2$. ((b) is from Takaki, etal.[15])

For $\gamma=0.02$, the amplitudes soon exceed unity showing divergence of solution. For $\gamma=0.1$, after a maximum of 5-mode, the 2- and 3-modes vary randomly, and then the 4- and 6-modes grow, but finally all modes decay. For $\gamma=0.2$, the 3-mode grows instead of 4- or 6-mode, and all modes decay also. The rapid rise of 6-mode, seen in the result for $\gamma=0.2$, occurs in any case at $t=37$ if other components remain small. It is due to the numerical instability, and the computation is stopped at this moment.

Timewise changes of the largest modes are interpreted as the mode transition observed in the experiment. However, we must be careful in interpretation of the noisy initial variations in the 3- and 2-modes. They are considered to have come from the unrealistic initial conditions (28), especially those for du_n/dt, because they appeared for any value of γ. In this period the system has been trying to recover natural distributions of u_n and $\dot{u}_n$. Therefore, these variations are not important for comparison with actual phenomena, and are neglected here. Growths of components in the simulation, which are physically reasonable, are those observed after certain periods. These growths, although they are modified by the nonlinear effect, are considered to come primary from instability of solution of the Mathieu equation for $\omega_0=\omega$, and this condition is satisfied at the time which is later for smaller n (see eq.(25)). Taking this situation into account, growths of modes are identified after $t=3$, 10 and 20 for 4-, 3- and 2-mode, respectively.

Variations of amplitudes are computed for several values of γ with a lot of combinations of (ε, α), but starting from the same initial conditions (28). The results are summarized as diagrams, two examples of which are shown in Fig. 14. In these diagrams large or small n-polygons indicate appearances of n-modes with large ($1>\mathrm{Max}(u)>0.2$) or small amplitudes ($0.2>\mathrm{Max}(u)>0.1$), respectively. The dots indicate unstable behavior, where amplitude of a certain mode exceeds unity before beginning of numerical instability. The slashes indicate decay of 5-mode without growth of other modes.

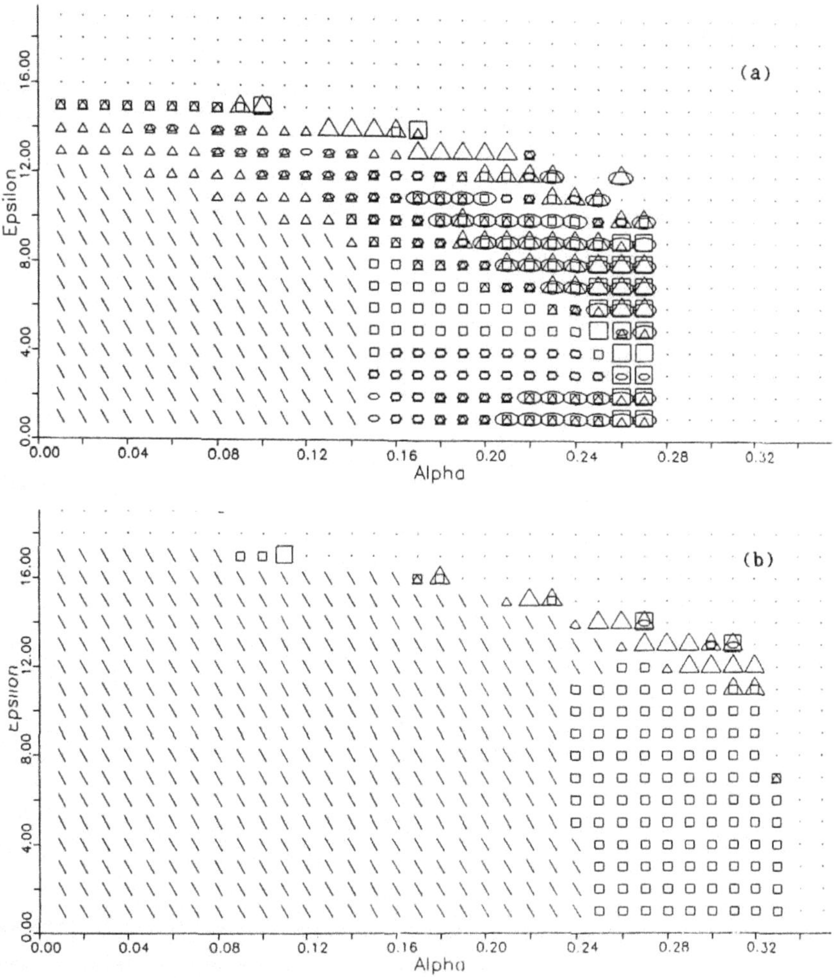

Fig. 14. Diagrams of mode appearances for (a) $\gamma=0$ and (b) $\gamma=0.2$, and for initial conditions (28). The large (small) n-polygons indicate appearances of n-modes with amplitudes larger than 0.2 (0.1). The dots indicate unstable behavior, and the slashes indicate simple decay of 5-mode.

From these diagrams one can extract the following conclusions.

(i) The condition in the parameter space (α, ε) for appearance of other modes after decay of 5-mode is limited to a narrow region, i.e. for $\gamma=0$ ($0.15<\alpha<0.27$, $\varepsilon<10$) or ($\alpha<0.15$, $10<\varepsilon<15$).

(ii) For α much larger than about 0.3 (stronger degree of inequilibrium) or $\varepsilon>17$ (stronger degree of nonlinearity), the system always becomes unstable. For opposite conditions, it tends to show a simple decay.

(iii) Appearance of successive modes are observed in the two regions. For $\gamma=0$, one is near the corner right-above at ($\alpha\sim0.24$, $\varepsilon\sim10$), and another is near the corner right-below at ($\alpha\sim0.24$, $\varepsilon\sim0$). For larger values of γ, the second region shrinks.

(iv) Effects of increasing γ are to shift the region in the diagram to the right direction and to suppress appearances of various modes.

Let us compare these results with experiment, by assuming that the parameter α stands for the ambient temperature and ε has a value of about 10. The instability of the system for larger α may correspond to tearing of a drop to small drops. This phenomenon was not observed yet but is expected to occur if the ambient temperature is too high and the drop suffers boiling. Small values of α with simple decay of 5-mode will correspond to conditions with low ambient temperature ($<-100^{\circ}$C) where no vigorous drop motion was expected to occur. Appearance of large amplitude vibration and mode transition for an intermediate region of α agrees well with experiment qualitatively. As is seen from this diagram, the mode transition process in the region ($\alpha\sim0.3$, $\varepsilon\sim10$) depends critically on α and ε. Physical meaning of the dependence on ε is not clear, because we do not have a correspondence of ε with physical quantities in a drop. From the observed dependence on material (see Sec. 2.1) and the result of simulation that 4-mode appears for smaller values of ε, it may be suggested that effect of material properties on transition process is understood in terms of changing ε.

After all, the present model simulation based on the Mathieu-type equation predicts observed phenomena well at least qualitatively. Quantitative prediction will become possible, if precise mechanism of drop vibration is taken into account in the theory.

5. Discussion

In this review article results of observation, linear analysis and model simulation are introduced, and discussions on their results are already given in respective sections. Here, in the last section, it will be meaningful to mention on the significance of investigating this phenomenon.

Patterns shown by an evaporating drop can be classified by the two parameters, the time and the degree of inequilibrium, as shown in Fig. 15. This diagram is partly based on the experimental results explained so far, but partly on speculation. In the region I the drop is in contact with the floor and makes a convective motion inside it owing to the Marangoni effect or the buoyancy, or it may show a weak surface wave. Processes in the regions II and III are what we investigated. Phenomena in the region IV have not been observed, but are expected to occur. Process in any region are terminated in the state of molecular chaos, and differ from each other in the route to this final state.

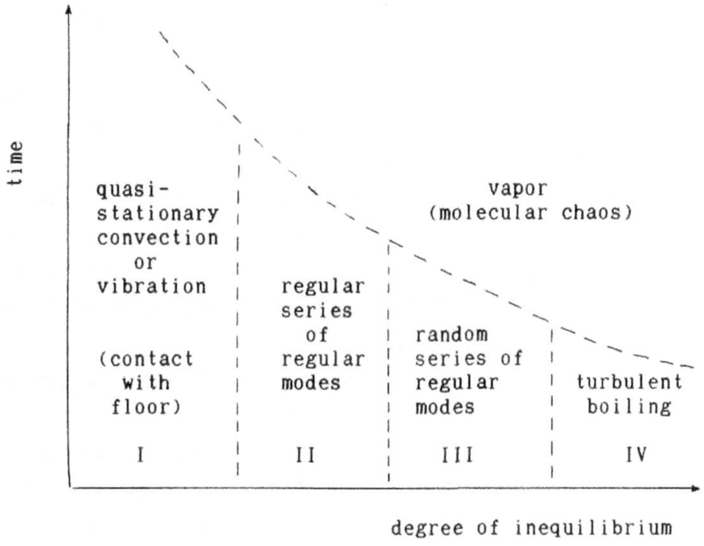

Fig. 15. Classification of dynamical processes.

Recently, formation of macroscopic order in systems, which is not in thermal equilibrium, is attracting interest of scientists with a name "dissipative structure".[16] Examples of such an order formation are the thermal convection, the BZ reaction, the diffusion-limited-aggregation, the crystal growth, the Taylor vortex, etc. Then, are the patterns observed in evaporating drop the dissipative structures ? The answer is yes in a sense that they occur in conditions out of thermal equilibrium. However, there is also an essential difference. In usual dissipative structures degrees of inequilibrium are not so large that macroscopic velocity of materials induced by the inequilibrium has relatively small value and the system shows in most cases only single mode for given values of parameters. The vibration of an overheated drop, on the contrary, is a phenomenon in a highly inequilibrium state. The drop shows not only a rapid motion, but also a successive transition to other modes.

Another characteristic feature which are not owned by other structures is the size of system, i.e. patterns are produced in a small drop. Because of a small heat capacity, patterns appeared in a short interval and made mode transitions until it reached thermal equilibrium. Such series of dynamical patterns can be looked upon as a pattern formation in the space-time framework. It may be an interesting concept to be investigated as a fundamental science.

Along with this aspect of phenomenon, there is also a practical significance in investigating the drop vibration. It affords a method to measure the surface tension coefficient of a surface far from equilibrium. As is mentioned in Sec. 2.3, the coefficients were smaller than those at boiling points by several percents. But, hydrodynamical analysis of vibration should be refined before it is confirmed.

The present status of study of drop vibration is still not satisfactory, and many problems are left unsolved. However, the present author believes that this phenomenon is a good topic for research, because it is quite interesting and it gives us a stimulation to develop various new ideas.

ACKNOWLEDGEMENTS

The present author would like to express his cordial thanks to Professor K. Adachi, Mr. N. Yoshiyasu, Mr. Y. Arai, Mr. A. Katsu and Professor O. Sano for their cooperations and helps throughout this series of works. These works are partly supported by Grant-in-Aids from the Ministry of Education, Culture and Science of Japan.

References

1) Holter,N.J. and Glasscock,W.R., J. Acoust. Soc. Am. 24, 682 (1952).
2) Adachi,K. and Takaki,R., J. Phys. Soc. Jpn. 53, 4184 (1984).
3) Stahler,A.N., ed., "The Earth Sciences", Harper & Row., (1971).
4) Wong,C.Y., Phys. Lett. 61B, 321 (1976).
5) Corrigan,T.M., Margetan,F.J. and Williams,S.A., Phys. Rev. C14, 2279 (1976).
6) Levich,V.G. and Krylov,V.S., Ann. Rev. Fluid Mech. 1, 293 (1979).
7) Valentine,P.S. Sather,N.E. and Heideger,W.J., J. Chem. Eng. Sci. 20, 719 (1965).
8) Trinh,E., Zwern,A. and Wang,T.G., J. Fluid Mech. 115, 453 (1982).
9) Nelson,A.R. and Gokhale,N.R., J. Geophys. Res. 77, 2724 (1972).
10) Takaki,R., Yoshiyasu,N. Arai,Y. and Adachi,K., "Proc. First Int. Symp. Sci. Form", KTK Scientific Publ. p.67 (1986).
11) Takaki,R., Katsu,A. and Arai,Y., Sci. Form 1, 89 (1986) [in Japanese].
12) Takaki,R., Katsu,A. and Arai,Y. and Adachi,K., J. Phys. Soc. Jpn. 58, 129 (1989).
13) Load Rayleigh, "Scientific Papers" Vol.3, Cambridge Univ. Press, (1902).
14) Load Rayleigh, Proc. R. Soc. 29 London, 71 (1879).
15) Takaki,R. and Adachi,K., J. Phys. Soc. Jpn. 54, 2462 (1985).
16) Glansdorff,P. and Prigogine,I., "Thermodynamic Theory of Structure, Stability and Fluctuations", Wiley-Interscience, (1971).
17) Wolfram,S., "Theory and Applications of Cellular Automata", World Scientific, (1986).

SIMULATION OF GROWTH PATTERNS IN A DIFFUSION FIELD

Yukio SAITO,
Chaouqi MISBAH, Heiner MÜLLER-KRUMBHAAR,†*
Tsuyoshi UETA and Makio UWAHA‡

Department of Physics, Keio University,
3-14-1 Hiyoshi, Kohoku-ku, Yokohama 223, Japan

* Groupe de Physique des Solides de l'Ecole Normale Supérieure, associé au Centre National de la Recherche Scientifique, Université Paris VII, place Jussieu, 75251 Paris Cedex 05, France

† Institut für Festkörper Forschung der Kernforschungsanlage, D-5170 Jülich, FRG

‡ Institute for Materials Research, Tohoku University, 2-1-1 Katahira, Aoba-ku, Sendai 980, Japan

ABSTRACT

Crystal shapes growing in a diffusion field are studied by various simulation methods, and universal behaviors in the selection of the growth rate and the growth patterns are found.

For a free dendritic growth dynamical evolution is simulated numerically by solving an integro-differential equation, and an anisotropy in the surface tension is found essential for stable steady growth. Scaling relation $VR^2=const$ between the velocity V and the tip radius R is obtained, and the sidebranch periodicity λ is found proportional to R. For directional solidification our simulation proves the same scaling relation $VR^2=const$ at sufficiently fast growth. In addition to a tip-splitting instability in the cellular growth regime, wavelength selection seems to occur via a tail-instability of sidebranches in the dendritic parameter range.

A microscopic approach by Monte Carlo simulation is applied to a lattice-gas model. As the chemical potential difference $\Delta\mu$ increases and the growth becomes fast, crystal morphology varies from polygonal to hopper-like, then to dendritic, and finally to homogeneous "spherulitic" structures. In the limit $\Delta\mu\rightarrow\infty$, the model is equivalent to that of a diffusion-limited aggregation (DLA) growing in a gas with finite density, n_g. At low n_g growth is controlled by a fractal structure in the short scale, and is characterized by the fractal dimension D_f as $V\propto n_g^{-1/(d-D_f)}$.

1. INTRODUCTION

In nature one often sees beautiful morphology of an interface separating two phases. This is, however, not the equilibrium phenomena, since the interface takes only simple patterns in equilibrium. For example, an equilibrium shape of a crystal is convex and monotonous.[1–3] The phase transition in the surface tension[3–7] induces a morphological change between faceted and non-faceted shapes, but the convexity of the profile remains. On the contrary, when two coexisting phases are out of equilibrium with each other, an interface between them propagates and shows a variety of complex patterns. Propagation velocity and interface morphology are the major subjects to be determined and to be understood in nonequilibrium statistical physics. Crystal growth is a typical field where the pattern formation and its propagation are readily observed.[8] There are various growth morphology of the crystal. It can take a simple form like a polyhedron with facets, but can be complex like a hopper and a dendrite or it can even look fractal. Variety of growth shapes comes from competition among the instability caused by the diffusion field in which the crystal is growing and the stabilizing effect of surface tension or surface kinetics.

In this work, we review some of our recent simulation studies on the growth morphology of the crystal.[9–12] In the next section a complete evolution of a free dendrite with sidebranches is simulated by integrating forwardly a non-local integro-differential equation of the front.[9] The still controversial problem of directional solidification is treated in a similar way in Sec. 3.[10] In Sec. 4 we describe a Monte Carlo study of a lattice-gas model for a crystal growing in a diffusion field with a finite density.[11] In the limit of an infinite chemical potential gain by the solidification, the model reduces to the growth of a diffusion-limited aggregation (DLA)[13] in a finite-density gas. Relation between the growth velocity and the fractal structure of the aggregate in this model is investigated in Sec. 5.[12]

In the rest of this introductory section we summarize the linear stability analysis of a planar interface growing from a diffusion field in terms of a solution growth.[8] Consider a two-dimensional binary-alloy growing from a solution at a temperature T_0. In equilibrium, the coexisting liquid and solid have concentrations $C_{L,eq}$ and $C_{S,eq}$, respectively, of one component. The partition coefficient is defined by $k=C_{S,eq}/C_{L,eq}$ and the miscibility gap by $\Delta C=C_{L,eq}-C_{S,eq}$.

Here we assume $C_{L,eq} > C_{S,eq}$.

With the local equilibrium assumption at the interface the liquid concentration there is obtained as

$$C_{L,i} = C_{L,eq} - \frac{T_M(\gamma + d^2\gamma/d\theta^2)}{\beta L}\kappa, \tag{1}$$

where $\gamma(\theta) = \gamma_0(1+\bar{\varepsilon}\cos m\theta)$ is the anisotropic surface tension depending on the orientation $\theta = \tan^{-1}(\partial z_i/\partial x)$ of the interface normal $\vec{n} = (-\partial z_i/\partial x, 1)/\sqrt{1+(\partial z_i/\partial x)^2}$ to the z-direction, $-\beta$ a liquidus slope $(dC_{L,eq}/dT)^{-1}$, T_M the melting temperature of the pure substance, L the latent heat, and $\kappa = -\partial^2 z_i/\partial x^2 / [1+(\partial z_i/\partial x)^2]^{3/2}$ is the curvature of the interface. Solid concentration at the interface is given as $C_{S,i} = kC_{L,i}$. By keeping the concentration in liquid infinitely far from the crystal front, C_∞, smaller than $C_{L,eq}$, the crystal grows due to a supersaturation $\Delta = (C_{L,eq} - C_\infty)/\Delta C$. When the solidification front, $z = z_i(x,t)$, advances, it repels solute molecules, which should be transported away from the interface to infinity by a chemical diffusion process

$$\partial C_{L,S}/\partial t = D_{L,S}\nabla^2 C_{L,S}. \tag{2}$$

Here $C_L(\vec{r},t)$ and $C_S(\vec{r},t)$ are the local concentration in liquid and solid respectively, and D_L and D_S are the corresponding chemical diffusion coefficients. Since the diffusion in solid is usually a very slow process, D_S is often set zero. This is called a one-sided model. Mass conservation leads to the continuity equation at the solidification front as

$$(C_{L,i} - C_{S,i})(\vec{v}\cdot\vec{n}) = (-D_L\vec{\nabla}C_L + D_S\vec{\nabla}C_S)_i\cdot\vec{n}, \tag{3}$$

where $\vec{v} = (0, \partial z_i/\partial t)$ is the local growth rate.

It is convenient to introduce a dimensionless diffusion field

$$u_{L,S}(\vec{r},t) = (C_{L,S}(\vec{r},t) - C_\infty)/\Delta C, \tag{4}$$

which satisfies the diffusion equation

$$\partial u_{L,S}/\partial t = D_{L,S}\nabla^2 u_{L,S}. \tag{5}$$

The local equilibrium condition at the solidification front reduces to

$$u_{L,i} = \Delta - d\kappa \qquad \text{and} \qquad u_{S,i} = k(u_{L,i} - \Delta) - (1-\Delta). \tag{6}$$

Here $d = d_0 \cdot A_m(\varepsilon)$ is the product of a usual capillary length

$$d_0 = T_M \gamma_0 / \beta L \Delta C, \tag{7}$$

and an m-fold anisotropy factor

$$A_m(\varepsilon) = 1 - \varepsilon \cos(m\theta). \tag{8}$$

The continuity condition reduces to

$$[1+(1-k)(u_{L,i} - \Delta)] v_n = -(D_L \partial_n u_L - D_S \partial_n u_S), \tag{9}$$

where v_n is the normal velocity $(\vec{v} \cdot \vec{n})$ and ∂_n represents the normal derivative $(\vec{n} \cdot \vec{\nabla})$.

When crystal grows from pure melt under supercooling, diffusion of temperatures in solid and liquid, T_S and T_L respectively, limits the growth. Dimensionless temperature fields

$$u_{L,S}(\vec{r},t) = (T_{L,S}(\vec{r},t) - T_\infty)/(L/c_p) \tag{10}$$

follow the same diffusion equation (5) with a few modifications in boundary conditions;

$$u_{L,i} = u_{S,i} = \Delta - d\kappa \tag{6a}$$

and

$$v_n = -(D_L \partial_n u_L - D_S \partial_n u_S). \tag{9a}$$

In this case Δ is the undercooling

$$\Delta = (T_M - T_\infty)/(L/c_p) \tag{11}$$

with c_p being the isobaric specific heat. D_S and D_L here correspond to the thermal diffusion coefficients in solid and liquid, respectively, which are of the same order. In case $D_S = D_L$ analysis becomes easy and the system is called a symmetric model.[14,15)]

For a planar interface moving steadily, $z_i(x,t) = Vt$, the diffusion fields are easily solved as

$$u_L(\vec{r},t) = \Delta \exp[-2(z - Vt)/l] \qquad \text{and} \qquad u_S(\vec{r},t) = const. \tag{12}$$

The spatial variation of the diffusion field in the liquid is characterized by the diffusion length defined by

$$l = 2D_L / V. \tag{13}$$

The continuity condition, (9), requires $\Delta = 1$ but V is arbitrary. The condition

$\Delta=1$ simply means that the planar interface can grow steadily only when the concentration in the solid is the same with that in the liquid matrix at infinity; $C_S(z=-\infty)=C_\infty$.

The planar interface is in fact unstable to the perturbation.[16] By neglecting the anisotropy ε for simplicity, an interface with a deformation, $z_i(x,t)=Vt+\delta\cos qxe^{\omega t}$, leads to a dispersion relation

$$\omega/2D_L=l^{-2}[(\sqrt{1+q^2l^2}-1)(1-\tfrac{1}{2}d_0lq^2)-\tfrac{1}{2}kd_0lq^2(1+\sqrt{1+r^2q^2l^2})], \tag{14}$$

in a quasi-stationary approximation. Here r is the ratio between the two diffusion coefficients; $r=D_S/D_L$. For the thermal model, k in Eq.(14) is replaced by 1. For slow growth ($l^{-1}\ll q$), the dispersion is simplified to

$$\omega/2D_L=(q/l)[1-\tfrac{1}{2}(1+kr)d_0lq^2]. \tag{15}$$

If there is no surface tension, $d_0=0$, ω is always positive and the interface is completely unstable. The growth rate increases indefinitely for shorter wavelengths. The capillary length d_0 stabilizes the short wavelength modes. The stability length is obtained from $\omega=0$ as

$$\lambda_s=2\pi/q_c=2\pi\sqrt{\tfrac{1}{2}(1+kr)d_0l}, \tag{16}$$

and we expect that a system developes a pattern, whose scale is characterized by this λ_s.

2. FREE DENDRITE[9]

2-1 Integral Equation

In experiments of crystal growth from a solution or a pure melt at $\Delta<1$, steady dendritic growth is observed with a unique and reproducible velocity and a shape.[17–19] In this section we discuss a dendritic growth from pure melt, and the diffusion field is assumed to satisfy the boundary conditions (6a) and (9a). In the absence of a surface tension, Ivantsov found a steady-state solution with a parabolic interface in two dimensions[20]

$$z_i(x,t)=Vt-x^2/2R_0. \tag{17}$$

The undercooling Δ determines the product of the tip radius R_0 and the growth rate V or the Peclet number

$$p = R_0 V / 2D_L = R_0 / l \tag{18}$$

as

$$\Delta = \sqrt{\pi p}\, e^p \mathrm{erfc}(\sqrt{p})$$
$$\sim \sqrt{\pi p} \qquad \text{for} \qquad \Delta << 1. \tag{19}$$

Here $\mathrm{erfc}(x) = 2\pi^{-1/2}\int_x^\infty e^{-t^2}dt$ is the error-function. The defficiency of the Ivantsov solution is that it selects only the product of the tip radius and the growth rate, but continuously many solutions are possible with different radii. On the other hand, in experiments both the radius and the rate take unique values under a given Δ. The infinite degeneracy is due to the lack of an intrinsic length scale in the system. When a finite surface tension is introduced, a capillary length d_0 plays the role of this intrinsic length scale. The diffusion length $l = 2D_L/V$ and the tip radius R should be related to d_0. The problem of mode selection is to find a mechanism to determine the relation as a function of supercooling and other material parameters.

Langer and Müller-Krumbhaar have performed a stability analysis of the Ivantsov parabola.[21,22] In the small Δ limit, a parabola with a small tip radius is stabilized by capillarity, while a parabola with a large tip radius is found unstable. The transition of stability is characterized by the tip radius R_0 in comparison to the stability length λ_s, or by the Langer-Müller-Krumbhaar (LMK) parameter

$$\sigma = d_0 l / R_0^2 = d_0 V / 2D_L p^2. \tag{20}$$

Since every dendrite with σ values larger than the marginal value σ^* is stable and is a candidate for a steadily growing dendrite, a unique working point of dendritic growth cannot be obtained. They then proposed the marginal stability hypothesis which claims that the velocity and pattern selected correspond to those at the point of marginal stability, $\sigma = \sigma^*$.

Thereafter, the marginal stability theory is further developed and sophisticated,[23] but new features have been revealed in the dendritic growth problem. By way of searching sound foundation of marginal stability, simplified models for dendritic growth have been invented, like a geometrical model[24], a boundary layer model[25] and so on. By numerical simulations and also later by theoretical analysis,[24] it is found that the LMK-model suffered from an oversimplification, and that an anisotropy in capillarity is essential for the stable growth. Without an anisotropy the tip of the dendrite always splits. With all these backgrounds it has become necessary again to study the full model of the dendritic growth with nonlocal interactions among surface elements.

When a crystal is growing steadily with a velocity V in z-direction, the diffusion field in the liquid satisfies a stationary equation in the moving frame of reference

$$\nabla^2 u_L + (2/l)\partial u_L/\partial\zeta = D_L^{-1}\partial u_L/\partial t = 0, \tag{21}$$

where ζ is a new coordinate; $\zeta = z - Vt$. Eq.(21) can be converted into an integro-differential equation on a solid-liquid boundary Γ;

$$\int d\Gamma'\, g_L(\vec{r}|\vec{r}\,')\partial u_L(\vec{r}\,')/\partial n' = \int d\Gamma'\, h_L(\vec{r}|\vec{r}\,')u_L(\vec{r}\,'). \tag{22}$$

Here $g_L(\vec{r}|\vec{r}\,')$ is the Green's function of an adjoint operator to Eq.(21),

$$[\nabla'^2 - (2/l)\partial/\partial\zeta']g_L(\vec{r}|\vec{r}\,') = -\delta(\vec{r}-\vec{r}\,'), \tag{23}$$

and it reads as

$$g_L(\vec{r}|\vec{r}\,') = (2\pi)^{-1}e^{-(\zeta-\zeta')/l}K_0(|\vec{r}-\vec{r}\,'|/l), \tag{24}$$

with the modified Bessel function K_0. The integral kernel h_L is given by

$$\begin{aligned} h_L(\vec{r}|\vec{r}\,') &= \partial_{n'} g_L - (2/l)n_z' g_L - \tfrac{1}{2}\delta(\vec{r}-\vec{r}\,') \\ &= (2\pi l)^{-1}e^{-(\zeta-\zeta')/l}[-n_z' K_0 - (\vec{n}\,'\cdot(\vec{r}-\vec{r}\,')/|\vec{r}-\vec{r}\,'|)K_1] - \tfrac{1}{2}\delta(\vec{r}-\vec{r}\,'), \end{aligned} \tag{25}$$

and satisfies the sum rule

$$\int_{-\infty}^{\infty} d\Gamma'\, h_L(\vec{r}|\vec{r}\,') = -1. \tag{26}$$

An equation similar to Eq.(22) holds on the solid side with a new Green's function g_S and a new kernel h_S, where l is replaced by $rl = D_S l/D_L$ and a sum rule is modified to

$$\int_{-\infty}^{\infty} d\Gamma'\, h_S(\vec{r}|\vec{r}\,') = 0. \tag{27}$$

In the symmetric model where $r=1$ the sum of Eq.(22) and the corresponding equation for the solid side reduces to a simple form;[14,15,27–38]

$$\Delta - d\kappa = \int_{-\infty}^{\infty} d\Gamma'\, g_L(\vec{r}|\vec{r}\,')v_n(\vec{r}\,')/D_L = (2/l)\int_{-\infty}^{\infty} dx'\, g_L(\vec{r}|\vec{r}\,'). \tag{28}$$

In the last equality crystal is assumed to have a needle shape, growing steadily with a local velocity $v_n(\vec{r}\,')=Vn_z{}'$. With the Ivantsov solution, $\zeta_{IV}(x)=-x^2/2R_0$, the r.h.s. of Eq.(28) reduces to the error function and the Ivantsov relation, Eq.(19), is reproduced for $d_0=0$. By scaling length in the unit of a Ivantsov radius R_0, Eq.(28) reduces to

$$-\tfrac{1}{2}\sigma A_m K=\int dX'\,[g_L(X,Z_i(X)\,|\,X',Z_i(X'))-g_L(X,Z_{IV}(X)\,|\,X',Z_{IV}(X'))],\quad(29)$$

with $x=R_0X$, $\zeta=R_0Z$, and $Z_{IV}=-\tfrac{1}{2}X^2$. Green's function in Eq.(29) is parametrized by the Peclet number p instead of l in Eq.(24), and a capillarity effect is parametrized by the LMK parameter σ with additional dependence on anisotropy, A_m.

The strategy of the method of solvability condition is to find first a solution $Z=Z_i(X)$ of Eq.(29) in a positive x-region under a given velocity V or σ. One always obtains a solution provided a finite cusp at the tip, $X=0$, is allowed. Among them the selected solutions of dendritic growth are those which have zero slope at the tip; $dZ_i(X=0)/dX=0$.[27–36] This is possible only for special values of V or σ. An important new finding is the essential role of a crystalline anisotropy ε in capillarity. Without an anisotropy in the surface tension, no solution is found for any σ. With a nonzero anisotropy a needle solution is possible at a discrete set of σ values. Further stability analysis[28,37–39] indicates that the fastest moving solution is weakly stable against tip splitting and also against sidebranching. Thus a needle crystal without sidebranches is a natural candidate for the pattern to be selected. Experiments, on the other hand, show always dendrites with sidebranches.[17–19] It thus seems not obvious whether the solvability condition corresponds to the actual selection mechanism.

We simulate a time evolution of the one-sided model[40] of a dendritic crystal, and determine the selected mode as a function of control parameters.[9] For the one-sided model the value of $u_{S,i}$ is irrelevant in the interface evolution. An initial velocity V is chosen arbitrary, and the crystal takes initially a shape of the corresponding Ivantsov parabola. On the interface values of the diffusion field are obtained from the local equilibrium condition (6a). The integral kernels, g_L and h_L, are calculable as functions of the steady growth rate V from Eqs.(24) and (25). Then the normal gradient, $\partial_n u_L$, is obtained as a solution of the linear equation (22). This yields the local velocity $v_n(\vec{r},t)$ by using the continuity condition (9a). The solidification front is shifted in the normal direction

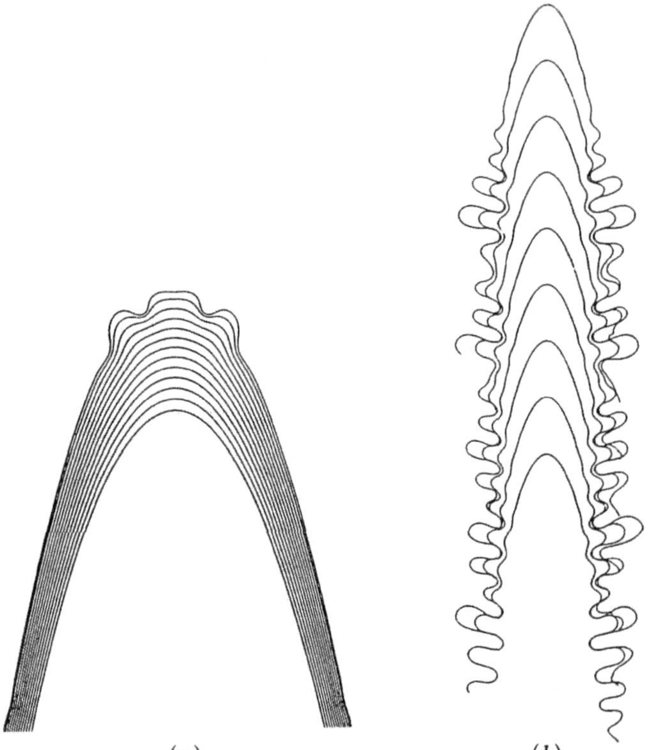

Fig.1 Stroboscopic plot of the evolution of a dendritic crystal. a) The supercooling is $\Delta=0.50$ and the anisotropy in capillarity is $\varepsilon=0$, and b) $\Delta=0.25$ and $\varepsilon=0.10$.

to a new configuration according to the local velocity $v_n(\vec{r},t)$. Since the front motion is generally different from that expected from the prescribed velocity V in z-direction, we have to adjust the velocity of the moving frame V to the growth rate of the crystal, or typically to the tip velocity $v_{tip}=v_n(x=0)$. The relaxation of V is represented by

$$dV/dt=-(V-v_{tip})/\tau \tag{30}$$

with an appropriate relaxation time τ. The new front takes no more a parabolic or even a needle shape. We iterate the procedure until the steady state is realized. The most characteristic point of our simulation is that the solidification front evolves to select its final profile spontaneously, and thus the possible degeneracy of the growth profile is removed automatically .

2.2 Scaling Relation

We consider two undercoolings, $\Delta = 0.25$ and 0.50, and the corresponding Peclet numbers are $p=0.0286$ and 0.187, respectively. Our length unit is so chosen that a capillary length is $d_0=10^{-3}$. In various simulations, the diffusion length l turns out to be of order $1\sim100$. The stability length $\lambda_s=\pi\sqrt{2d_0 l}$ which characterizes the crystal structure is thus expected to be of order $1\sim10$. Without an anisotropy, $\varepsilon=0$, the tip of a parabolic crystal flattens and finally it splits as is shown in Fig.1a. With a small four-fold anisotropy ε, the tip stabilizes to form a dendrite as is shown in Fig.1b.

The ε dependence of the growth rate V or $\sigma=d_0V/2Dp^2$ for different undercoolings is shown in Fig.2. It seems that σ depends on an anisotropy but is almost independent of the undercooling Δ. The solvability condition for the symmetric model[32-36] predicts the relation

$$V\sim\Delta^4\varepsilon^{7/4} \tag{31}$$

at an infinitesimal Δ. Since $p\sim\Delta^2/\pi$ for small Δ, the Δ dependence of V in Eq.(31) is compatible with our finding of the universal curve of σ. For large Δ numerical solution of the solvability condition yields σ as a function of ε as is shown by a solid line in Fig.2.[36] Scaling behaviors in our simulation of one-sided model and in the solvability analysis of the symmetric model agree with each other with a difference in the absolute factor by about 2. The factor 2 comes from the difference in the models and has been explained by Misbah.[41]

Growth takes place in the direction of the small capillarity, and the actual tip radius R of the dendrite is sharper than the Ivantsov value R_0 for the corresponding Δ and V. The ratio R/R_0 depends weakly on an anisotropy but is almost independent of Δ as is shown in Fig.3; $R/R_0=\tau(\varepsilon)$. Combining these two results, we reach the scaling relation between the velocity V and the actual tip radius R such that

$$\overline{\sigma}\equiv d_0 l/R^2=2Dd_0/VR^2=\sigma(\varepsilon)/\tau^2(\varepsilon) \tag{32}$$

depends only on an anisotropy parameter ε but is independent of the

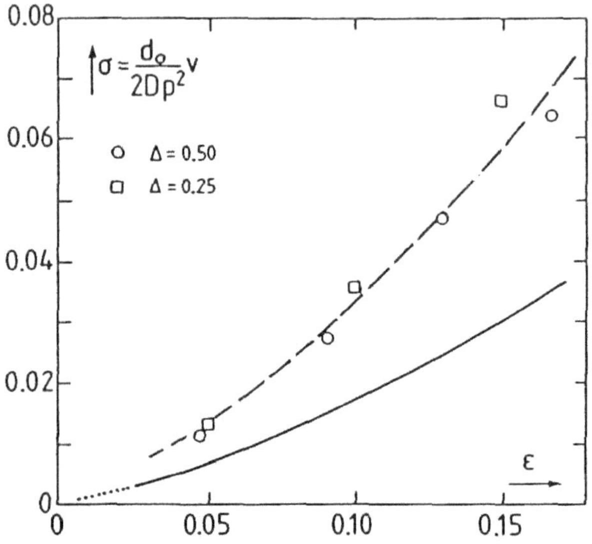

Fig.2 LMK parameter $\sigma = d_0 V/2Dp^2(\Delta)$ versus ε for two different undercoolings, Δ=0.25 and 0.50. Solid line is obtained by the solvability condition of a needle crystal for the symmetric model,[33,36] and dashed line is twice of that value, which is shown to correspond to the one-sided model.[41]

undercooling Δ.

We always obtain dendrites with sidebranches in our simulation, in contrast to the prediction of the solvability analysis[27] but in agreement with usual experiments.[17–19] The wavelength λ of the side branch is defined in the simulation as the growth rate V divided by the number of side branches produced per unit time. The ratio of the sidebranch periodicity λ to the stability length λ_s defined in Eq.(16) is easily calculated to be proportional to $\sigma^{-1/2}(\lambda/R_0)$. This ratio is plotted in Fig.4, and is found independent of both the undercooling Δ and the anisotropy ε. It suggests that the sidebranching is driven by the diffusional instability with the stabilizing effect of the capillarity. Since the actual tip radius R is proportional to the Ivantsov radius R_0 as $R/R_0=\tau(\varepsilon)$, the scaling between λ and R_0 can be read as the scaling relation between λ and R as

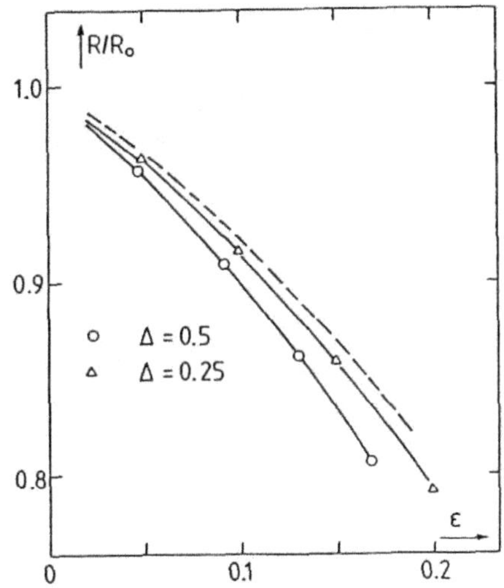

Fig.3 Ratio of the tip radius R to the Ivantsov value R_0 as a function of anisotropy parameter ε. Dashed curve represents the value for symmetric model.[41)]

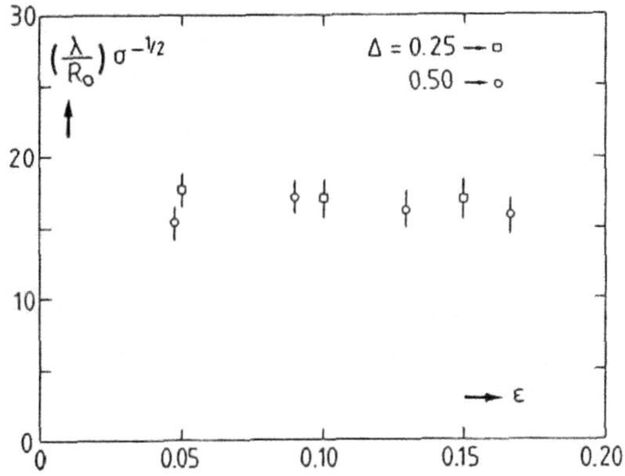

Fig.4 Scaled wavelength $(\lambda/R_0)\sigma^{-1/2}$ vs anisotropy ε for two different undercoolings.

$$\lambda/R \propto \sigma^{1/2}/\tau = \overline{\sigma}^{1/2}. \tag{33}$$

We summarize remarkable results as follows:

(1) Anisotropy is necessary for stable dendrites against tip-splitting instability.

(2) The system converges to a unique quasisteady state, which has sidebranches produced almost periodically.

(3) The selected velocity V and the tip radius R satisfy the scaling relation so that the product VR^2 depends on ε through the function $\overline{\sigma}(\varepsilon)$ but is independent of Δ, precisely as the solvability condition has predicted for the needle crystal.

(4) The sidebranch periodicity λ is proportional to the tip radius R. More generally it is proportional to the stability length λ_s (see Fig.4) independent of Δ and ε. Note that these scaling relations appear to hold both in two and three dimensions as long as supercooling is expressed through the Peclet-number p which absorbs the dependence on Δ in different dimensions.

As for the first point, the effect of anisotropy has also been demonstrated in the viscous finger problem.[42–45] In an isotropic system viscous fingers show tip-splitting instability,[43] whereas by imposing an anisotropy fingers form a stable dendritic pattern with regular sidebranches.[44,45] The third point indicates that the selection of the velocity and the tip radius is largely governed by the immediate neighborhood of the tip, and the velocity selection and the side-branching can be treated as separate problems. As for the sidebranching current belief is the selective amplification of broadband external noise near the tip.[26,37–39] The random character of sidebranches is also observed in the experiment.[46] A systematic study of noise at the tip in our simulation seems promising to clarify this point further.

3. DIRECTIONAL SOLIDIFICATION[10]

3.1 Fundamentals and Integral Representation

In the previous section the simulation of interface dynamics shows that the selected velocity of the free dendrite agrees with the prediction by the solvability condition. More subtle is the selection mechanism in the directional solidification.[8,16] When a cell containing a binary solid and its liquid solution is pulled at a constant speed along a temperature gradient, a solidification front remains planar until a critical pulling speed V_c is exceeded. By increasing the speed above V_c, the planar interface becomes unstable and forms various periodic structures like a parallel array of cellular shapes, of needle-like cusps, and of dendrites.[47-50] In this section we describe briefly results of our numerical study on the dynamical evolution of the solidification front.[10]

In a simplified model of the directional solidification[8] temperature is assumed to vary linearly in the growth direction as $T=T_0+G\zeta$ with a constant temperature gradient G. The local equilibrium assumption determines the interfacial temperature T_i as a function of the liquid concentration there $C_{L,i}$;

$$T_i=T_M-\beta C_{L,i}-T_M(\gamma+d^2\gamma/d\theta^2)L^{-1}\kappa=T_0+G\zeta_i. \tag{34}$$

Eq.(34) modifies the interfacial boundary condition of the liquid from Eq.(6) to

$$u_{L,i}=\Delta-d\kappa-\zeta_i/l_T \tag{35}$$

with a thermal length $l_T=\beta\Delta C/G$. When the system is pulled with a slow rate, steady growth of a planar interface is possible at $\Delta=1$. The amplitude of a deformed interface $\zeta_i(x,t)=A(t)\cos qx$ is shown[51-53] to evolve for small A as

$$dA/dt=a_0A+a_1A^3+O(A^5), \tag{36}$$

with

$$\begin{aligned} a_0/2D_L=l^{-2}[(\sqrt{1+q^2l^2}-1)(1-½l/l_T-½d_0lq^2)-\\ -½k(l/l_T+d_0lq^2)(1+\sqrt{1+r^2l^2q^2})]. \end{aligned} \tag{37}$$

The explicit form of a_1 is too complicated and is omitted.[51-53]

The condition $a_0=0$ determines the neutral curve, which separates the $V-q$ phase space into a region where a planar interface is stable to an infinitesimal deformation ($a_0>0$), and the other where it is unstable ($a_0<0$). An extremum condition, $\partial a_0/\partial q=0$, determines the most unstable mode, $q_m(V)$, which has the highest amplification rate of deformation when $a_0<0$. The critical velocity V_c and the critical wave number q_c are determined as the cross point of the neutral curve with the most unstable mode:

$$a_0(V_c,q_c)=\partial a_0/\partial q=0. \tag{38}$$

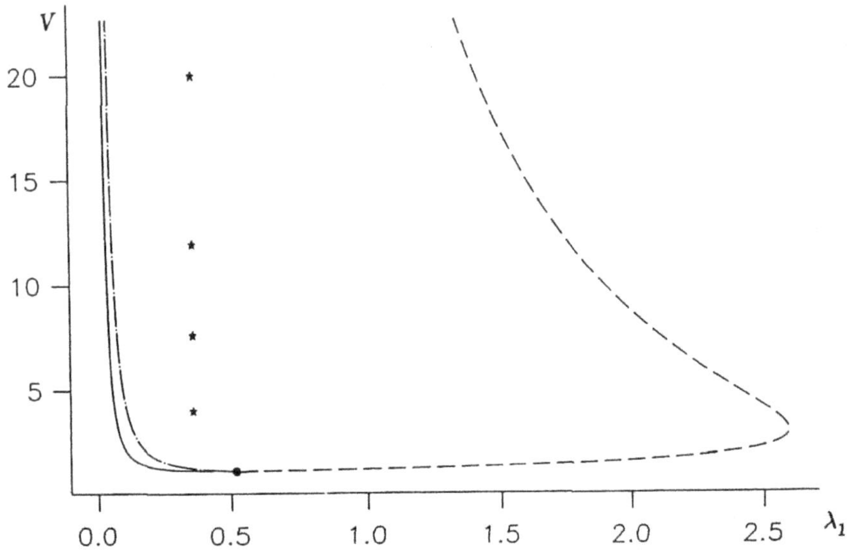

Fig.5 Neutral curve (solid and dashed curves) and the most unstable mode (dot-dashed curve) of the steel with 18.5wt% of Cr or Ni impurity near the critical point. Parameters are l_T=1, D_L=1, k=0.9 and d_0=2.95$\times10^{-4}$. Stars denote parameters where simulations are performed.

Fig. 5 depicts the neutral curve and the most unstable mode for control parameters corresponding to steel with 18.5wt% of Cr or Ni ingredients.[54)] (See Table 1.) λ_1 is a primary spacing or the wavelength; $\lambda_1=2\pi/q$. By taking length and time units such that $l_T= D_L= 1$, the capillary length is d_0=2.95 $\times10^{-4}$. The segregation coefficient is k=0.9.

The critical point determined from Eq.(38) depends on various material and control parameters. For fixed d_0 and r, the critical point value of a_1 changes its sign according to the values of k. For large k or small l, $a_1>0$ and the bifurcation from the planar interface to cellular one is normal (or supercritical), whereas for small k or large l, $a_1<0$ and the bifurcation is inverted (or subcritical). In the case of steel with Cr or Ni ingredients, $a_1>0$ and the

Table 1. Material parameters and typical control parameters for directional solidification. Lower part describes values in dimensionless units where l_T=1 and D_L=1.

	CBr_4-Br_2	$Steel-Cr,Ni$
$T_M[K]$	366.5	1776
$L[J/cm^3]$	34.5	1650
$\gamma[\ 10^{-7}\ \ J/cm^2]$	7.00	260
$\beta[K/wt\%]$	2.90	0.67
k	0.16	0.90
$D_L\ [10^{-5}\ cm/s^2]$	1.2	1.8
$C_\infty[wt\%]$	0.12	18.5
$G[K/cm]$	140	20
$V[\mu m/s]$	12.0	20
$\lambda[\mu m]$	43	250
$l_T[\mu m]$	130	689
$d_0[\mu m]$	0.041	0.203
$d_0\ \times 10^4$	3.22	2.95
$r=D_s/D_s$	0	0
V	1.31	7.65
λ	0.328	0.363
V_c	1.042	1.136
λ_c	0.941	0.514

bifurcation is normal. If the bifurcation is inverted or the pulling rate exceeds far from the stability limit, deformation is no more small and the perturbation expansion method like Eq.(36) is not valid. More detailed and complete analysis such as numerical simulation is required. Previous numerical studies were performed in a weakly nonlinear regime, far from the dendritic growth regime.[55-58]

We extend our numerical simulation method developed in the previous section for the free dendrite now to the directional solidification. Since the solidification front shows various periodic structures above the critical velocity, we assume from the beginning that it takes a structure with a fixed primary spacing or wavelength λ_1. This means that we still have difficulties to address the most interesting problem of wavelength selection.[59-61] By assuming an axial symmetry about z-axis, the integro-differential equation in the liquid corresponding to Eq.(22) now reads as

$$\int_0^{\lambda_1/2} d\Gamma' \sum_{m=-\infty}^{\infty} [g_L(x,\zeta|x'+m\lambda_1,\zeta')+g_L(x,\zeta|-x'+m\lambda_1,\zeta')]\partial u(x',\zeta')/\partial n'$$

$$=\int_0^{\lambda_1/2} d\Gamma' \sum_{m=-\infty}^{\infty} [h_L(x,\zeta|x'+m\lambda_1,\zeta')+h_L(x,\zeta|-x'+m\lambda_1,\zeta')]u(x',\zeta') \quad (39)$$

where both $\zeta=\zeta(x)$ and $\zeta'=\zeta(x')$ lie on the solidification front. Since Green's function g_L as well as h_L decay exponentially fast, summation over $|m|\leq 5$ periods is sufficient for most numerical simulations. An equation similar to (39) holds on the solid side with appropriate modifications. Starting from a sinusoidal initial shape with a small amplitude, we follow the dynamics of the front just like in the free dendrite case. Of course, since the pulling velocity is here given from the outside, we omit the velocity relaxation of the moving frame Eq.(30).

3.2 Cellular and Dendritic Structures

First of all we study the stability limit of the planar interface and the character of the bifurcation near the critical point. With parameters $k=1$, $D_S=0$ and $d_0=10^{-3}$, the bifurcation at $\lambda_{1,c}=2\pi/q_c=0.72$ and $V_c=1.24$ is expected supercritical with a positive Landau coefficient a_1. We found by simulation that the critical point is reproduced within 1 percent range. At $V=1.25$ an interface grows steadily with a sinusoidal modulation of a small amplitude. Thus the bifurcation is proven to be supercritical.

At $k=0.1$, the critical point located at $\lambda_{1,c}=1.24$ and $V_c=1.07$ is expected subcritical with a negative a_1. Simulation at $\lambda_1=\lambda_{1,c}$ and $V=1.00$ reveals that evolution of the interface deformation depends on the initial amplitude $A(t=0)$.

With $A(0)=0.10$ the deformation dies out, whereas with $A(0)=0.20$ it developes to an array of cusps, whose tips flatten and split ultimately. Therefore, the transition from the planar interface to the modulated structure has a subcritical character with a hysteresis in this case. Even though V is below the stability limit, the modulation amplitude is not small and nonlinear coupling between higher harmonics leads to period halving. The same tip splitting can be observed at $k=1$ but at higher growth rate, $V>2V_c$. When the crystal with $k=1$ grows faster than $V\gtrsim 15V_c$, the tip recovers stability.

In order to make comparison with experiments, simulations are performed with realistic parameters. Parameters corresponding to CBr_4-Br_2 alloy are listed in Table 1. Profiles numerically obtained are compared with those obtained in experiments.[62] For velocities $V/V_c=$ 3.1, 5 and 5.7 simulations produce stationary stable cells, and the overall shape with deep grooves is in reasonable agreement with experiments.[62] However, cells in simulation always have sharper tips than in experiments. The discrepancy may be attributed to the third dimension in the real experiments.

We also perform simulations with parameters corresponding to steel with 18.5 $wt\%$ of Cr and Ni impurities,[54] listed also in Table 1. The anisotropy parameter is rather arbitrarily chosen to be $\varepsilon=0.1$. Velocities studied are $V=4$, 7.65, 12 and 20 and the width of the system is fixed at $\lambda_1=0.36$(Fig.5). At $V=4$ the front takes a cellular shape without sidebranching as is shown in Fig.6. Deep grooves have been chopped off for plotting reasons. The width of the system λ_1 is more than a factor 4 larger than the wavelength of the most unstable mode $\lambda_{1,m}=2\pi/q_m$ and the short wavelength limit of the neutral curve. Therefore, the tip could show splitting instability. However, due to the anisotropy in the surface tension, the tip is stabilized with this rather long wavelength. Here sidebranching is suppressed, probably due to the overlapping of the neighboring diffusion field, as the diffusion length $l=2D_L/V=0.5$ is of the same order of the system width $\lambda_1=0.36$. At $V=12$, the diffusion length $l=0.17$ is shorter than the system width, and sidebranches appear as in Fig.6. At $V=$ 20, sidebranching becomes more evident.

From these simulations we find that the tip radius follows precisely the scaling relation Eq.(32), even for velocities as low as $V=4$. This is consistent with the observation from the free dendrite case, that details in the tail region

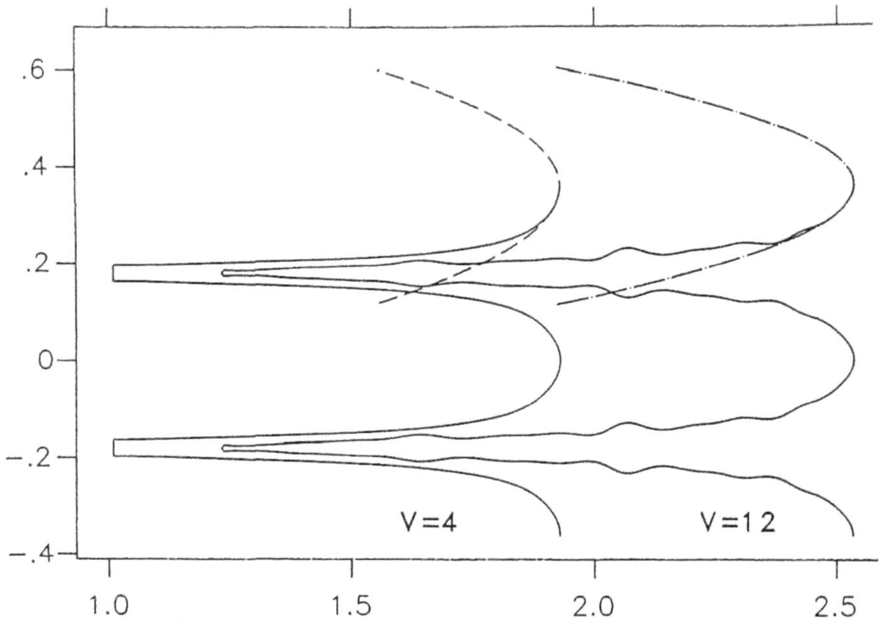

Fig.6 Solidification front at V=4 and V=12. Dased and dot-dased curves represent parabolas fitted at the tip for two velocities.

are not so important for the shape selection of the tip region, as long as the global conservation laws are satisfied. In contrast, if one compares the results with the Ivantsov relation Eq.(18,19) for the free dendrite without surface tension, satisfactory agreement is only reached for velocities larger than about 8. For smaller velocities the Ivantsov radius R_0 calculated through Eq.(18) by using the actual supercooling Δ at the tip to obtain a Peclet number p by Eq.(19) turns out too large. This is easily understandable as the Ivantsov relation assumes global conservation to be fulfilled by a single (dendritic) cell, but for small velocities several cells are competing for the diffusion field within a diffusion length l.

3.3 Scaling Relations

Our attention is restricted to the conventional experimental situation where the diffusion length is large compared to the capillary length. We have made a quantitative comparison with experiments at small segregation coefficient where the instability of a plane interface is of inverse type. We found very good agreement if the experimental pattern was effectively two-dimensional.[62] In most cases, however, the experimental setup unfortunately seems to produce three-dimensional structures. Detailed quantitative comparison there is no more possible, but scaling relations should still apply. Furthermore, important quantities like kinetic coefficients or anisotropy of surface tension are not known experimentally. For comparison with results on free dendrites we took $\varepsilon=0.1$. In order to continuously reach the curve of neutral stability from the cellular side we most intensively looked at cases of normal bifurcation corresponding e.g. to steel-alloy. Nevertheless we expect the basic scaling results for the dendritic region to be of general validity, as described in the following.

(1) Close to the neutral curve we find precise agreement with the predictions from linear stability analysis, using an almost flat interface as initial condition.

(2) The cellular-dendritic region inside the neutral curve can be divided into two areas, the dividing line being $\lambda_1=l$. For velocities below that line we find cells without sidebranches and without an obvious parabolic tip shape, whereas for velocities higher than the dividing line (at fixed λ_1) we find dendrites. In the latter case the scaling of the tip radius with respect to velocity, wavelength of the sidebranches etc. corresponds quite precisely to the free dendritic growth.

(3) This dividing line intersects the short-wavelength branch of the neutral stability curve at some velocity V_i. For larger velocities, therefore, only dendritic patterns should be observed in the sense of scaling like in the free dendrite case.

(4) For velocities large enough ($V>2V_c$) to develop deep cells but below the dividing line (point 2) we reach the *Saffman-Taylor* [42] -limit upon reduction of the cellular spacing λ_1 towards the short wavelength branch of the neutral curve. We there find cells separated by wide grooves, such that the cell width near the tip is only slightly larger than half the basic

periodicity (spacing) of the cells. This region may not easily be reached experimentally, as an Eckhaus-instability may occur and the cell spacing is not directly controllable. Approaching that region, one has the diffusion length becoming increasingly larger than the cell spacing. The tip radius R there accordingly should scale with the cell spacing λ_1 only. This is found numerically in agreement with recent experiments[62)], giving a relation of $\lambda_1 \approx 5R$ in the experimentally accessible range.

(5) In the cellular range we do not yet have a selection criterion for the primary spacing. On the other hand, even experimentally one observes a rather broad scatter of wavelengths in that region.

(6) In the dendritic region the selection of cell spacings appears to occur by a 'tail-instability' of the sidebranching dendrite (at nonzero crystalline anisotropy). For cells which are much further apart than the diffusion length (as observed experimentally and numerically), the free dendritic pattern is a stable mode of operation of each individual cell since the direct interaction of the tips via the diffusion field becomes exponentially weak. This is clearly the case in the limit of infinite thermal length where free dendrites correspond to infinite spacing. Even for finite thermal length we obtained stationary dendrites at for example $V=20$ with spacings larger by a factor 14 compared to the short wavelength limit of the neutral curve at that velocity.

The notion of 'tail-instability' means that at high speeds the supersaturation in the tail region due to the nonzero thermal gradient becomes so high that one of the sidebranches develops a protrusion in forward direction which then rapidly grows out of the groove until it reaches the average position of the growth front. This reduces the primary spacing locally by a factor of two. The effect depends on the anisotropy such that for smaller anisotropy the instability occurs closer to the tip. An analytical treatment seems to be rather difficult as it seems to be a collective mechanism among fully developed sidebranches.

4. LATTICE GAS MODEL[11)]

4.1 Model and Monte Carlo Simulation

In the previous sections growth morphology of a crystal are treated macroscopically, and the anisotropy in the surface tension is found essential in selecting the growth rate and the pattern. There the surface tension is treated as an input parameter given phenomenologically from outside. However, the surface tension depends on the temperature and leads to phase transitions in thermodynamic equilibrium: a roughening transition of a particular interface[4–6)] and a faceting transition of an equilibrium crystal shape.[4,7)] For these phase transitions, a microscopic treatment of the surface tension is neccessary.

Also in the previous sections surface kinetics is assumed to be fast enough to maintain a local equilibrium at the solidification front. This assumption is not always valid. It is possible to include the effect of surface kinetics into the boundary condition and to solve a modified integro-differential equation.[63)] Instead, here we propose a simple lattice-gas model of a crystal growing in a diffusion field. This includes the microscopic effect of surface tension as well as surface kinetics, and allows a unified treatment of the equilibrium and the growth forms of the crystal.

In order to make a model as simple as possible, we interprete that the gas and the solid phases consist of their respective atoms: A gas atom is mobile to give an entropy contribution, whereas a solid atom is immobile but its chemical potential has the contribution from the energy. Further simplification is brought about by considering a lattice system, where the space is divided into discrete lattice points. Each lattice site can be occupied by a solid atom, by a gas atom or can be empty. Double occupancy of a lattice site is forbidden. We consider here a square lattice. Although there is no interfacial phase transition at finite temperatures for a square lattice, various exact results known is useful to check our simulation algorithm. Taking advantage of the four-fold symmetry of a square lattice, we consider only an upper-right quarter of a crystal in the positive x and y region. A skewed boundary condition where the x- and y-axes are contiguous to each other fixes the center of mass at the origin.

To each solid atom is associated an energy gain $\mu_s(<0)$, and the effect of the surface tension is incorporated in an energy cost of the value $2J>0$ to break a solid-solid nearest neighbour bond. These energetics are summerized in a Hamiltonian

$$H=2J\sum_{<i,j>}[C_i(1-C_j)+(1-C_i)C_j]+\mu_s\sum_i C_i \tag{40}$$

where the first summation runs over all the nearest neighbor pairs, and the crystallization order parameter C_i on an i-th lattice site is unity when it is occupied by a solid atom and vanishes otherwise. In terms of an Ising spin variable $S_i \equiv 2C_i - 1 = \pm 1$, the Hamiltonian reduces to that of a ferromagnetic Ising model in a field.

We *simulate the equilibrium and nonequilibrium morphologies of this* model system by the Monte Carlo method.[64] Every gas atom hops freely from a lattice site to one of its nearest neighbors, unless it is occupied. This exchange process mimics the diffusion in a gas phase, and gives entropy contribution to the equilibrium free energy of the gas. When a gas atom diffuses to come into contact with solid atoms, it crystallizes with a certain probability. Inverse to this crystallization process is an evaporation, where a solid atom at an interface tries to turn into a gas atom. These crystallization and evaporation processes at the interface are associated with energy changes and should satisfy a detailed balance condition. By adapting the heat-bath algorithm, transition probability W for a process with an energy change ΔE at a temperature T is set to be $W(\Delta E) = [1+\exp(\Delta E/T)]^{-1}$.

Our process neglects evaporation in the bulk crystal and solidification in the bulk of the gas phase, since our main concern is on the crystal morphology at low temperatures where these excitations in bulk phases are quite rare. This is the same assumption for the solid-on-solid(SOS) model which simulates the thermal behavior of a singular interface.[5] The equilibrium shape of the SOS model has been simulated by the Monte Carlo method previously.[65,66] However, in order to study growth forms like dendrite or fractal patterns, the SOS model is inappropriate since it does not allow overhanging of the interface.

We start a simulation with a solid nucleus situated in a closed vessel. Some solid atoms evaporate to a gas phase and diffuse away from the interface, but some gas atoms condense back. From this simulation we get the equilibrium form of the crystal. Instead, if the system is expanded steadily by adding an atom reservoir with a prescribed gas density at the periphery of the system, crystal continues to grow and we get a growth form. First we check our simulation algorithm by comparing equilibrium shapes with the exact results, and then study variation of the growth morphology as the growth rate varies.

4.2 Equilibrium Shape and Size

In Fig.7 we show averaged interface profiles obtained by simulations at temperatures a) T/J=0.3 and b) T/J=0.6. The average is performed over $2\sim3\times10^4$ Monte Carlo steps per site (MCS) for T/J=0.6 and over $3\sim6\times10^4$ MCS for T/J=0.3, disregarding the initial 10^4 MCS for equilibriation in every simulation. The temperatures simulated, T/J=0.3 and 0.6, are far below the Ising critical temperature $T_c/J=2/\ln(1+\sqrt{2})\approx 2.27$, and holes are not created in the crystal. On heating the crystal becomes rounded. At each temperature an increment in the chemical potential $|\mu_s|/T$ leads size increase, but the shape preserves similarity.

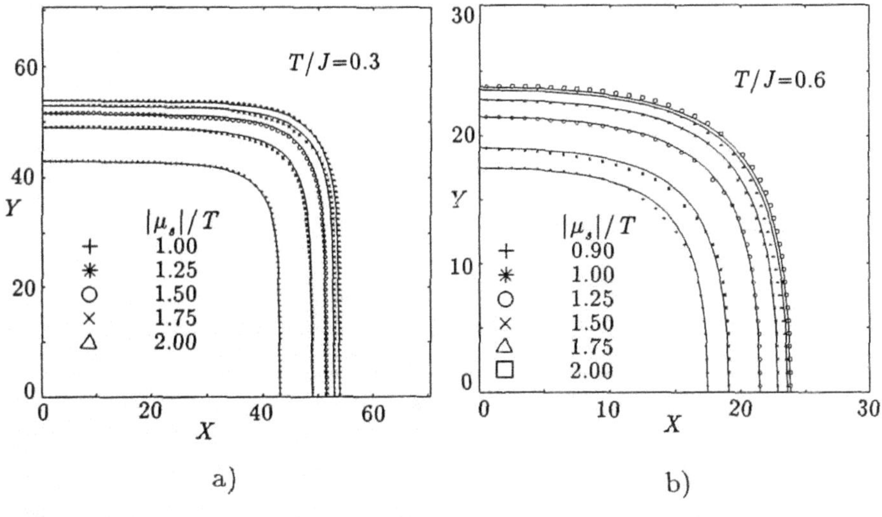

Fig.7 Equilibrium crystal shapes at a) T/J=0.3 and b) T/J=0.6 with various chemical potentials. Curves are the exact profile of the two-dimensional Ising model, where the top heights are fitted to the simulated profiles.

Theoretically, the equilibrium shape of a crystal can be constructed from the orientation-dependent surface tension $\gamma(\theta)$ by the Wulff's theorem.[1-3] As stated previously, our model, Eq.(40), corresponds to the Ising ferromagnet on a square lattice, and $\gamma(\theta)$ is known exactly.[3] Theoretically calculated profiles are drawn by curves in Fig.7 with an adjustment in the top height of the crystal to the simulated profiles. The shapes agree at both temperatures and at various chemical potentials with the simulation results.

Size of the crystal increases on increasing $|\mu_s|$. It is explained by a minimization of the free energy cost

$$\Delta F(N_s) = =2\overline{\gamma}(T)\sqrt{N_s} + \mu_s N_s - T\ln[(\Omega - N_s)!N!/(N-N_s)!\Omega!] \tag{41}$$

to form a crystal with a size N_s in a vessel of volume Ω. In Eq.(41) the first term on the r.h.s is the interfacial energy cost proportional to a linear size $L=\sqrt{N_s}$, and $\overline{\gamma}(T)$ is an averaged surface tension,

$$\overline{\gamma}(T) = \int\gamma(\theta(s))ds/2[\int ydx]^{1/2}. \tag{42}$$

Here the integration is performed along the periphery s of the crystal in the equilibrium form. The second term represents the bulk energy gain by crystallization, and the third is the entropy change of gas atoms. The extremum of ΔF determines the critical radius as

$$\sqrt{N_s} = \Delta\mu/\overline{\gamma}(T). \tag{43}$$

Here $\Delta\mu \equiv \mu_g - \mu_s$ is the chemical potential difference between the gas,

$$\mu_g = T\ln(N-N_s)/(\Omega - N_s) = T\ln n_g, \tag{44}$$

and the solid μ_s. ΔF has two extrema for sufficiently large $|\mu_s|$; one maximum at a small $N_s^{(1)}$ and one minimum at a large $N_s^{(2)}$ as is shown in Fig.8a. In a closed and small system, the crystal takes a size $N_s^{(2)}$ which corresponds to the minimum of the free energy cost ΔF, if $\Delta F(N_s^{(2)})<0$. As $|\mu_s|$ decreases, $N_s=0$ becomes the absolute minimum position of the free energy, and $N_s^{(2)}$ corresponds only to a metastable state. Finally when $|\mu_s|$ becomes smaller than the critical value $|\mu_{s,c}|$ where $\partial\Delta F/\partial N_s = \partial^2\Delta F/\partial N_s^2 = 0$, only the gas phase is stable and the crystal should evaporate. The relation between the crystal size N_s and the chemical potential $|\mu_s|$ at $T/J=0.3$ is shown in Fig.8b. The data points obtained by simulation in Fig.8b lie quite close to the theoretical curve given by Eq.(43).

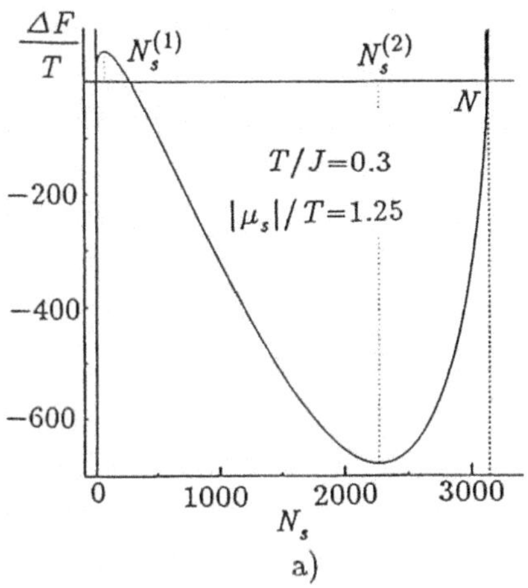

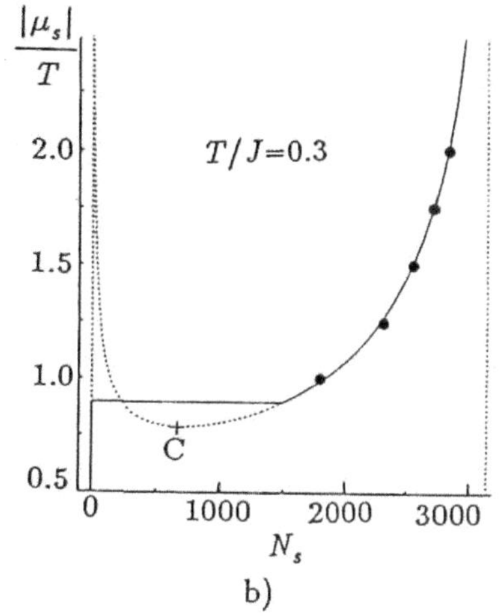

Fig.8a) Free energy cost ΔF for the formation of a crystal nucleus with a size N_s. The system size is $\Omega=70\times70$, the total atom number $N=56\times56$, the temperature $T/J=0.3$, and the chemical potential is $|\mu_s|/T=1.25$.

b) Chemical potential $|\mu_s|/T$ versus the equilibrium crystal size N_s. Closed points represent simulation results, a continuous curve the stable branch, and a broken curve the metastable and the unstable branch. C represents a critical point.

When one increases the system size Ω and the total number of atoms N so as to keep n_g constant, $N_{s,c}=(\overline{\gamma}/\Delta\mu)^2$ remains the extremum of ΔF. However, its stability changes when N exceeds the threshold N_{th} determined from the relation $\partial\Delta F/\partial N_s=\partial^2\Delta F/\partial N_s^2=0$. For a large system with $N>N_{th}$, it becomes unstable. Thus for an open system with infinite Ω and N with a fixed $n_g=N/\Omega$, a crystal nucleus with a size N_s larger than a critical size $N_{s,c}=(\overline{\gamma}/\Delta\mu)^2$ should grow.

4.3 Growth Shape

In an open system where Ω and N are infinite the crystal grows. For the dynamics we choose a time unit such that a single diffusion trial of a gas atom among N_g of them corresponds to a time increment of $\Delta t=(4N_g)^{-1}$. By choosing the lattice parameter as a unit of length the diffusion constant D reduces to unity. We have to consider two independent processes in the simulation; diffusion and solidification-evaporation processes. In this paper both processes are tried with equal weights. First, one selects randomly an atom from a list of gas and interfacial solid atoms. If it is a gas atom diffusion is tried, and if its new position is next to the solid atom solidification is tried. If the selected is a solid atom, evaporation is tried. The shape of the vessel is assumed circular in order to lessen an anisotropy in the diffusion field.

Simulation is performed at a very low temperature as $T/J=0.3$. The gas density is fixed rather high to be $n_g=0.5$ in order to make the scales of length and time short enough for practical simulation. When a chemical potential μ_s is set smaller than $\mu_g=T\ln n_g=-0.69T$, the crystal is in supersaturation. By setting an initial size of a crystal larger than the critical one $(\overline{\gamma}/\Delta\mu)^2$, and the initial system size large enough, the crystal starts to grow.

To keep a continuous growth of the crystal, we provide a particle reservoir at the periphery of the vessel, which is a thin region of a fixed gas-density, $n_g=0.5$. The reservoir should be set far enough from the interface in order not to disturb the density distribution of the gas. For a growth velocity V, variation of the diffusion field is characterized by a diffusion length $l=2D/V$ as in Eq.(12). Therefore, it is desirable to shift the particle reservoir outwardly so as to keep its separation from the crystal tip larger than l. This is possible at $|\mu_s|/T\geq 8$ where V is large enough. At small $|\mu_s|/T\leq 8$, on the other hand, crystal grows very slowly, and the diffusion length $l=2D/V$ is very large. In order to keep the system small enough to perform simulation, we choose an alternative criterion to shift the reservoir: It is shifted outwardly only when the density in front is depressed more than the level of random fluctuation. Since the growth at these small $|\mu_s|$'s is strongly governed by the interface kinetics, we expect that our criterion of the reservoir shift may not influence the growth kinetics essentially. At $|\mu_s|/T=8.0$, simulations with different criteria in shifting the reservoir produce similar shape evolutions. Growth velocity also continues smoothly at $|\mu_s|/T=8$. These observations confirm the validity of our shift criteria of the reservoir.

Resulting growth shapes are shown in Fig.9 at various energy gains $|\mu_s|/T$ associated with the solidification. At small $|\mu_s|/T=4.0$ and 5.0 crystal grows

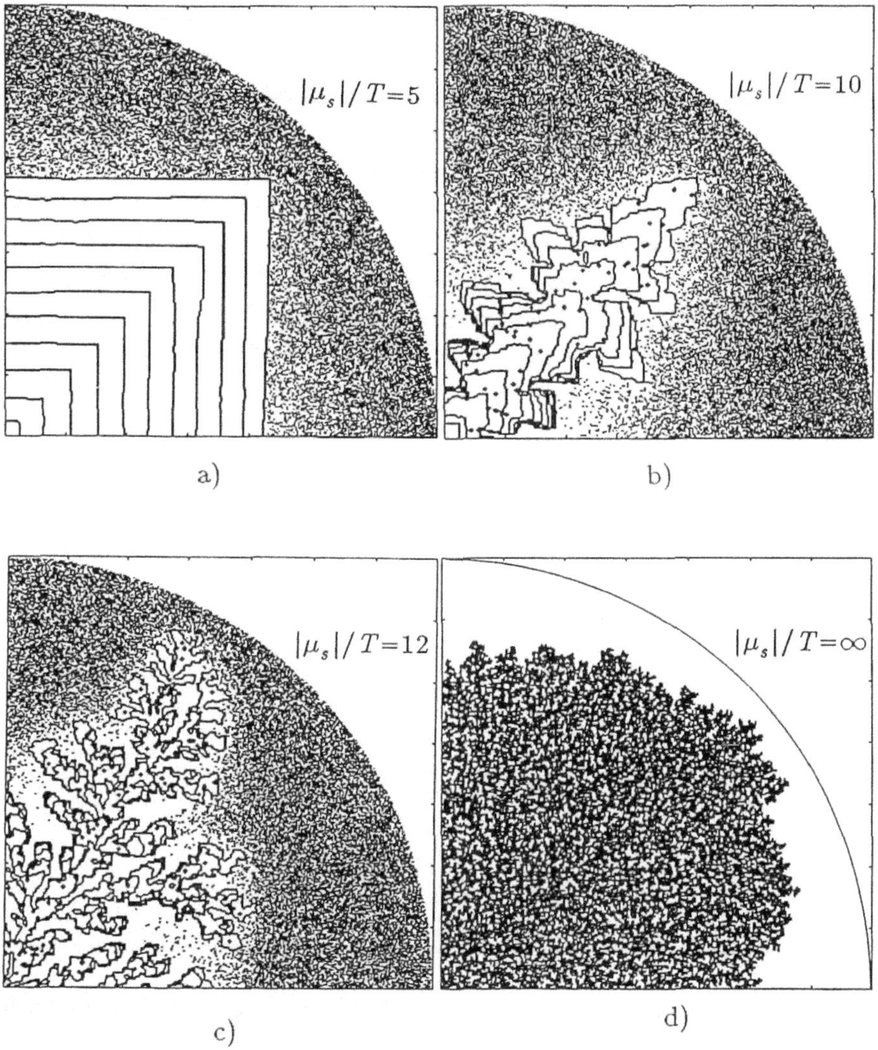

Fig.9 Stroboscopic growth shapes of the crystal and the final arrangement of gas atoms at various μ_s's. Chemical potential $|\mu_s|/T$ is a) 5.0, b) 10.0, c) 12.0 and d) ∞. For d) only the final crystal shape is shown in black, and gas atoms are omitted. Radius of the circular vessel is 350.

slowly and its shape is polygonal up to our maximum simulation size with a vessel radius of 350.(Fig.9a) On increasing $|\mu_s|/T$ from 6.0 to 10.0, the crystal grows initially in a square form, but becomes unstable at the corner, and turns into a dendritic form ultimately. (Fig.9b) Further increase in $|\mu_s|/T$ leads to a splitting of dendrite tips and many irregular branches are observed.(Fig.9c) Until $|\mu_s|/T$=12, the structure is rather open, while for $|\mu_s|/T \geq 13$ the width and the spacing between dendritic branches become very fine and the whole structure is in a dense-branching morphology.(Fig.9d)[44] It resembles to a spherulitic crystal. Similar morphological transitions from a round compact structure to a dendritic and finally to a homogeneous spherulitic structure have been found experimentally in a melt-growth by increasing the undercooling.[67]

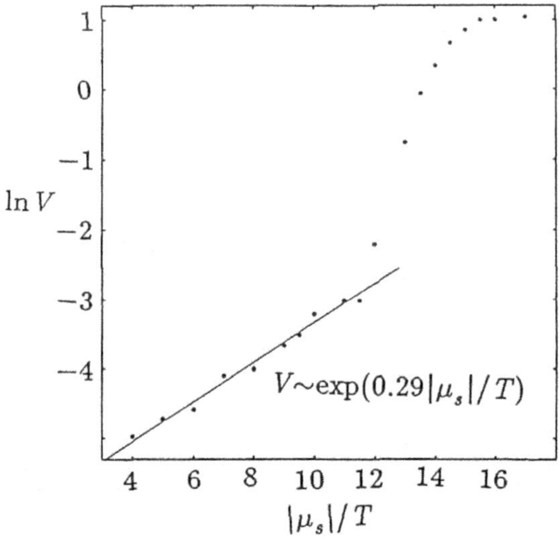

Fig.10 ln V versus $|\mu_s|/T$. A solid line shows the least square fit for small $|\mu_s|$'s.

Growth rate is monitored from the time evolution of the crystal tip. The distance r_{tip} of the farthest tip of the crystal from the origin increases asymptotically in proportion to the time t. Growth rate is defined by the time variation of r_{tip} as $V=\Delta r_{tip}/\Delta t$. When the crystal grows in a square form or in an open dendritic form, the growth rate is very small. On the other hand, when the crystal grows in a compact dense-branching morphology,[44] its growth rate increases drastically. The logarithmic plot of the rate V in Fig.10 reveals that for small $|\mu_s|$'s V depends exponentially on $|\mu_s|/T$ as

$$\ln V \propto -(0.29\pm0.05)|\mu_s|/T. \tag{45}$$

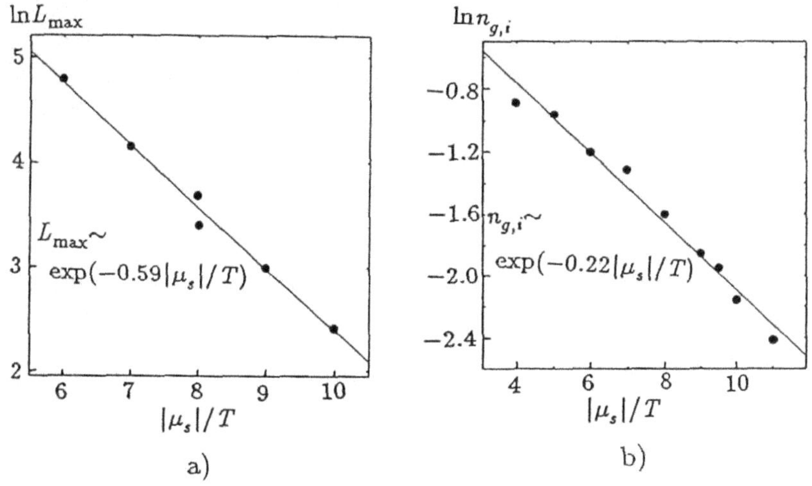

Fig.11 a) Logarithm of the stability limit L_{max} of the polygonal growth.
b) Logarithm of the gas density $n_{g,i}$ at the crystal tip.

In order to characterize the crystal structure at small $|\mu_s|$'s, we estimate the stability limit of the square crystal, L_{max}. When a crystal grows to a size N_s in a square form, the number of surface crystal atoms $N_{s,i}$ is proportional to the linear dimension $L=\sqrt{N_s}$. When a crystal becomes dendritic, the interface becomes complex and the ratio between $N_{s,i}$ and L increases. The transition of L-dependence in $N_{s,i}$ determines the stability limit of the square crystal, L_{max}, as is shown in Fig.11a. It depends also exponentially on the chemical potential $|\mu_s|$ as

$$\ln(L_{max}) \propto -(0.59\pm0.07)|\mu_s|/T. \tag{46}$$

When the crystal grows larger than L_{max}, it becomes dendritic. The tip size and the sidebranch periodicity of the dendritic structure still seem to be of order L_{max}, and thus L_{max} seems to characterize the structure of the crystal.

These behaviors in V and L_{max} may be interpreted by the nucleation and growth kinetics on the edge of a polygonal crystal. For a small $|\mu_s|$ a crystallization probability of an isolated atom on an edge site is very low as

$$P_1 \approx n_{g,i} \exp[(|\mu_s| - 4J)/T]. \tag{47}$$

Here $n_{g,i}$ is the gas density at the interface, and $4J$ represents the energy cost to nucleate an isolated solid atom on a flat edge with two excessive broken bonds. Once a solid atom is nucleated on the flat edge, further spreading costs no energy and a layer growth takes place quickly. Since the total probability of nucleation on an edge of the length $2L$ is $2LP_1$, the growth rate of an edge limited by the surface kinetic process is obtained to be

$$V_K = 2LP_1. \tag{48}$$

In a steady state gas atoms consumed by the nucleation and growth at the interface should be supplied through the diffusion from the bulk of gas. The surface kinetic rate V_K should be balanced with the diffusional rate V_D, which is proportional to the gradient of the diffusion field $q = \partial_n n_g$ at the interface as

$$V_D = Dq \tag{49}$$

for small supersaturation. Another relation among the gas density $n_{g,i}$, the growth rate V and the edge length $2L$ involves the Berg effect.[68–70] When the crystal grows steadily in the polygonal form, the density difference is produced between the corner and the center of the edge. Since the density variation is characterized by the diffusion length $l = 2D/V$, the difference between the density at the corner, $n_g(L)$, and at the center, $n_g(0)$, is of order L/l. A precise calculation[69] in the static approximation to the diffusion field leads to the difference as

$$n_g(L) - n_g(0) = 0.28qL \approx 0.4L/l\,. \tag{50}$$

When the density at the center $n_g(0)$ is suppressed and reaches to the corresponding equilibrium density $n_{g,eq} = e^{-|\mu_s|/T}$ (<<1), central part cannot grow any more.[71] This determines the limit of polygonal growth for our model system. The maximum linear size L_{max} of the polygonal growth is obtained from Eq.(48) and (50) to be

$$L_{\max} \sim \sqrt{n_{g,i}/P_1} \propto \exp[(4J-|\mu_s|)/2T], \tag{51}$$

and the velocity is

$$V \sim V_D \sim V_K \sim \sqrt{n_{g,i}P_1} \propto n_{g,i}\exp[(|\mu_s|-4J)/2T], \tag{52}$$

where we have approximated $n_g(L)=n_{g,i}$ and $n_g(0)=0$.

From Eq.(51) one expects $\ln L_{\max}$ is proportional to $-|\mu_s|/2T$, which differs not so much from the simulated slope -0.59. From the simulation the interfacial gas density at the tip $n_{g,i}$ is found to depend on $|\mu_s|$ exponentially as is shown in Fig.11b and is fitted in the form

$$\ln n_{g,i} \propto -(0.22\pm0.03)|\mu_s|/T. \tag{53}$$

It may be fortuitous, but the ratio $n_{g,i}/V$ at small $|\mu_s|/T$ satisfies the relation (52).

Another characteristic of the growth shape is its openness in the dendritic structure. Up to the chemical potential $|\mu_s|/T\leq10$, the system consists of a single primary dendrite, whereas for high $|\mu_s|/T\geq11$ multiple branches are observed. This structural openness may be parametrized by the fractal dimension D_f defined by

$$N_s \propto R_g^{D_f}. \tag{54}$$

Here N_s is the mass of the crystal and R_g is the radius of gyration defined by

$$R_g^2=(1/N_s)\sum_{i=1}^{N_s} r_i^2, \tag{55}$$

with r_i being the distance of the i-th solid atom from the origin. We have obtained the minimum value of $D_f\approx1.7$ around $|\mu_s|/T=10$.

For $|\mu_s|/T$ higher than $12\approx4J/T-1$, the energy barrier against nucleation is of the order or less than the temperature as is shown in Eq.(47). Most gas atoms solidify instantaneously when they attach to the solid. This causes the drastic increase of the growth rate V around $|\mu_s|/T=12$. Also random shot noise by solidification will be frozen and it drives the system into an irregular structure via tip splitting, as is observed at $|\mu_s|/T\geq11$. Similar effect of strong randomness which leads to an irregular structure is observed in a dendrite experiment.[72)]

When $|\mu_s|$ is infinite evaporation never takes place, and the problem is the same as the diffusion-limited aggregation (DLA)[13] growing in a gas with a finite density.[12] With the present high density of gas, $n_g=0.5$, the growth rate is so fast that the diffusion length is less than unity. The growth is not controlled by the diffusion but by the surface kinetics, and the resultant structure is compact with $D_f=d=2$. However, at low densities diffusion becomes important. The growth rate and the structure of DLA grown from a finite-density gas depend on the supersaturation or the gas density n_g. This point will be studied in the next section.

5. AGGREGATION GROWTH FROM A FINITE-DENSITY GAS[12]

5.1 Velocity Selection via Length Matching

The DLA is a model for a cluster growth limited by diffusion of incoming particles, which are assumed to stick rigidly on an aggregate.[13] This rigid sticking introduces random shot noise. Without evaporation process the bond-breaking energy $2J$ considered in the previous section cannot influence the growth kinetics, and the resulting profile is independent of the surface tension. Without the effect of the surface tension, system loses the characterisitc capillary length d_0 defined in (7), and thus the stability length λ_s disappears. The shot noise is amplified to form an irregular structure. In the Witten-Sander(WS) model,[13] the diffusion field is replaced to its static limit, i.e. the Laplacian field, and the system loses its dynamics. The diffusion length also disappears from the problem. Absence of any length scale in the system leads to a self-similar or fractal structure of the aggregate. Its fractal feature is reflected in the relation between the cluster size R and the mass or the number of solid atoms N_s which it contains as in Eq.(54). The average density of the cluster

$$n_{DLA}(R)\propto R^{D_f-d} \tag{56}$$

then vanishes in the limit of an infinite cluster size, $R\rightarrow\infty$. This means that the DLA is the growth in the low-density limit of the diffusion field, where the density n_g of the gas atoms is zero.

Fractal aggregates are often said[13] to bear resemblance to critical objects at a thermodynamic phase transition point.[73] Fractal dimension is compared to the critical exponents or anomalous dimensions. The latter describe not only structures at criticality, but also critical behaviors near the transition point. Therefore, it is natural to imagine that the fractal dimension D_f describes also scaling behaviors near criticality. For that purpose, however, we have to know relevant physical quantities to drive the system off critical. For thermodynamic phase transitions, the temperature is a relevant quantity.[73] For polymer conformation, the degree of polymerization is the one.[74] We show in the following that in DLA problem the gas density n_g seems to be the relevant quantity.

When an aggregate grows *steadily* from a gas with a finite density n_g, it should have an average density $n_s = n_g$ infinitely far from the solidification front due to the mass conservation. The supersaturation Δ is thus unity. The finite n_s means that the aggregate cannot ultimately be fractal as its size increases indefinitely. In contrast, in the proximity of the aggregate the gas density decays to zero, since all the gas atoms are absorbed to the aggregate. The characteristic length of the density suppression in the gas, ξ_g, is about half the diffusion length $l = 2D/V$ according to Eq.(12). In a boundary layer of thickness ξ_g around the aggregate diffusing particles are rare just as in the WS model, and the aggregate results in a fractal and self-similar structure. The solid density decreases as in Eq.(56) when the scale R increases. However, the density decrement should stop and saturate to n_s at a characteristic length ξ_s. The crossover length ξ_s is determined from the relation $n_g \sim \xi_s^{D_f - d}$ to be

$$\xi_s \sim n_g^{-1/(d-D_f)}. \tag{57}$$

Fractal structure is possible only over a scale where the gas density is sufficiently suppressed. Therefore, it is natural to assume that ξ_s be of the same order as ξ_g. This assumption leads to the density dependence of the growth velocity of the aggregate in the steady state as

$$V \sim \xi_g^{-1} \sim \xi_s^{-1} \sim n_g^{1/(d-D_f)}. \tag{58}$$

The result is remarkable since the fractal dimension of the DLA cluster D_f controles the steady-state growth velocity V, even though the DLA itself has no dynamics and no scale in its structure.

A simulation of the DLA growth from a gas of finite density has been initiated previously,[75,76] but no attention was paid to the relation between the steady growth rate and the fractal dimension. We perform Monte Carlo simulations of the lattice-gas model similar to that in the previous section with $|\mu_s|=\infty$, and test the new mechanism of the velocity selection via the matching of the dynamical length ξ_g and the structural length ξ_s.

5.2 Scaling of Velocity and Width

The vessel here takes a rectangular box shape, which is filled initially with gas atoms of a density n_g. A linear seed of the aggregate is provided at the bottom of the box. This one-dimensional growth geometry is chosen to realize a steady growth easily.[77] When a randomly chosen gas atom sits or diffuses next to the aggregate, it solidifies. If an aggregate grows to a predetermined level from the top of the box, a gas reserver which is a thin layer of a constant gas-density is added at the top, so as to keep the density far from the top of the aggregate constant.

Figs.12 show morphology of aggregates grown at various gas densities $n_g=$ 0.08, 0.10, 0.15, and 0.40. The width of the system L at low densities ($n_g\leq$ 0.15) is 1024, and that at $n_g=$ 0.40 is 256. The height increases in the simulation up to a maximum of 1500. We have also performed a series of simulations with other densities and also with other box sizes, whose maximum values are 400×4100 and 100×10100.

For low density samples the branching structure at short scales looks quite similar to that of a fractal cluster. The number of main trees of the aggregates is small, indicating the long correlation length ξ_s. On increasing the gas density the fractal structure becomes less obvious, since the characteristic length ξ_s decreases to the lattice constant. At very high densities the solid is dense and uniform, only with many vacancies.

Number of solid atoms N_s in the aggregate increases almost linearly in time after the solid density reaches the steady state value. The velocity of the aggregate height is defined by

$$V=\frac{\Delta N_s}{n_g L\Delta t}, \tag{59}$$

when the number of solid atoms increases ΔN_s in a time interval Δt. Density n_g-dependence of the velocity is shown in Fig.13 with $m=1$ being the present

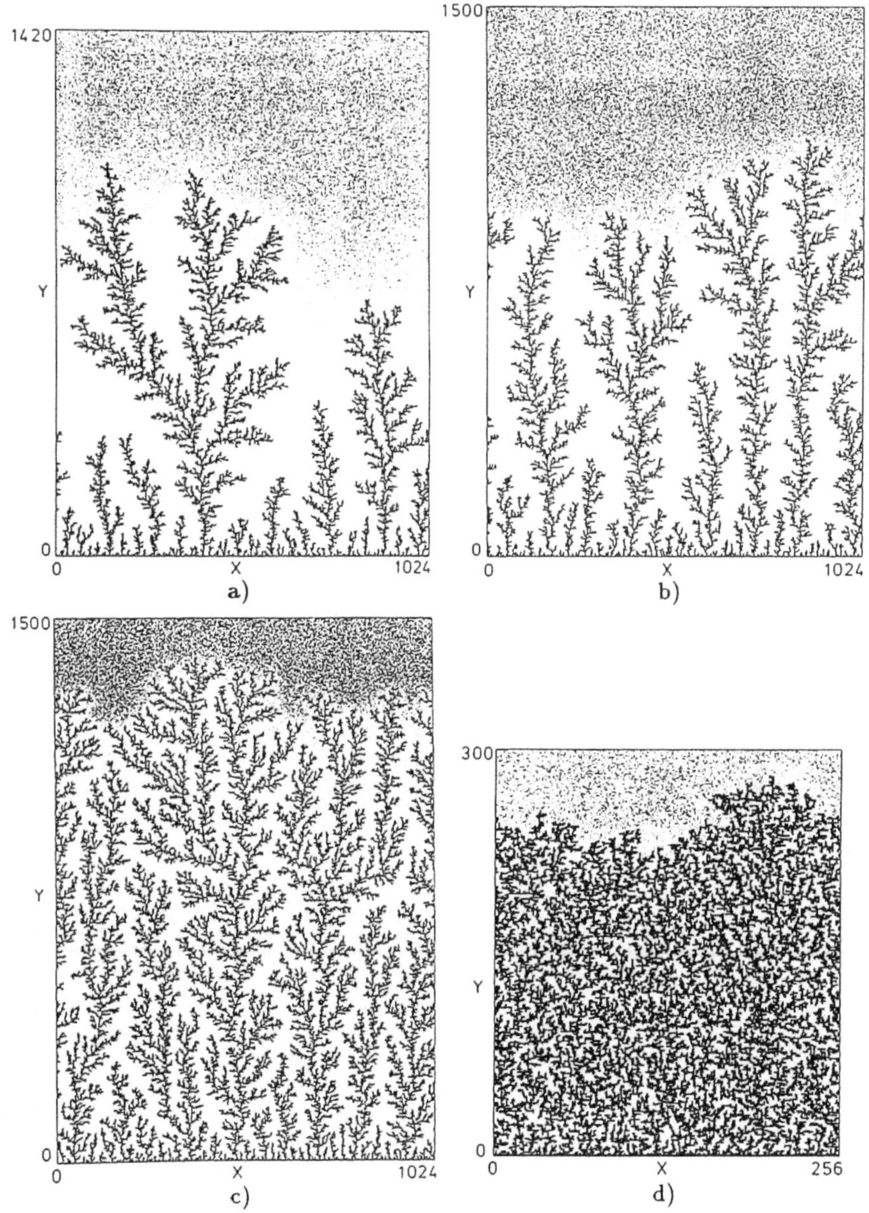

Fig.12 Aggregations from various gas densities, n_g= a) 0.08, b) 0.10, c) 0.15 and d) 0.40.

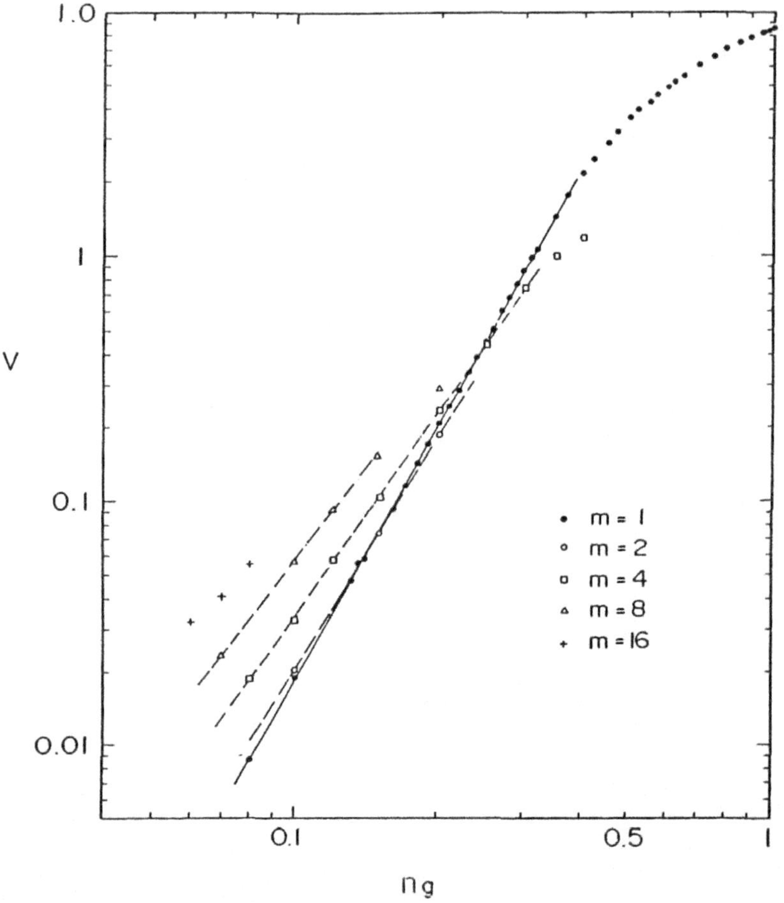

Fig.13 Log-log plot of the growth rate V versus gas density n_g for different mesh sizes m.

case. At low densities, $0.08 \leq n_g \lesssim 0.35$, the velocity satisfies the relation

$$V = 58 n_g^{\ 3.50}, \tag{60}$$

with an accuracy of 0.10 for the exponent. Combining with Eq.(58) we get the fractal dimension

$$D_f=1.71\pm0.01, \tag{61}$$

which agrees with the one obtained previously.[78)]

We now consider the growth at high densities. At high densities where $\xi_s \lesssim 1$ the concentration of gas atoms at the interface becomes finite, and the growth is mainly limited by the interface kinetics. In the high density limit of this model ($n_g=1$), gas atoms are packed dense and cannot diffuse at all. Crystallization progresses via a random choice of surface gas atoms. This is nothing but the growth process of the Eden model.[79,80)] The steady growth velocity approaches the limiting value 8.64±0.08 for $n_g=1$. Since each gas atom tries to solidify 4-times in our unit time, the growth rate of the Eden model is 4-times the average number of gas atoms next to the solid per width (i.e. the ratio of the perimeter length to the width L). From the steady state value of the velocity at $n_g=1$ the interface length is about 2.16 times longer than that of the flat interface. This enhancement factor agrees with the value 2.180±0.007, obtained previously for the very large Eden model simulation. [79,80)]

After the confirmation of the conjecture (58) concerning the velocity, we study related quantities, ξ_g and ξ_s, at low densities. ξ_g characterizes the exponential relaxation of the local gas density $\langle n_g(r)\rangle$ around the growth points. It is calculated by averaging the gas configuration about the most recently grown 100 points. ξ_g is found proportional to the power of the gas density n_g, and is in fact almost equal to l. For a plane interface ξ_g is expected to be $\frac{1}{2}l$, and the enhancement factor ~2 may be attributed to the interfacial structure.

The characteristic length scale in solid is more difficult to measure. The solid density correlation function $g(\vec{r})=\langle n_s(\vec{0})\cdot n_s(\vec{r})\rangle$ is expected to decrease from 1 at $r=0$ in a power law as $g(r)\sim r^{D_f-d}$ for distances r up to the crossover length ξ_s, and then saturates at the value n_s for larger r. Instead of calculating $g(r)$ directly, we divide the system into square cells of a linear dimension b, and count the number of solid atoms in a cell, $N_s=\sum_{i\in \text{cell}} n_s(i)$. The quantity defined from the second moment of N_s by

$$n(b)=b^{-d}\{\langle N_s^2\rangle_b/\langle N_s\rangle_b-1\}\approx b^{-d}\int_0^b g(r)\,d^dr \tag{62}$$

is essentially an average of $g(r)$ in a cell of the dimension b. The subtraction of

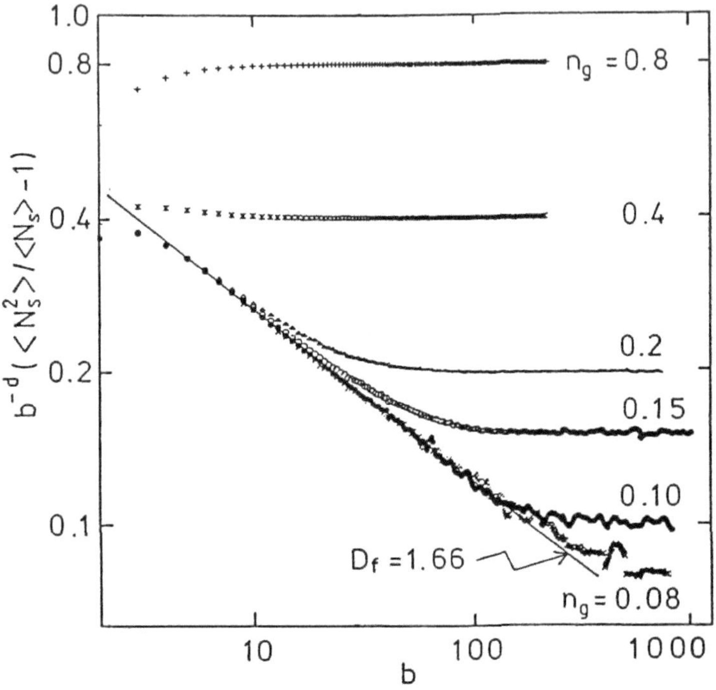

Fig.14 Log-log plot of correlation of solid atom numbers in a cell $n(b)$ versus the cell dimension b. The power law decrease of $n(b)$, which characterizes the DLA, changes into a constant at a crossover length $\sim\xi_s$.

unity in Eq.(62) removes the strong auto-correlation. Reflecting the crossover of $g(r)$ at $r=\xi_s$, $n(b)$ is expected to be proportional to $b^{(D_f-d)}$ for $b\lesssim\xi_s$, and $n(b)=n_s$ for $b\gtrsim\xi_s$. Fig.14 depicts $n(b)$ as a function of the cell size b, taken at a centeral part of the aggregates. Crossover of $n(b)$ in b-dependence is apparent. From the data at $b\lesssim\xi_s$, the fractal dimension is calculated to be D_f=1.66±0.02, a little smaller than the expectation 1.71. The discrepancy may be attributed to the insufficient thickness of the central solid region to get enough statistics for $n(b)$. At low densities diffusing gas particles remain deep in the solid, and the completely solidified region is rather small. However, considering the unsettled controversy[81] on the fractal dimension of DLA, we do not exclude the possibility that the discrepancy is real.

Another feature of the aggregation structure is its interface width. Interface can be defined as the location where the growth has just taken place. Provided the system is strongly correlated within the correlation length ξ_s but uncorrelated for large separations, the interface can be interpreted to consist of blobs with a size ξ_s and these blobs perform a random walk in the "time" of L/ξ_s steps. The value of the averaged interface width w is estimated from the analogy to the Brownian motion as

$$w^2 \propto (\text{step unit length})^2 \times \text{time} \sim \xi_s^2 (L/\xi_s)$$

or

$$w \propto (\xi_s L)^{1/2} \propto n_g^{-1/2(d-D_f)} L^{1/2}. \tag{63}$$

The width of the interface in simulation is defined as the height fluctuation of newly grown atoms from the average. It is defined over every M growing points as

$$w^2(\overline{t}) = \frac{1}{M} \sum_{i=1}^{M} (y_{j+i} - \frac{1}{M} \sum_{i=1}^{M} y_{j+i})^2 - (M/Ln_g)^2/12 \tag{64}$$

at an average time $\overline{t} = \sum_{i=1}^{M} t_{j+i}/M$. The last term on the r.h.s. is the correction due to the systematic drift of the average height. The width of the growing region w is shown in Fig.15 for systems with various widths L=100, 400 and 1024.

At high densities ($n_g \gtrsim 0.3$) one can clearly observe the size dependence $\sim L^{1/2}$ for L=100 and 400. In the limit of Eden model ($n_g = 1$) our results agree with a previous extensive simulation study[82] of the interface width which predicts $w \approx 0.42\sqrt{L}$. In order to reach this asymptotic behavior, however, the system should reach as high as $\overline{y} \sim 0.4L^{3/2}$. For narrow systems with L=100 and 400 we have taken the maximum box height at $h_{\max}(L=100)=10100$ and $h_{\max}(L=400)=4100$, which satisfies the above criterion. For a wider system with L=1024, the above criterion requires the height of more than 13000 which is out of scope of our computational ability. The data for L=1024 shown in Fig.15 are obtained only at height 1500 and have only illustrutive meaning.

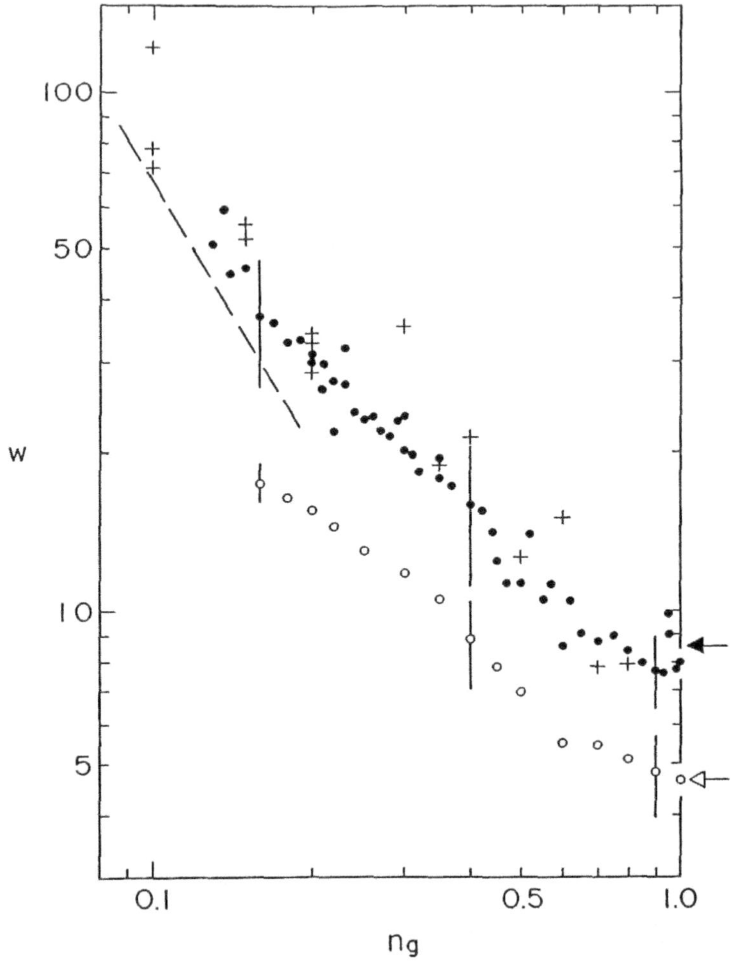

Fig.15 Average width w of the interface versus gas density n_g for L=1024 (+), L=400 (○) and L=100 (●). The bars are typical fluctuation of the width. The expected slope $w \propto n_g^{-1/2(d-D_f)}$ is shown by the dashed line. Arrows indicate the values given in Ref.82) for the Eden model (n_g=1.0).

As density n_g decreases the width w increases. But our data does not seem to be sufficient for quantitative discussions. The data for narrow system as L=100 is strongly influenced by the periodic boundary condition in the horizontal direction. The widest system with L=1024 again suffers from the shortage of height and the data values have very large error bars. With all these limitations, however, one can observe that the slope in w versus n_g becomes steeper as the system width L increases. For L=400 and L=1024, simulation results of the interface width w at low densities are compatible with the hypothesis Eq.(63) with D_f=1.71.

5.3 Anisotropy Effect

In Sec.2 and 3 we have seen that an anisotropy in the system is essential to stabilize a dendrite. For DLA it is also shown that an anisotropy of the underlying lattice induces dendritic shapes but at very large system size.[83] In order to obtain a dendrite in a system with a reasonable size, one has to suppress somehow the shot noise and to enhance relatively the anisotropy effect. Among various methods applied to obtain dendritic DLA,[84–86] we use the one originally proposed by Matsushita and Kondo.[86] The solid growth takes place only on every m-th row and column, while gas atoms diffuse on whole of the original lattice. Aggregates thus formed have more regular structures as the mesh size m increases, as is shown in Fig.16. This feature is the same with that obtained in the usual DLA simulation.[84–86]

Table 2. Apparent fractal dimension D_f versus mesh size m. For large m we have fewer data points than for m=1, and the values of D_f are less accurate.

m	1	2	4	8
D_f	1.71	1.69	1.64	1.61

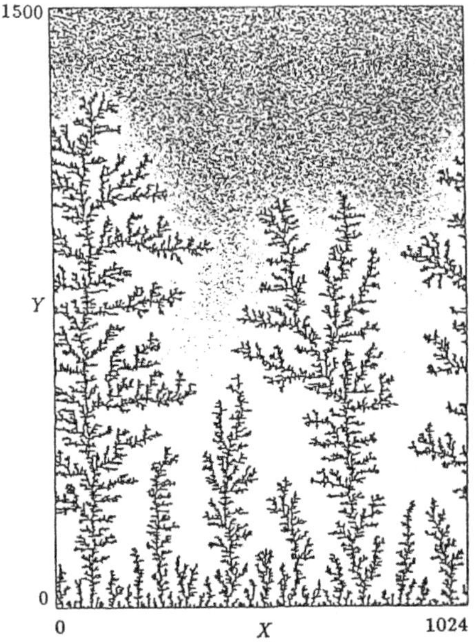

a)

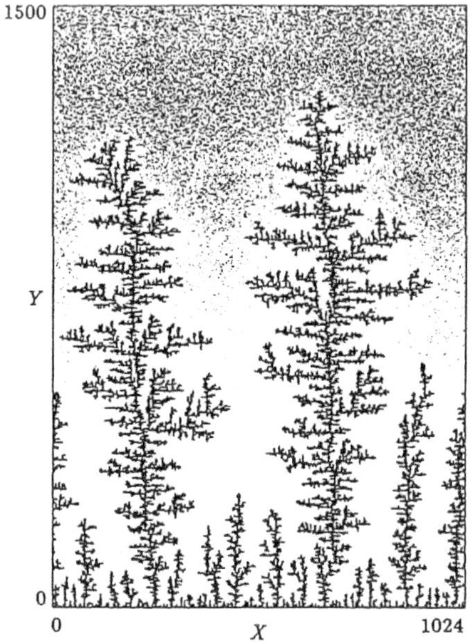

b)

c)

Fig.16

Aggregation configurations grown with various mesh sizes m and at various gas densities n_g. Parameters are a) m=2, n_g= 0.10, b) m=4, n_g= 0.08, and c) m=8, n_g= 0.07.

As the dendrite looks regular, the fractal dimension D_f is claimed to decrease possibly down to D_f=1.5.[85] If this holds, it should be reflected in our $V-n_g$ relation. Our simulation with various m's, in fact, proves this expectation, as is shown in Fig.13. The apparent fractal dimension D_f is obtained from the slope of $\ln V$ versus $\ln n_g$, and it decreases as the mesh size m increases, as is summarized in Table 2.

6. SUMMARY AND DISCUSSIONS

We have presented above our studies on morphology and dynamics of the growing solidification front in diffusion field by means of two complementary approaches; one is macroscopic and the other is microscopic.

Macroscopic approach of solving diffusion equation is applied to problems of free dendrite and directional solidification. In these cases, surface kinetics are assumed fast enough to attain a local equilibrium at the interface. The stabilizing effect comes from the surface tension. For the free dendrite and also for an array of dendrites in directional solidification, the tip radius R and the growth rate V satisfy a scaling relation such that a value $VR^2/2Dd_0=1/\sigma$ depends only on an anisotropy in capillarity and takes a universal value for a fixed anisotropy.

Microscopic approach is realized in terms of a lattice-gas model. In addition to the diffusion process of gas atoms, solidification and evaporation processes are included in a Monte Carlo simulation. Latter processes contain the effect of surface tension as well as surface kinetics. On increasing the driving force $\Delta\mu$, the crystal shape varies from a polygonal through dendritic to irregular structures. For polygonal growth, surface kinetics by nucleation is the limiting process. When crystal grows so fast that the rearrangement of solid atoms are practically impossible, random shot noise is frozen in and the crystal profile becomes irregular.

Random shot noise by solidification is strongly enhanced when $\Delta\mu$ is infinite. In this case evaporation is forbidden, and surface tension plays no role in determining morphology. Front instability leads to an irregular structure, which is fractal in the region where gas density n_g is suppressed. Fractal nature is represented by the fractal dimension D_f. The extent of the fractal region $\xi_s \sim n_g^{-1/(d-D_f)}$ matches with the thickness of the density suppression layer, which is of the order of the diffusion length $l=2D/V$. This results in a velocity selection as a fuction of the gas density: $V \sim n_g^{-1/(d-D_f)}$.

Two approaches so far have dealt with different regimes of growth kinetics. To obtain a perspective view one has to compare both approaches in the same region. Inclusion of the surface kinetics in the macroscopic approach is now being undertaken by Yokoyama and Kuroda.[63] Noise effect on the dendrite is another interesting problem to be investigated by the macroscopic approach: Sidebranches are said to be caused by the noise with a finite amplitude,[29] but strong noise is said to split dendrite tips and makes them fractal.[72] Although numerical simulations always contain round-off errors which play in a sence a noise source, a study with a well controlled noise is necessary. As for the extension of the microscopic approach, inclusion of a fast surface diffusion process or elevation of the system temperature may help to achieve a local equilibrium at the crystal interface. Then the universality in the growth rate and the tip radius of the dendrite can be investigated on a microscopic level. All these are still open problems.

References

1) G. Wulff, Z. Kristallogr. Mineral. 34, 449 (1901)

2) A.F. Andreev, Sov. Phys. -JETP 53, 1063 (1982).

3) C. Rottman and M. Wortis, Phys. Rev. B29, 328 (1984).

4) H.J. Schulz, J. Phys. (Paris) 46, 257 (1985).

5) J.D. Weeks, in *Ordering in Strongly Fluctuating Condensed Matter Systems* ed. T. Riste (Plenum, New York and London, 1980), p.293.

6) Y. Saito and H. Müller-Krumbhaar, in *Applications of the Monte Carlo methods in statistical physics*, ed. K. Binder (Springer, Berlin, 1984), p.223.

7) C. Rottman and M. Wortis, Phys. Rev. B24, 6274 (1981).

8) J.S. Langer, Rev. Mod. Phys. 52, 1 (1980).

9) Y. Saito, G. Goldbeck-Wood and H. Müller-Krumbhaar, Phys. Rev. Lett. 58, 1541 (1987); Phys. Rev. A38, 2148 (1988).

10) Y. Saito, C. Misbah and H. Müller-Krumbhaar, Nucl. Phys. B5A, 225 (1988), and in preparation.

11) Y. Saito and T. Ueta, Phys. Rev. A (1989) accepted.

12) M. Uwaha and Y. Saito, J. Phys. Soc. Jpn. 57, 3285 (1988), and Phys. Rev. A (1989) accepted.

13) T.A. Witten and L.M. Sander, Phys. Rev Lett. 47, 1400 (1981); Phys. Rev. B27, 5686 (1983). See also a recent review by P. Meakin in *Phase Transitions and Critical Phenomena* Vol.12 eds. C. Domb and J.L Lebowitz (Academic Press, London, 1988), and references therein.

14) J.S. Langer and L.A. Turski, Acta Met. 25, 1113 (1977).

15) J.S. Langer, Acta Met. 25, 1121 (1977).

16) W.W. Mullins and R.F. Sekerka, J. Appl. Phys. 35, 444 (1964).

17) M.E. Glicksman, R.J. Schaefer and J.D. Ayers, Metall. Trans. 7A, 1747 (1976).

18) S. C. Huang and M.E. Glicksman, Acta Metall. 29, 701 (1981); 29, 717 (1981).

19) H. Honjo, S. Ohta and Y. Sawada, Phys. Rev. Lett. 35, 841 (1985).

20) G.P. Ivantsov, Dokl. Akad. Nauk SSSR 58, 567 (1947).

21) J.S. Langer and H.Müller-Krumbhaar, Acta Metall. 26, 1681 (1978); 26, 1689 (1978); 26, 1697 (1978).

22) H. Müller-Krumbhaar and J.S. Langer, Acta Metall. 29, 145 (1981).

23) G. Dee and J.S. Langer, Phys. Rev. Lett. 50, 383 (1983), and W. van Saarloos, Phys. Rev. Lett. 58, 2571 (1987).

24) D. Kessler, J. Koplik and H. Levine, Phys. Rev. A30, 3161 (1984).

25) E. Ben-Jacob, N. Goldenfeld, J.S. Langer and G. Schön, Phys. Rev. Letters 51, 1930 (1983), Phys. Rev. A29, 330 (1984).

26) R. Pieters and J.S. Langer, Phys. Rev. Lett. 56, 1948 (1986).

27) D.I. Meiron, Phys. Rev. A33, 2704 (1986).

28) D.A. Kessler and H. Levine, Phys. Rev. Lett. 57, 3069 (1986).

29) D.A. Kessler and H. Levine, Phys. Rev. A33, 7867 (1986).

30) D.A. Kessler, J. Koplik and H. Levine, Phys. Rev. A33, 3352 (1986).

31) D.A. Kessler, J. Koplik and H. Levine, in *Pattern, Defect and Microstructures in Nonequilibrium Systems,* edited by D. Walgraef (Martinus

Nijhoff, Dordrecht,1987).

32) P. Pelce and Y. Pomeau, Studies Appl. Math. 74, 1283 (1986).

33) M. Ben Amar and Y. Pomeau, Europhysics Lett. 2, 307 (1986).

34) A. Barbieri, D.C Hong and J.S. Langer, Phys. Rev. A35, 1802 (1987).

35) B. Caroli, C. Caroli, C. Misbah, and B. Roulet, J. Phys. (Paris) 48, 547 (1987).

36) M. Ben Amar and B. Moussallam, Physica 25D, 155 (1987),

37) D.A. Kessler and H. Levine, Europhys. Lett. 4, 215 (1987).

38) M.N. Barber, A. Barbieri and J.S. Langer, Phys. Rev. A36, 3340 (1987).

39) B. Caroli, C. Caroli and B. Roulet, J. Phys.(Paris) 33, 1423 (1987).

40) B. Caroli, C. Caroli, B. Roulet and J.S. Langer, Phys. Rev. A33, 442 (1986).

41) C. Misbah, J. Phys.(Paris) 48, 1265 (1987).

42) R.G Saffman and F.R.S.G. Taylor, Proc. R. Soc. London, Ser. A245, 312 (1958).

43) L. Paterson, J. Fluid Mech. 113, 513 (1981).

44) E. Ben-Jacob, R. Godbey, N.D. Goldenfeld, J. Koplik, H. Levine, T. Mueller and L.M. Sander, Phys. Rev. Letters 55, 1315 (1985).

45) M. Matsushita, to be published.

46) A. Dougherty, P.D. Kaplan and J.P. Gollub, Phys. Rev. Lett. 58, 1652 (1987).

47) K. Somboonsuk, J.T. Mason and R. Trivedi, Metal. Trans. 15A, 967 (1984).

48) R. Trivedi, Metal. Trans. 15A, 977 (1984).

49) S. de Cheveingné, C. Guthmann and M.M. Lebrun, J. Phys.(Paris) 47, 2095 (1986).

50) J. Bechhoefer and A. Libchaber, Phys. Rev. A35, 1393 (1987).

51) D.J. Wollkind and L.A. Segel, Phil. Trans. R. Soc. 268, 351 (1970).

52) B. Caroli, C. Caroli and B. Roulet, J. Phys.(Paris) 43, 1767 (1982).

53) C. Misbah, J. Phys.(Paris) 47, 1077 (1986).

54) G. Lesoult, private communication.

55) G. Dee, Physica 12D, 270 (1984).

56) M. Kerzberg, Phys. Rev. B27, 3909 (1983), ibd. B 27, 6796 (1983).

57) G.B. McFfadden and S.R. Coriell, Physica 12D, 253 (1984).

58) L.H. Ungar and R.A. Brown, Phys. Rev. B29, 1367 (1984), ibd. B30, 3993 (1984), ibd. B31, 5923 (1985), ibd. B31, 5931 (1985).

59) A. Karma, Phys. Rev. Lett. 57, 858 (1986), and Phys. Rev. B34, 4353 (1986).

60) T. Dombre and V. Hakim, Phys. Rev. A36, 2811 (1987).

61) M. Ben Amar and B. Moussallam, Phys. Rev. Lett. 60, 317 (1988).

62) S. de Cheveigné and C. Guthmann, private communication.

63) E. Yokoyama, Ph.D. Thesis to Hokkaido Univ. 1989, and E. Yokoyama and T. Kuroda, unpublished.

64) *Monte Carlo Methods in Statistical Physics,* ed. K. Binder (Springer, Berlin, 1979).

65) Y. Saito and M. Uchiyama, in *Proc. Oji Int. Seminar of Morphology and Growth Unit of Crystals,* 1985, Yamagata, Japan.

66) Y. Akutsu and N. Akutsu, J. Phys. Soc. Jpn. 56, 9 (1987).

67) D.E. Ovsienko, G.A. Alfintsev and V.V. Maslov, J. Cryst. Growth 26, 233 (1974).

68) W.F. Berg, Proc. Roy. Soc. A164, 79 (1938).

69) A. Seeger, Phil. Mag. Ser. 7, 44, 1 (1953).

70) T. Kuroda, T. Irisawa and A. Ookawa, J. Cryst. Growth, 42, 41 (1977).

71) J.W. Cahn, in *Crystal Growth,* ed. H.S. Peiser, (Plenum, Oxford, 1966) p.681.

72) H. Honjo, S. Ohta and M. Matsushita, J. Phys. Soc. Jpn. 55, 2487 (1986).

73) L.P. Kadanoff, W. Götze, D. Hamblen, R. Hecht, E.A.S. Lewis, V.V Palciauskas, M. Rayl, J. Swift, D. Aspens and J. Kane, Rev. Mod. Phys. 39 395 (1967).

74) P.G. de Gennes, *Scaling Concepts in Polymer Physics* (Cornell, Ithaca and London, 1979).

75) R.F. Voss, J. Stat. Phys. 36, 861 (1984).

76) P. Meakin and J.M. Deutch, J. Chem. Phys. 80, 2115 (1984).

77) For the same geometry, a simulation of DLA and an experiment of electrodeposition have been done: P. Meakin, Phys. Rev. A27, 2616 (1983); M. Matsushita, Y. Hayakawa and Y. Sawada, Phys. Rev. A32, 3814 (1985).

78) P. Meakin, Phys. Rev. A33, 3371 (1986).

79) H.P. Peters, D. Stauffer, H.P. Hölters and K. Loewenich, Z. Phys. B34, 399 (1979).

80) R. Hirsch and D.E. Wolf, J. Phys. A19, L251 (1986).

81) F. Argoul, A. Arneodo and G. Grasseau, Phys. Rev. Lett. 61, 2558 (1988), and references therein.

82) D.E. Wolf, J. Phys. A20, 1251 (1987).

83) P. Meakin, Phys. Rev. A33, 3371 (1986).

84) J. Kertesz and T. Vicsek, J. Phys. A19, L257 (1986).

85) J. Nittmann and H.E. Stanley, Nature 321 663 (1986).

86) M. Matsushita and H. Kondo, J. Phys. Soc. Jpn 55, 2483 (1986).

PATTERN FORMATION OF GROWING SNOW CRYSTALS
-DENDRITIC PATTERN AND HABIT CHANGE-

TOSHIO KURODA AND ETSURO YOKOYAMA

INSTITUTE OF LOW TEMPERATURE SCIENCE
HOKKAIDO UNIVERSITY
SAPPORO, 060 JAPAN

ABSTRACT

Snow crystals are ice crystals growing from the vapor. A spherical single ice crystal is first produced by freezing of a supercooled water droplet in the cloud, then it grows into a hexagonal prism bounded by two basal {0001} and six prism {10$\bar{1}$0} faces and develops into various patterns such as plate like, columnar, dendritic, and needle like patterns according to the growth conditions given by temperature, supersaturation and so on. The important processes relevant to the growth of snow crystals are the surface kinetic process for incorporating water molecules into the ice crystal lattice at the surface and the diffusion process for supplying the molecules to the surface.

First we qualitatively discuss how these processes contribute to pattern formation in growth of snow crystals. Then we show the results of simulation of two-dimensional pattern formation at -15°C based on a new model in which both the processes mentioned above are taken into account. The results are classified into three patterns: 1) hexagonal pattern, 2) dendritic pattern and 3) circular pattern due to kinetic roughening.

Finally we treat the habit change occurring with decreasing temperature and the surface melting and roughening of ice.

1. INTRODUCTION

Crystals of minerals such as diamond, calcite and quartz grow in magma in the earth. On the other hand, ice crystals grow from water vapor during their fall through the atmosphere covering the earth. The latter mentioned are snow crystals which show remarkable variation of morphology or patterns with growth conditions given by supersaturation, temperature and so on.[1,2)]

Late Prof. Ukichiro Nakaya said "A snow crystal is a letter to us from the sky", since meteorological information from the inside of clouds is recorded on the patterns of snow crystals. Later, Sir Charles Frank[3)] answered him in a Haiku:

Diamonds are letters
still better worth the reading:
We can read the sky

The variety of beautiful patterns of crystals stimulates a curiosity of scientists.

In 1936 Nakaya succeeded in growing artificial snow crystals in a cold room and started to investigate the relation between patterns of snow crystals and growth conditions to read the meteorological information written on the letter. Since then, experimental studies of growth of snow crystals have been carried out in the field of meteorology and cloud physics.[1,4-9)] In contrast to the large number of experimental studies, the theoretical treatment of pattern formation of snow crystals has been a few and consequently there remains theoretically unclarified problems, e.g. how a definite pattern as obtained by experiment is formed under the given growth conditions and why the pattern sensitively changes with slight variation of the conditions.

When we discuss the external form of crystals, we distinguish the equilibrium form from the growth form. The former represents the form of crystals in equilibrium with an ambient phase, i.e. thermodynamically most stable form, which is determined by minimizing the total free energy of a crystal surface.[10)] On the other hand, the latter represents the form of crystals growing in non-equilibrium state, e.g. in supersaturated vapor, supercooled melt and so on. The growth form is continuously produced by motion of each position of the crystal surface and a pattern is formed as a whole after the crystal has grown up. Since the growth rate of the surface depends on the growth conditions, various patterns are formed in response to them.

The rate of vapor growth is controlled by following three processes in principle:

i) Surface kinetic process for incorporating water molecules into ice crystal lattice, the rate of which depends on the crystallographic orientation of the surface,

ii) Diffusion process for supplying the molecules in air to the surface, and

iii) Transport process of latent heat of sublimation.

Therefore, we must self-consistently solve the motion of growing surface by taking into account these processes to understand the mechanism of pattern formation of snow crystals. It is to be noted that the structure of hexagonal ice crystal is reflected to the growth patterns of snow crystals (Figs.1 and 2) through the surface kinetic process. The process is necessary for understanding why six prism faces develop into facets or why six primary branches of a dendritic snow are oriented to a-axes.

Several models for pattern formation of snow crystals are recently presented.[11-13] However, there seems a lack of knowledge of actual growth processes and time evolution of pattern of growing snow crystals. For an example, it is never noticed that pattern formation starts from a small spherical ice single crystal which is produced by the freezing of supercooled water droplet in cloud and hexagonal symmetry appears in the patterns during growth through the surface kinetic process. Thus, we briefly describe nucleation and growth of snow crystals in cloud and the relation between their various patterns of snow crystals and growth conditions in section two. Then, we explain the processes relevant to growth of snow crystals in section three and qualitatively discuss the factors controlling the pattern formation in section four. In section five, we summarize the previous theoretical studies on dendritic pattern formation developed from polyhedral snow crystals growing from vapor and that developed from disc-shaped ice crystals growing from melt. In section six, we show the results of simulation[14-16] of development of facets and dendritic pattern in which the surface kinetic process

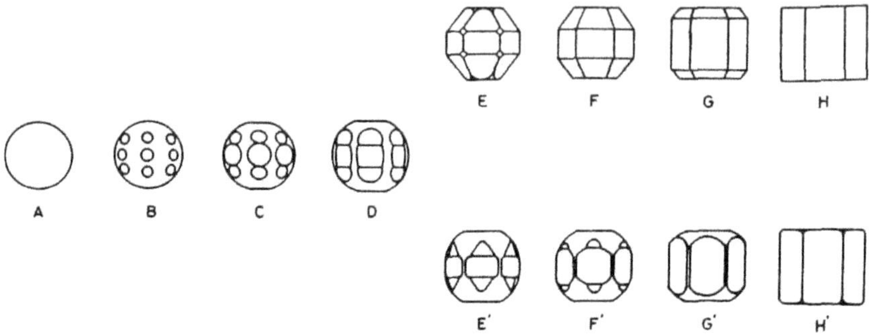

Fig. 1 Development of facets starting from a spherical ice crystal at -15°C and at humidities above water saturation (E～H) and below water saturation (E'～H') (Gonda and Yamazaki[69]).

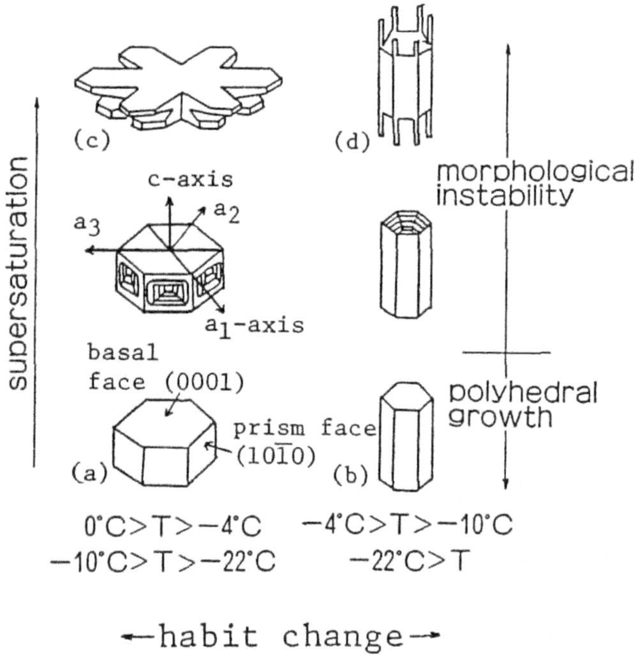

Fig. 2 The schematic representation of variation of growth patterns of snow crystals with growth conditions.

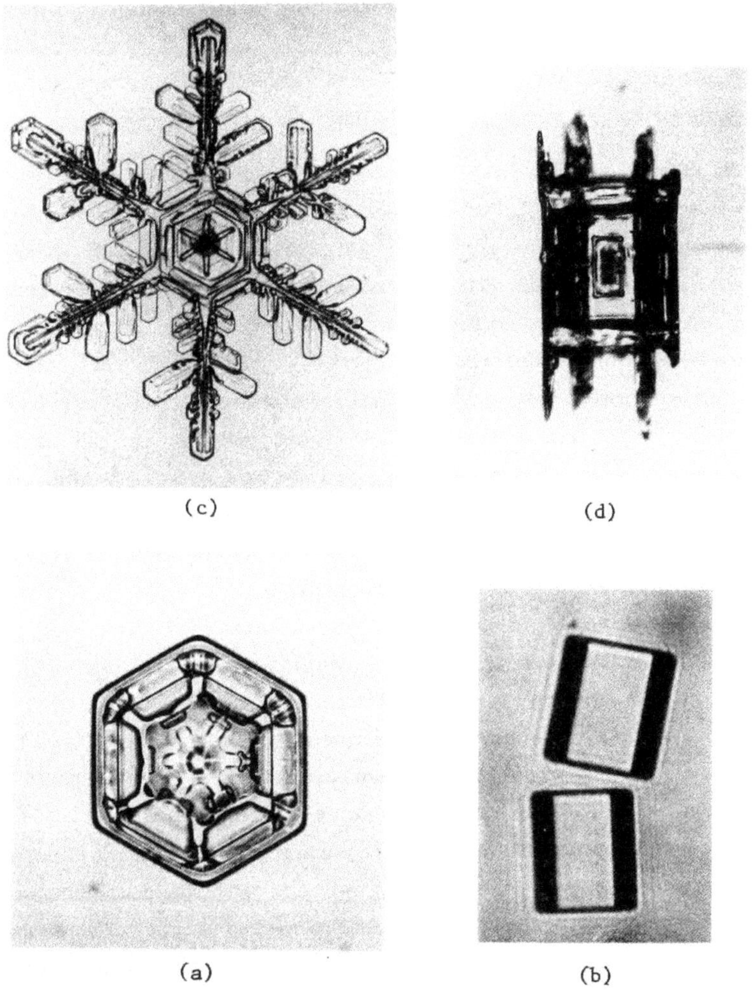

Fig. 3 Photographs of natural snow crystals corresponding to Fig. 2 (provided by Y. Furukawa).

and the diffusion process are taken into account. Furthermore, habit change associated with surface melting and roughening of ice is treated in section seven.

2. GROWTH OF SNOW CRYSTALS AND FEATURES OF THEIR PATTERNS

An air mass containing water vapor cools down to the temperature below freezing point $0^{o}C$ of ice because of adiabatic expansion during its elevation, and equilibrium or saturation pressure of water vapor exponentially decreases with decreasing temperature. Therefore, the water vapor becomes supersaturated and cloud droplets, i.e. supercooled liquid droplets, are formed, even though the temperature falls considerably below $0^{o}C$. This phenomenon is called Ostwald's step rule,[17,18] i.e. a supercooled liquid droplet in the metastable state appears at first in the nucleation process and then it transforms into stable crystal over a considerable range of temperature below the freezing point, since the specific surface free energy of liquid is smaller than that of crystal.

When water vapor is consumed by nucleation and growth of a large number of the cloud droplets, it is equilibrated with the droplets in a range from a few μm to several ten μm in radius and no further nucleation occurs. Namely, the pressure of water vapor in cloud is nearly equal to the equilibrium vapor pressure p_e^w with respect to the supercooled water. This state is called water saturation.

Since the equilibrium vapor pressure p_e^w of water is larger than the equilibrium vapor pressure p_e^i of ice at the temperature below $0^{o}C$, water vapor in cloud is in supersaturated state with respect to ice. Therefore, ice crystals continue to grow from the vapor under the conditions of water saturation, once they are formed by freezing of supercooled water droplets in cloud. This is the growth of snow crystals.

Although the form of an initial ice crystal produced by freezing of droplet is simply sphere, it grows into a hexagonal prism bounded by two basal $\{0001\}$- and six prism $\{10\bar{1}0\}$- faces (Fig.1) and then develops into various patterns according to the growth conditions.

Fig.2 is a schematic diagram representing the relation between patterns of snow crystals and their growth conditions which is based on the results of experimental studies.[1,4-9] The photographs of snow crystals corresponding to Fig.2 are shown in Fig.3. The diagram shows that three conversions occur in the basic pattern or habit at -4°C (plates to columns), -10°C (columns to plates), and -22°C (plates to columns). It has recently been reported that habit at temperatures below -20°C is not definitely columnar.[19,20] Interpretation[21] of the habit change in terms of surface melting and roughening of ice crystal is shown in section seven.

The diagram shows also that increasing supersaturation causes transitions from very thick plates to thin plates to sector plates to dendrites, or from solid prismatic columns to hollow prisms to needles. In other words, the surfaces of polyhedral ice crystals become morphologically unstable by way of preferred growth of edges and corners of the crystal with increasing supersaturation. In section six, we shall discuss the formation of dendritic pattern of the polyhedral snow crystals on the basis of recent simulation.

It is to be noted that the crystals shown in Figs.2 and 3 are categorized as single crystals. In nature a variety of spatially developed polycrystalline snow crystals have been observed. Their shapes include the twin prism, twelve-branched crystal, combination of bullets (Fig.4a), plane assemblage of the spatial type (spatial dendrite Fig.4b), plane assemblage of the radiating type (radiating dendrite), and crystals of the so-called "peculiar" type. It has been observed that the distribution of the angle between the c-axes of each component of them is predominantly concentrated at a definite value.[22-26] Studies on structure and formation mechanism of snow polycrystals have been actively carried out by Kobayashi et al.[24-28] For the details, the reader may refer to the review article.[2]

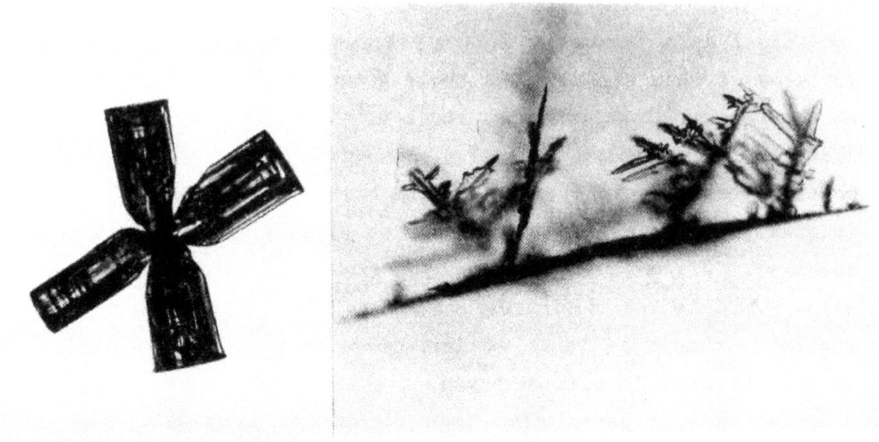

(a) (b)

Fig. 4 Snow polycrystals of twin type. (a) Combination of bullets and (b) spatial dendrite (provided by Y. Furukawa).

3. PROCESSES RELEVANT TO GROWTH OF SNOW CRYSTALS AND A MODEL FOR SIMULATION OF PATTERN FORMATION

In this section we summarize the surface kinetic process for incorporating water molecules into crystal lattice and the diffusion process for supplying the molecules in air to the surface, and introduce a model for simulation of pattern formation of growing crystals.

3.1 Surface Kinetic Process

Water molecules which impinge on the surface stay there for time τ_s and migrate in an area x_s^2 in τ_s. These molecules are incorporated into the ice crystal, if they reach a step during surface diffusion. On the other hand, they are evaporated into the air, if they do not meet the step in τ_s. The process for incorporating the molecules into crystal mentioned above is called the surface kinetic process.

The growth rate V_k in the direction normal to the surface which is inclined by angle θ to a singular surface, e.g. prism surface (Fig.5) is expressed as

$$V_k = \beta(\theta, \sigma_s)\sigma_s, \quad (1)$$

where σ_s is the local supersaturation at the surface and $\beta(\theta, \sigma_s)$ is the kinetic coefficient which represents the surface activity for incorporating the molecules into crystal and depends on θ or the distance λ of neighboring steps acting as sink for molecules and also on σ_s. The surface supersaturation σ_s is defined as

$$\sigma_s = \frac{p_s - p_e^i}{p_e^i}, \quad (2)$$

where p_s and p_e^i are the pressure of water vapor at the surface and the equilibrium vapor pressure of ice respectively. According to a surface diffusion model,[14,29] $\beta(\theta, \sigma_s)$ is given by

$$\beta(\theta, \sigma_s) = \beta_{max} \frac{s}{s_1} \tanh \frac{s_1}{s} . \quad (3)$$

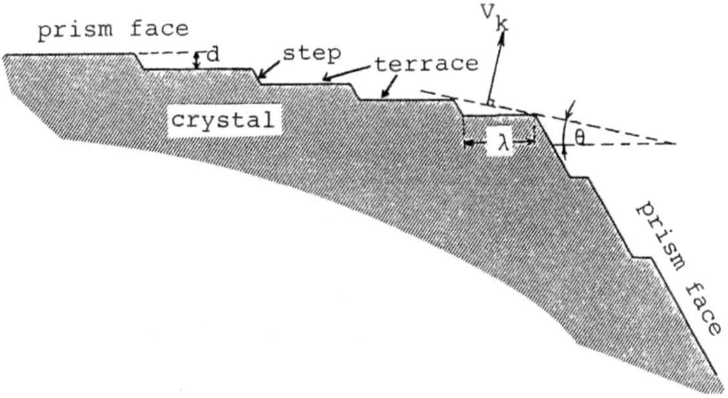

Fig. 5 The schematic representation of the growing surface composed of terraces and steps.

In eq.(3) s and s_1 are the slopes defined as follows:

$$s = \tan\theta = d/\lambda \ , \quad s_1 = d/2x_s \ , \qquad (4)$$

and

$$\beta_{max} = \alpha v_c \, p_e^{i}/\sqrt{2\pi mkT} \ , \qquad (5)$$

where λ is the mean step distance, d the step height, x_s the mean surface diffusion distance of a molecule on the surface, α the sticking coefficient, v_c the volume of a molecule in the crystal, m the mass of a molecule, k the Boltzmann constant, and T the absolute temperature.

Fig.6 qualitatively represents the dependence of $\beta(\theta, \sigma_s)$ on θ which is a rotation angle about c-axis of the hexagonal ice crystal. It shows minima at 0°, $\pm60^o$, $\pm120^o$ and 180° corresponding to six prism faces on which there is no step in equilibrium state. However, the value of $\beta(0, \sigma_s)$ at the prism faces is not zero, but positive in actual supersaturated state, since growth hillocks with a local slope $s_0=\tan\theta_0$ are formed by step generation with the aid of screw dislocations or two-dimensional nucleation (Fig.7).

In the case of spiral growth[29] with the aid of screw dislocations, the local slope s_0 of a growth hillock is given by

$$s_0 = \frac{d\sigma_s}{2x_s\sigma_1} \ , \qquad (6)$$

$$\sigma_1 = \frac{9.5f_0\kappa}{kTx_s} \ , \qquad (6')$$

where κ is the edge free energy per unit length along a step and f_0 the surface area occupied by a molecule. Thus, $\beta(0, \sigma_s)$ at prism faces linearly increases with increasing σ_s for $\sigma_s \ll \sigma_1$ and approaches β_{max} for $\sigma_s \gg \sigma_1$ in the case of spiral growth (eqs. (3), (4), (6) and Fig.8). On the other hand, $\beta(0, \sigma_s)$ in the case of two-dimensional nucleation growth is equal to zero at small σ_s, then exponentially

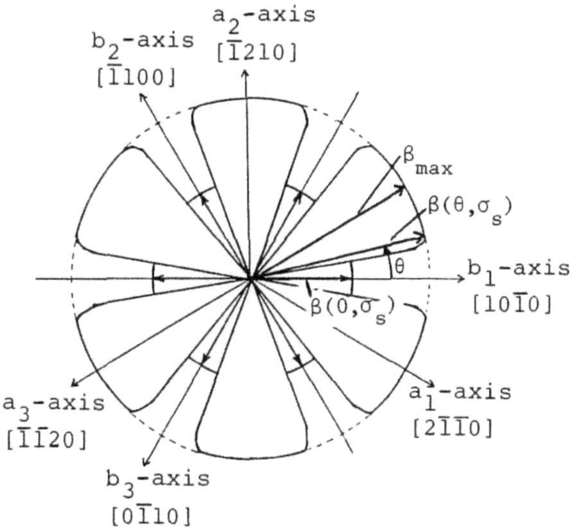

Fig. 6 Polar plot of the kinetic coefficient $\beta(\theta,\sigma_s)$ as a function of the crystallographic orientation θ of the surface.

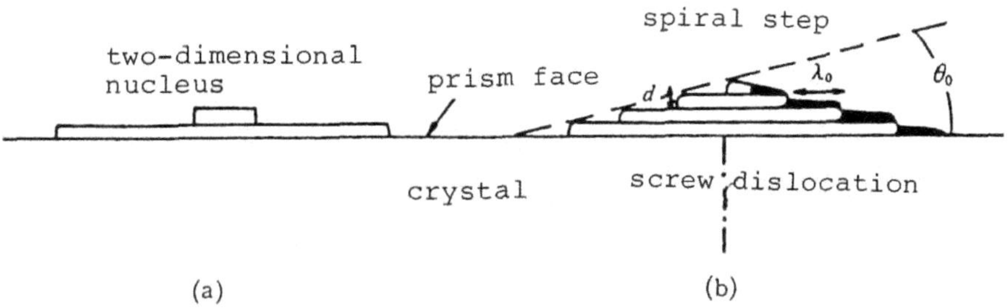

Fig. 7 Generation of the steps with the aid of (a) two-dimensional nucleation and (b) screw dislocation.

increases with increasing σ_s above the critical value σ_{sc} and approaches β_{max} (Fig.8).

In both cases, the local slope s_0 of the growth hillock as well as the kinetic coefficient $\beta(0, \sigma_s)$ of the prism faces increase with increasing σ_s. Thus, the anisotropy, i.e. the dependence on θ of $\beta(\theta, \sigma_s)$ decreases with increasing σ_s, and $\beta(\theta, \sigma_s)$ becomes isotropic when $\beta(0, \sigma_s)$ reaches β_{max}. It is to be noted that the growth vate V_k (eq.(1)) of the prism faces, i.e. singular faces in general, behaves non-linearly with σ_s, since $\beta(0, \sigma_s)$ depends on σ_s. This non-linearity tends to change the patterns of growing crystals drastically in response to the slight change of the growth conditions. Furthermore, the surface kinetic process is coupled with the diffusion process through σ_s, since the σ_s which determines $\beta(\theta, \sigma_s)$ is governed by the diffusion equation.

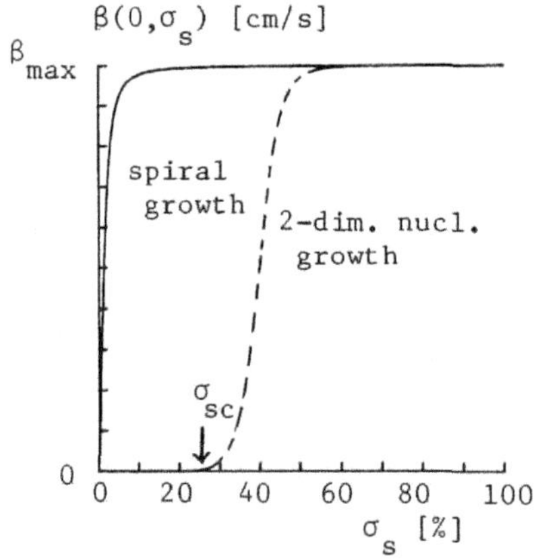

Fig. 8 The dependence on the surface supersaturation σ_s of the kinetic coefficient $\beta(0,\sigma_s)$ for prism face ($\theta=0^o$) in the case of spiral growth with the aid of screw dislocation (a solid line) and in the case of two-dimensional nucleation growth (a dashed line).

3.2 Diffusion Process

The growth rate V_d determined by the diffusion process is given by

$$V_d = \frac{v_c \, p_e^i}{kT} D \, q_s, \qquad (7)$$

where D is the diffusion coefficient of water molecules in air and $q_s = (\partial\sigma/\partial n)_{surface}$ is the normal gradient of supersaturation on the surface which is unknown.

From the viewpoint of mass balance at the crystal surface, the growth rate V_k determined by the surface kinetic process should be equal to the growth rate V_d determined by the diffusion process under steady state conditions [30]:

$$V_k = V_d = V. \qquad (8)$$

By inserting eqs.(1) and (7) into eq.(8), we obtain

$$\beta(\theta, \sigma_s)\sigma_s - \frac{v_c \, p_e^i}{kT} D \, q_s = 0 \, . \qquad (9)$$

This is an important relation for determining self-consistently the growth rate V controlled both by diffusion process and by surface kinetic process.

The supersaturation σ in the region Ω (Fig.9) surrounding a crystal is determined by solving the quasi-static diffusion equation

$$\Delta\sigma = 0 \qquad (10)$$

subject to the following boundary conditions: the supersaturation is specified by

$$\sigma = \sigma_\infty \qquad (11)$$

on the boundary Γ_1 which is apart from the crystal by R, and the mass balance equation (9) is satisfied on the boundary Γ_2 representing the crystal surface. The problem characterized by the boundary condition

(9) including both σ_s and its derivative q_s is a so-called third boundary value problem.

3.3 Method of Simulation of Pattern Formation[14,15,16)]

We solve the problem by means of the boundary element method as follows. At first, we obtain the boundary integral equation from eq.(10) using Green's theorem and boundary conditions (eqs.(9) and (11)):

$$\sigma_{si}/2 + \int_{\text{along}\Gamma_2} \left(q^* + \frac{\sigma^*\beta(\theta,\sigma_s)kT}{v_c P_e^{i} D}\right)\sigma_s d\Gamma = \int_{\text{along}\Gamma_1} (q_\infty \sigma^* - \sigma_\infty q^*)d\Gamma, \qquad (12)$$

where σ_{si} is the supersaturation at a point i on Γ_2, q_∞ the normal gradient of supersaturation on Γ_1, σ^* satisfies $\nabla^2\sigma^*+\delta_i=0$ (δ_i: the Dirac delta function), and q^* is $\partial\sigma^*/\partial n$. We approximately replace q_∞ with following equation under the assumption that Γ_1 is far from Γ_2.

$$q_\infty \simeq \sigma_\infty / R\ln(R/r_c), \qquad (13)$$

which is obtained for diffusion field surrounding a circular crystal with a radius r_c which is half of mean length along a-axis and b-axis and with perfect sink ($\sigma_s=0$). The value R/r_c is so chosen that the growth rate of a circle of r_c in radius is equal to that of sphere of r_c in radius. Then, we change the eq.(12) to the algebraic equations with respect to the surface supersaturation σ_s by dividing Γ_2 into N elements.

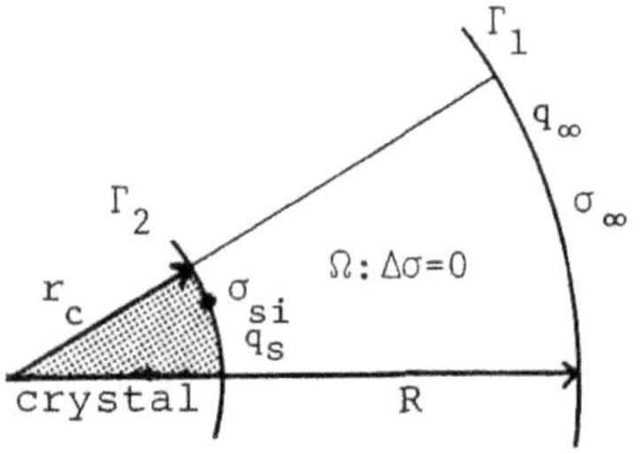

Fig. 9 Diffusion field surrounding a crystal.

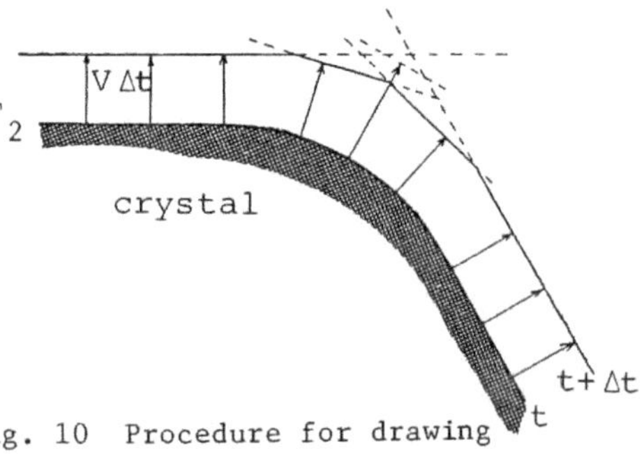

Fig. 10 Procedure for drawing a growth form after Δt second.

By solving the N simultaneous equations of σ_{si} for all points i on Γ_2, we obtain σ_s, consequently the growth rate V at each position on the surface at a certain moment and then determine a shape of the growing surface after Δt second. The surface growing faster is cut by neighboring surfaces growing slower, while the surface growing slowest develops its area largest. Therefore, a growth form is determined by inner envelope of the lines perpendicular to each vector V Δt at its point (Fig.10). By repeating this procedure we simulate the time evolution of patterns in growth of snow crystals. The results are shown in section six.

4. FACTORS CONTROLLING PATTERN FORMATION

From the features of growth led by the surface kinetic process and the diffusion process, four factors (4-i)-iv)) are expected to contribute to pattern formation of growing crystals from vapor or solution.

4-i) Faceting due to Anisotropy of Kinetic Coefficient $\beta(\theta, \sigma_s)$

If the kinetic coefficient $\beta(\theta, \sigma_s)$ has deep cusps, i.e. singularities, at the definite orientations θ of the surfaces, the surfaces develop into facets. In other words, the anisotropy of $\beta(\theta, \sigma_s)$ works as a stabilizing factor of polyhedral pattern of the crystals. The prism $\{10\bar{1}0\}$ facets seen at tips of primary and secondary branches of a dendritic snow crystal (Fig.3) are due to storong anisotropy of $\beta(\theta, \sigma_s)$.

4-ii) Preferred Growth of Corners and Edges of a Crystal due to Non-Uniformity in Supersaturation σ_s over the Crystal Surface

When a polyhedral crystal caused by the anisotropy of $\beta(\theta, \sigma_s)$ grows, the supersaturation is not uniform over its surfaces, being largest at the corners and smallest at the centers of the surfaces (Fig.11). This non-uniformity has been theoretically pointed out by Seeger[31] for the first time and then experimentally confirmed by many authors using a interference method etc. in case of solution growth.[32-41]

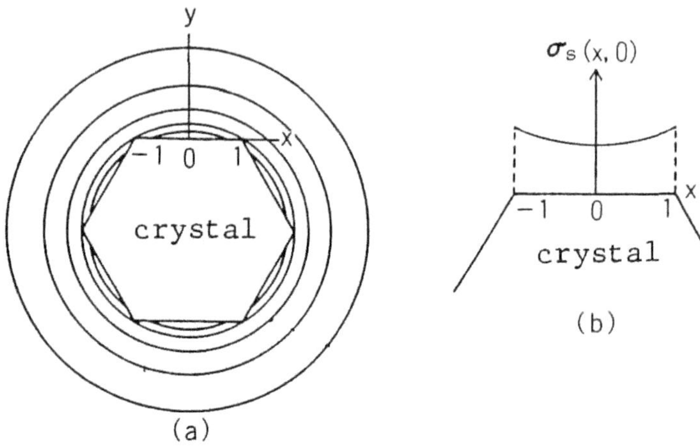

Fig. 11 The schematic representation of (a) equi-supersaturation curves and (b) non-uniform supersaturation $\sigma_s(x,y)$ over the surface.

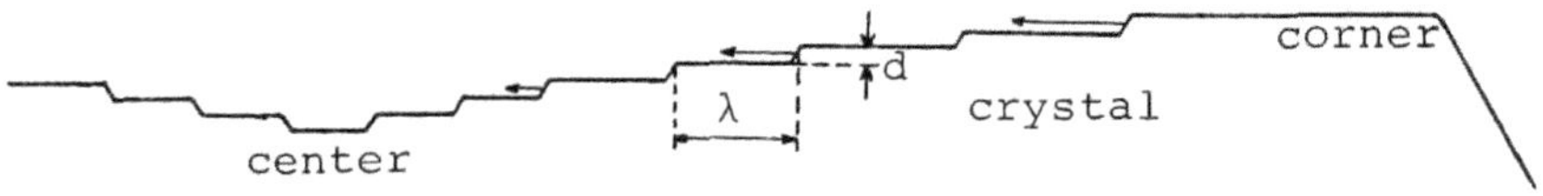

Fig. 12 The schematic representation of step distribution for compensation of non-uniformity in surface supersaturation σ_s.

The excess supersaturation at corners tends to cause preferred growth there,[30,42-49] which is the onset of dendritic growth. Thus, the non-uniformity of surface supersaturation σ_s is the destabilizing factor of polyhedral pattern of crystals.

On the other hand, Chernov argued that the non-uniformity of σ_s can be compensated by the variation of local kinetic coefficient $\beta(\theta, \sigma_s)$ by adjustment of step distribution.[42,43] For an example, the step nuleated at corners slow down as they approach the center of the surface because of a decrease in σ_s and consequently the step density increases near the center (Fig.12). Since an increase in the density of the steps at which the molecules are incorporated into the crystal causes a larger kinetic coefficient, the growth rate V_k can be constant over the whole surface. Therefore, the destabilizing factor of a polyhedral pattern is the non-uniformity in supersaturation over the surface which is caused by the volume diffusion process, while the stabilizing factor is the variation of the local kinetic coefficient by adjustment of step distribution. The competition of these two factors determines wheather the polyhedral pattern is stable or not.

4-iii) Development of Concavo-Convex Pattern due to Diffusion Field

If a convex part is formed at crystal surface by fluctuation or preferred growth caused by non-uniformity of σ_S, diffusion flux is concentrated toward the convex part. Thus, the convex part tends to grow faster than the other flat parts.

On the other hand, the concave part tends to grow slowly, since the diffusion flux toward there is smaller. Thus, the diffusion field contributes to development of concavo-convex pattern of the crystals. This is a driving force for the famous Mullins-Sekerka instability. [50,51]

4-iv) Suppression of Concavo-Convex Pattern by Surface Tension - Gibbs Thomson Effect-

On the other hand, the equilibrium vapor pressure $p_e(K)$ or equilibrium concentration $c_e(K)$ of the convex part with a curvature K becomes larger than that $p_e(0)$ or $c_e(0)$ of flat surface (K = 0) because of surface tension $\gamma(\theta)$, i.e.

$$p_e(K) = p_e(0)[1 + \frac{v_c}{kT}(\gamma(\theta) + \frac{\partial^2\gamma}{\partial\theta^2})K] \qquad (14)$$

in two dimensional case.[52] Such effect is called Gibbs-Thomson effect.

Since an increase in $p_e(K)$ reduces the surface supersaturation $\sigma_S=(p_S-p_e(K))/p_e(K)$, the convex part tends to grow slowly by the Gibbs-Thomson effect. On the other hand, the concave part (K<0) tends to grow faster because of a decrease in $p_e(K)$. Thus, the surface tension suppresses the formation of concavo-convex pattern.

If the molecules are quickly incorporated into crystal at the surface and the supply of the molecules to the surface is rather difficult on the contrary, the growth rate is controlled only by the diffusion process. Since equilibrium condition holds at the surface in such diffusion controlled case, the vapor pressure at the surface is set to be equal to the equilibrium vapor pressure (eq.(14)) on solving the diffusion equation. Thus, the anisotropy of the surface tension $\gamma(\theta)$ contributes to anisotropic pattern formation in the diffusion controlled case, while the anisotropy of the kinetic coefficient $\beta(\theta, \sigma_s)$ is important in the case that the surface kinetic process plays a role in the growth. For an example, the anisotropy of interfacial tension $\gamma(\theta)$ between ice and water contributes to the selection of definite orientations, i.e. a-axes (Fig.13), of the dendrites developed from the disc-shaped ice crystals growing from the supercooled water, the growth rate of which is controlled by thermal diffusion.[61]

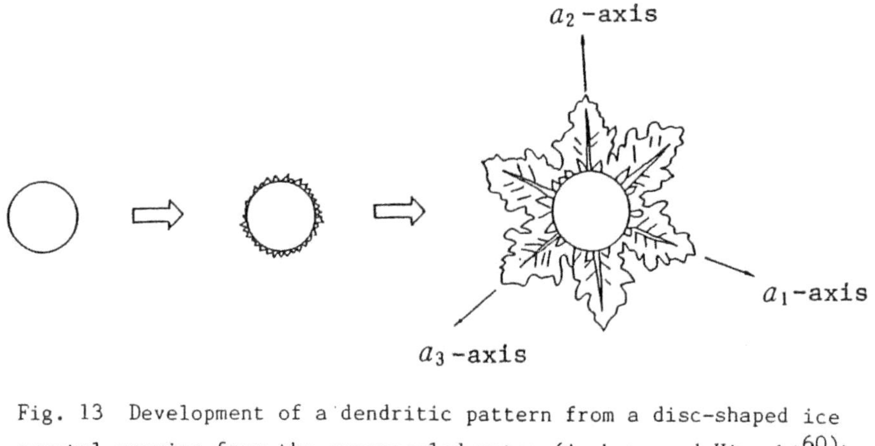

Fig. 13 Development of a dendritic pattern from a disc-shaped ice crystal growing from the supercooled water (Arakawa and Higuchi[60]).

5. PREVIOUS THEORETICAL STUDIES ON DENDRITIC PATTERN FORMATION DEVELOPED FROM POLYHEDRAL SNOW CRYSTALS GROWING FROM VAPOR AND THAT DEVELOPED FROM DISC-SHAPED ICE CRYSTALS GROWING FROM MELT

As mentioned in sub-sections 4-i) and 4-ii), a hexagonal prism bounded by two basal {0001} and six prism $\{10\bar{1}0\}$ faces can be developed from an initial spherical ice crystal during growth from vapor because of deep cusps of the kinetic coefficient $\beta(\theta, \sigma_s)$ at <0001>- and $<10\bar{1}0>$- orientations. Thus, the supersaturation becomes non-uniform over the crystal surface as a result of diffusion field surrounding a polyhedral crystal. If the degree of the non-uniformity of σ_s exceeds a critical value, preferred growth of corners where the supersaturation σ_s is highest occurs. This is the onset of dendritic pattern formation of snow crystals. It should be noticed that the primary branches of the snow dendrites (Figs.2 and 3) are necessarily oriented to the directions which bisect an angle between neighboring prism $\{10\bar{1}0\}$ faces, i.e. a-axis ($<11\bar{2}0>$-axis), at the temperatures -10 to -22°C where the growth of basal face is much smaller than that of prism face (Fig.2).

Let us summarize theoretical studies on the conditions for transition from polyhedral to dendritic patterns. Cahn[44] required as his stability condition for polyhedral pattern that the supersaturation should remain positive at the center of a surface, and discussed the maximum size of a crystal with polyhedral pattern. Chernov[42,43] pointed out that a catastrophic loss of stability occurs when the slope at the center of the surface reaches a critical value corresponding to the loss of anisotropy of the kinetic coefficient.

Kuroda, Irisawa and Ookawa[30] elaborated the characteristics of the crystal growth of cube from solution at rest, satisfying a consistent compromise between the three-dimensional diffusion field surrounding the crystal, the interface growth kinetics, and the step distribution over the surface. For a specified condition of the supersaturation σ_∞ at infinity and the crystal size L, the growth rate in a stable region is determined. Instability occurs when the local kinetic coefficient at the face center reaches the maximum value. The limiting curve of instability, σ_∞^{**} versus L, was obtained quantitatively, where the slope at the face center tends to exceed a value of the order of d/x_s. Here, d is the step height and x_s is the diffusion distance of growth units on the surface.

Formation of the three-dimensional external shape characteristic of snow crystals, the so-called lacunary (skeletal) growth, was discussed qualitatively by Frank[45,46] who suggested that steps nucleated at crystal corners were propagated toward the face centers.

Irisawa, Kuroda and Ookawa[48,49] extended the treatment of the cube crystals to the morphological instability of the hexagonal prism-shaped crystals bounded by basal and prism faces. In this case, we have a question: On which face does instability occur first for specified supersaturation σ_∞ at infinity, ratio l_c/l_a of the length l_c along c-axis and the length l_a along a-axis and crystal size l_c? It was shown that the instability occurs first on the prism face in the case of $l_c/l_a \simeq V_b/V_p < 0.7$, where V_b and V_p are the growth rate of basal and prism faces respectively, while the instability occurs first on the basal face in the case of $l_c/l_a \simeq V_b/V_p > 0.9$. The theoretical results are in qualitative agreement with experimental results by Gonda and Koike[53] at -30°C and by Nenow, Stoyanova and Ganadiev[54] at -2 to -12°C.

Kuroda[55] qualitatively discussed the factors other than supersaturation and crystal size that effect the morphological stability, e.g. i) diffusion coefficient of growth units, ii) thermal conductivity and iii) flow velocity of the environment. For the experimental studies on these factors, the reader may refer to the papers: (i) Gonda[56,57] and Nenow and Stoyanova,[58] ii) Gonda[56] and iii) Keller and Hallett.[59]

The previous studies mentioned above treat only the conditions for transition from polyhedral to dendritic patterns. However, we stand still in need of understanding time evolution of patterns of growing snow crystals. Recent results of simulation of pattern formation in growth of snow crystal are shown in section six.

It should be noticed that the morphological instability of the polyhedral crystals bounded by smooth surface in the case of vapor or solution growth should be distinguished from the Mullins-Sekerka instability[50,51] for rough crystal/melt interface. In the latter problem, the destabilizing factor is supercooling in the melt, while the stabilizing factor is a decrease in melting temperature associated with interfacial tension and interface curvature. For an example, Fig.13 shows the morphological instability of an ice crystal growing from the slightly supercooled water (Arakawa and Higuchi[60]). The external shape of the snow dendrite growing from vapor (Fig.3(c)) is similar to that of ice dendrites growing from the melt (Fig.13) in the sense that the primary branches of the dendrites grow along a-axes in both cases. However, the reason for the definite orientation of the former dendrites is quite different from that of the latter dendrites. The primary branches of the snow dendrites are necessarily oriented to the directions which bisect an angle between neighboring prism faces,i.e. a-axis, since the morphological instability of the hexagonal prism-shaped crystal is caused by an excess supersaturation along edges where the neighboring prism faces intersect. On the other hand, the orientation of the ice dendrites growing from the supercooled water is attributable to the anisotropy of the interfacial tension $\gamma(\theta)$ between ice and water.[61] The effect of anisotropy of $\gamma(\theta)$ on dendritic growth of rough interface was pointed out by Komabayashi[62] for the first time. Recent studies on the effect of anisotropy of $\gamma(\theta)$ on the growth of dendrites will be treated in the chapter written by Saito et al.

6. RESULTS OF SIMULATION OF PATTERN FORMATION OF GROWING SNOW CRYSTALS

In this section, we show the results[14-16] of simulation of pattern formation of two-dimensional snow crystals starting from circle perpendicular to c-axis. We consider only the spiral growth to occur with the aid of screw dislocations and assign $\kappa = 2.0\times10^{-6}$ erg/cm for the edge free energy, since the growth rates obtained by eqs.(1) and (3)-(6') well agree with experimental values[63]. Furthermore, the steps are supplied from the screw dislocations emerging at the centers of six prism faces until a circular crystal develops into a perfect hexagon, and they are generated at its corners where the supersaturation is largest, after a perfect hexagon has been developed.

The heat conduction process is not taken into account in this simulation, since its contribution to the growth rate is relatively small in comparison with the diffusion process and the surface kinetic process, unless the growth rate is too large[64]. Thus, it is assumed that the temperature of the whole system is uniform at $-15^{o}C$ which is typical temperature for development of dendritic pattern of snow crystals.

The numerical values used for simulation are as follows: equilibrium vapor pressure of ice $p_e^{\ i}=1.66x10^3$ dyn/cm^2=$1.66x10^2$ Pa corresponding to $-15^{o}C$, molecular volume of ice crystal $v_c=3.25x10^{-23}$ cm^3, surface areas occupied by a water molecule $f_0=8.3x10^{-16}$ cm^2, mass of a water molecule $m=3.0x10^{-23}$ g, step height or lattice constant $d=4.5x10^{-8}$ cm, mean surface diffusion distance of an admolecule $x_s=400d$[65] and radius of initial crystal r_{c0} $=5x10^{-3}$ cm.

The following three patterns of snow crystals have been obtained by simulation[15,16]: (1) Hexagonal pattern: An initial circular crystal develops into the perfect hexagon bounded by six prism faces, and then the hexagonal crystal can grow retaining its flat surface macroscopically (Figs.14(a),15 and 16(a)). (2) Dendritic pattern: An initial circular crystal develops into the perfect hexagon, and then hexagonal crystal can not grow retaining its flat surface but develops into dendritic crystal (Figs.14(b) and (c)). (3) Circular pattern due to kinetic roughening: Prism faces can be hardly developed because of the vanishment of anisotropy of growth rate, i.e. kinetic roughening, and a circular pattern remains during growth (Fig.16(b)).

Figure 14 shows produced patterns for smaller diffusion coefficient $D=0.2cm^2/s$ (corresponding to $1x10^5$ Pa of air pressure, i.e. latm) and $\alpha=0.01$[66]. When supersaturation σ_∞ is small, e.g. 1%

(Fig.14(a)), an initial circular crystal develops into a perfect hexagon, and then the hexagonal crystal grows retaining its flat surfaces in spite of non-uniformity in surface supersaturation, since the degree of the non-uniformity is not large. For example, the surface supersaturation σ_s is 0.47% at the corners and σ_s is 0.41% at the centers of surfaces at 7000s.

On the other hand, for larger σ_∞, e.g. 16% (Fig.14(b)) and 32% (Fig.14(c)), the growth rate at corners is larger than that of center of the surface because of large non-uniformity in surface supersaturation, and consequently preferred growth of corners occurs. In Fig.14(b) σ_s=2.4% at the corners and σ_s=1.7% at the center of surface at 800s, and in Fig.14(c) σ_s=3.6% at the corners and σ_s=2.3% at the center of surface at 800s. It should be noted that the facets are developed at the tips of primary branches and the macrosteps are formed because of bunching of monomolecular steps on the facets. Such macrosteps may act as trigger of formation of secondary branches. The width of a primary branch formed at σ_∞=32% (Fig.14(c)) is smaller than that formed at σ_∞=16% (Fig.14(b)), since the position where the growth rate can never reach the growth rate at corners moves toward the corner from the center of the surface with increasing non-uniformity in surface supersaturation.

It should be noticed that the dendritic patterns shown in Fig.14(c) is obtained for σ_∞(=32%) much larger than the maximum supersaturation 16% at -15°C which is realized when water vapor is equilibrated with supercooled water droplets in clouds, while the similar patterns are seen in natural snow crystals grown during their fall in clouds at $\sigma_\infty \leqq 16\%$. The difference between σ_∞ in this simulation and σ_∞ in natural growth can be explained by a decreases in the thickness of diffusion boundary layer due to relative air flow to the snow crystals[55].

Figure 15 shows produced patterns for large diffusion coefficient D=10 cm^2/s (corresponding to $2x10^3$ Pa of air pressure) and α = 0.1[67,68]. An increase in D leads to an increase in growth rate and a decrease in anisotropy of the growth rate. Thus the size of a

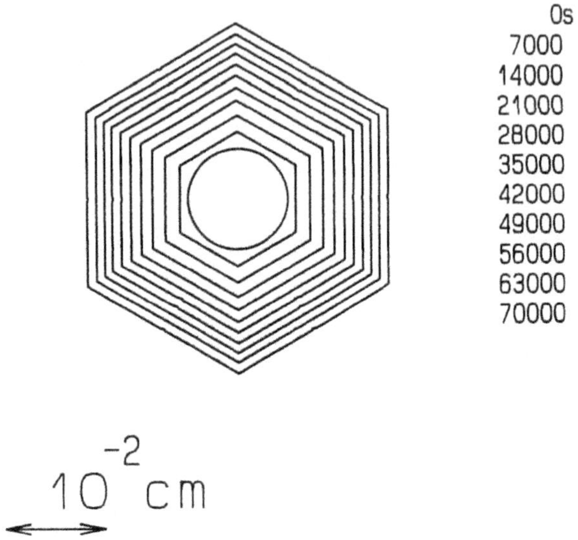

Fig. 14 Produced patterns for smaller $D=0.2cm^2/s$ (corresponding to $1x10^5$Pa of air pressure, i.e. 1atm) and for various σ_∞. (a) Hexagonal pattern at $\sigma_\infty=1\%$.

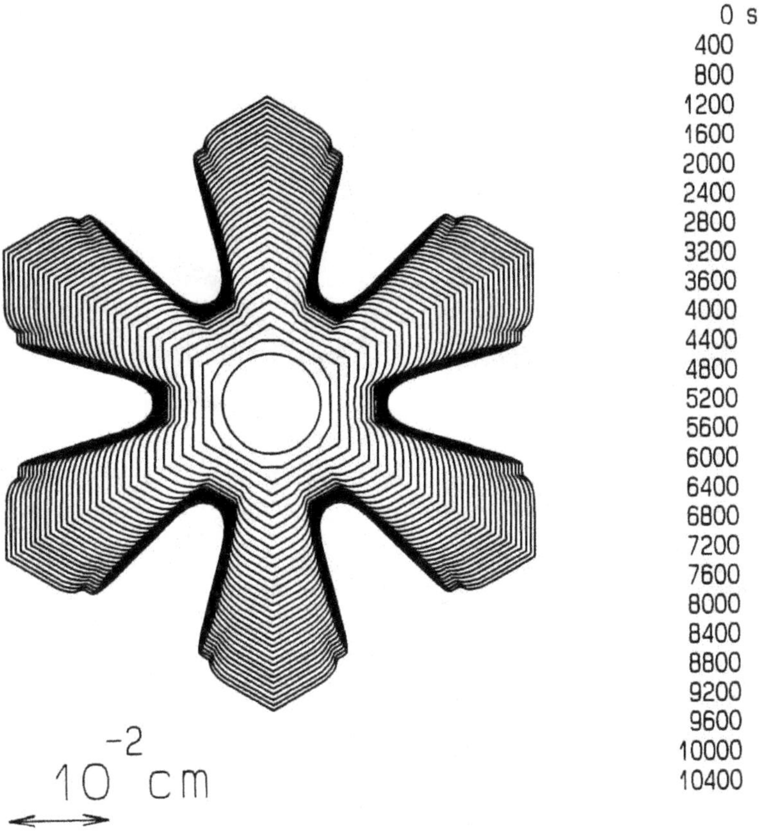

Fig. 14(b) Dendritic pattern at σ_∞=16%.

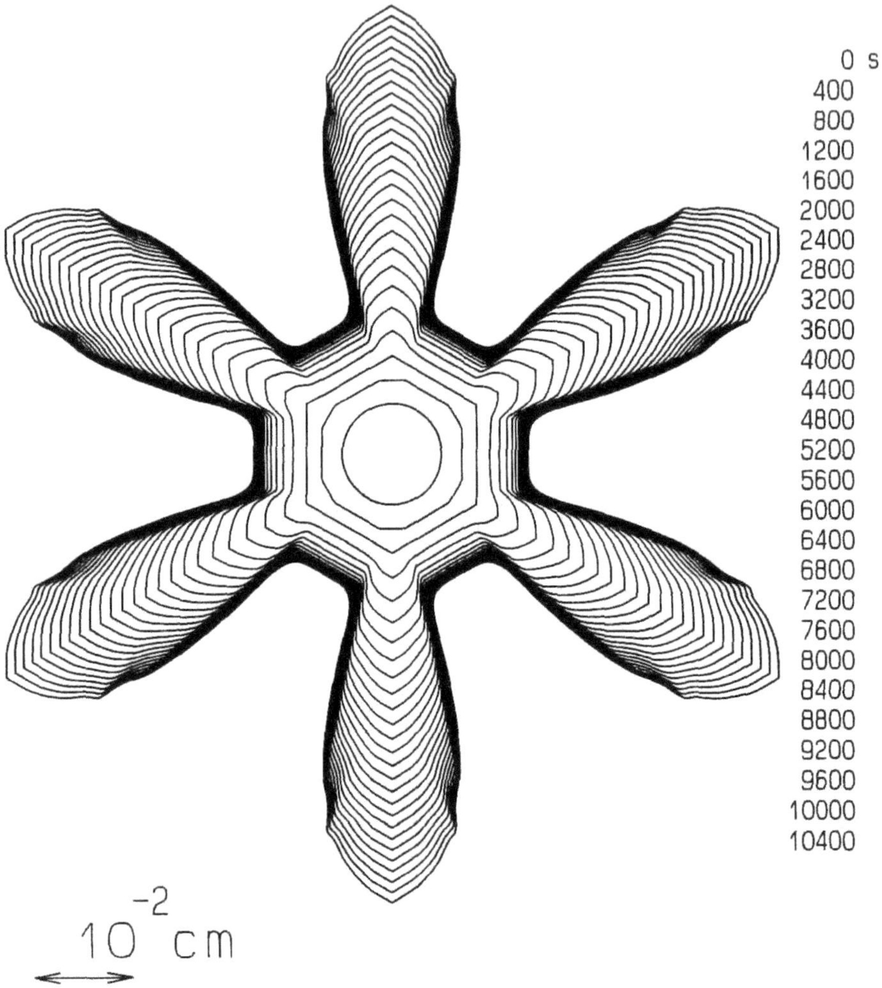

Fig. 14(c) Dendritic pattern at σ_∞=32%.

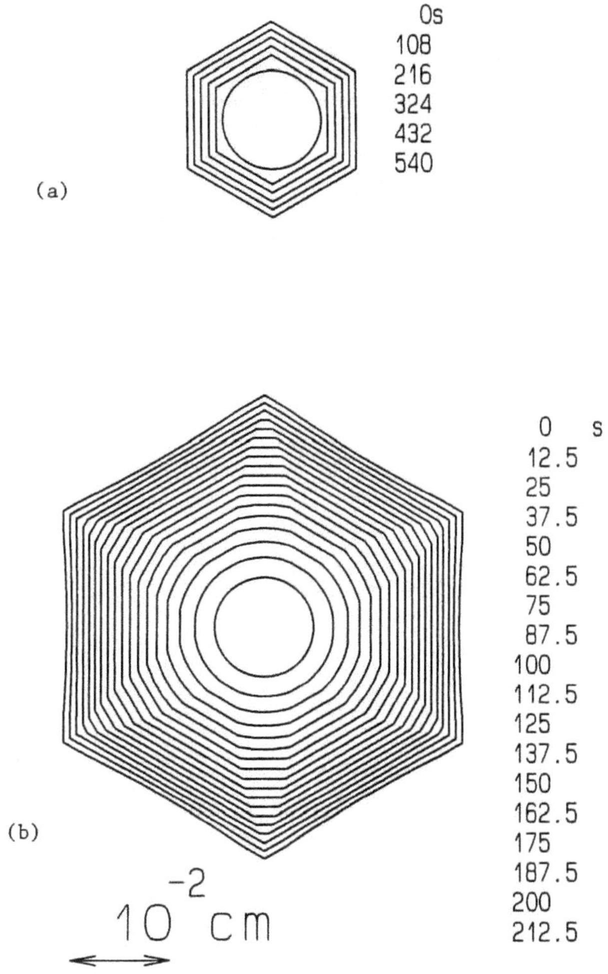

Fig. 15 Produced patterns for large $D=10cm^2/s$ (corresponding to $2x10^3$Pa of air pressure). (a) Hexagonal pattern at $\sigma_\infty=1\%$. (b) Hexagonal pattern at $\sigma_\infty=16\%$.

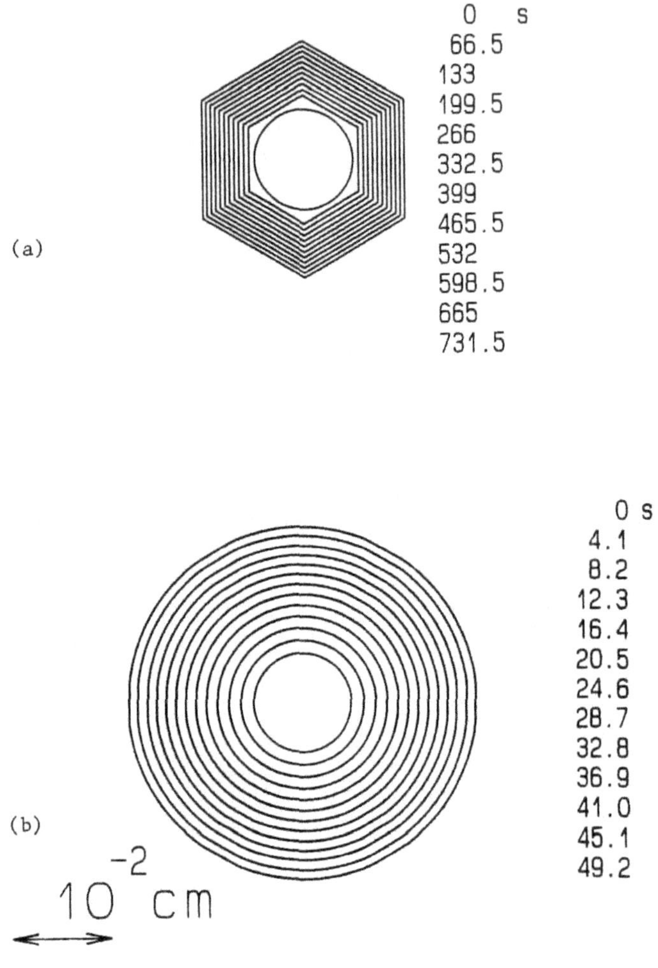

Fig. 16 Produced patterns for larger D=500cm^2/s (corresponding to 40Pa of air pressure). (a) Hexagonal pattern at σ_∞=1%. (b) Circular pattern at σ_∞=16%.

regular hexagon for σ_∞ =16% (Fig.15(b)) is larger than that for α_∞ =16% and D=0.2 cm^2/s (Fig.14(b)), when it has been formed. A hexagonal crystal of small size grows for σ_∞ =1% (Fig.15(a)), since anisotropy of growth rate is small because of small σ_∞.

Figure 16 shows produced patterns for larger diffusion coefficient D=500 cm^2/s (corresponding to 40 Pa of air pressure) and α=0.1[67,68]. In this case σ_s is nearly equal to σ_∞ because of very large D. Thus, a hexagonal pattern composed of six prism faces can be developed only at small σ_∞ (=1%, Fig.16(a)). On the other hand, the prism faces can be hardly developed at large σ_∞ (=16%, Fig.16(b)), since the anisotropy of growth rate vanishes because of large σ_s ($\simeq \sigma_\infty$) (see subsection 3.1). The vanishment of anisotropy of growth rate due to kinetic reason is called kinetic roughening. It should be noted that small facets begin to develop as the crystal grows further. This is due to a decrease in σ_s with increasing crystal size, since the supply of water molecules to the surface becomes insufficient because of increase in the area of surface acting as sink for molecules.

The three simulated patterns mentioned above have been quantitatively analyzed[15,16]. It has been shown that the conditions for kinetic roughening and the conditions for transition from hexagonal to dendritic pattern are given by the supersaturation σ_∞ and the dimensionless crystals size $\mathcal{L}$ with reference to the diffusion coefficient and the mean speed of water molecules in air (Fig.17).

In this simulation, we treat the growth of snow crystals only at -15°C where the basic habit is a hexagonal plate (Fig.2). In the same way, further simulation for growth in another temperature region where the basic habit is a hexagonal column (Fig.2) will be carried out and a diagram showing relation between patterns and conditions (σ_∞, $\mathcal{L}$) will be found in the future.

It had been considered that the pattern of snow crystals were determined only by the temperature and the supersaturation so far.[1,4-9] However, it has been shown that the patterns of snow crystals also depend on the dimensionless crystal size $\mathcal{L}$ [15,16]. A three-

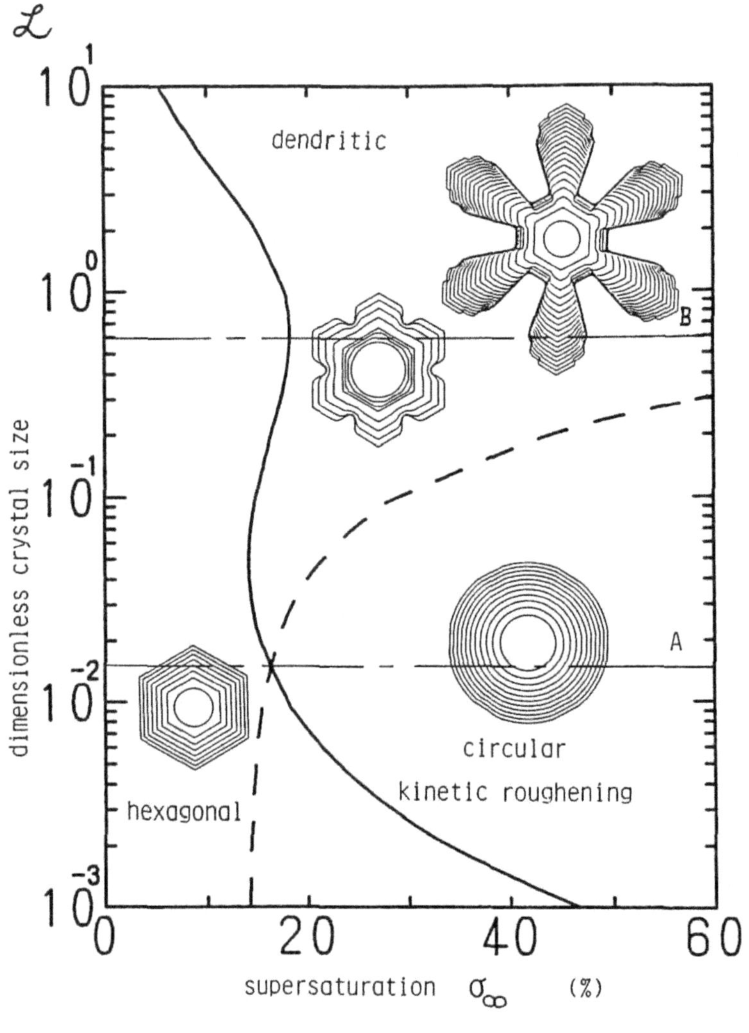

Fig. 17 Diagram showing the relation between produced patterns and conditions (σ_∞, $\mathcal{L}$). σ_∞:supersaturation at the position far from a crystal, $\mathcal{L}$:dimensionless crystal size defined by $\mathcal{L} = 0.022\alpha L \bar{v}/D$, where L is the side length of hexagon, α the sticking coefficient, $\bar{v}$ the mean speed of water molecules and D the diffusion coefficient.

dimensional (T, σ_∞ and $\mathcal{L}$) diagram, therefore, is necessary in order to classify accurately the patterns of snow crystals.

7. HABIT CHANGE AND SURFACE MELTING OF ICE

The basic habit of snow crystals bounded by two basal {0001} and six prism {10$\bar{1}$0} faces changes with decreasing temperature: plates to columns at -4°C, columns to plates at -10°C and plates to columns at -22°C (Figs.2 and 18).

When we consider the growth of snow crystals, we should notice that their growth temperature is very high in the sense that it is not far below the melting point of ice. Thus, Kuroda and Lacmann[21]) theoretically investigated the dependence on temperature of the structures of basal and prism surfaces of ice as well as their growth mechanisms, and proposed a following interpretation of habit change of snow crystals.

The surface is covered with a quasi-liquid layer just below 0°C, whose thickness d or coverage ϑ(>1) decreases with decreasing temperature and becomes equal to the thickness of the monolayer (ϑ=1) at a transition temperature $T_{I/II}$. The surface at temperatures $T<T_{I/II}$ may be rough on a molecular level because of adsorption of water molecules themselves with converge $\vartheta<1$. The number of adsorbed molecules decreases with decreasing temperatures and becomes very small, e.g. $\vartheta<0.02$ at the next transition temperature $T_{II/III}$. The surface at temperatures $T<T_{II/III}$ is molecularly smooth as is assumed in previous theories of crystal growth from vapor[29]). The growth mechanism depends on the surface structure. Therefore, three growth mechanisms take place on one surface with decreasing temperature (Fig.18):

I) Vapor-Quasi-liquid-Solid growth mechanism on the surface covered with a quasi-liquid layer (0°C > T > $T_{I/II}$; ϑ > 1),

II) Adhesive growth mechanism on the rough surface ($T_{I/II}$ > T > $T_{II/III}$; 1 > ϑ > 0.02) and

III) Two-dimensional nucleation growth mechanism on the smooth surface ($T_{II/III}$ > T ; 0.02 > ϑ).

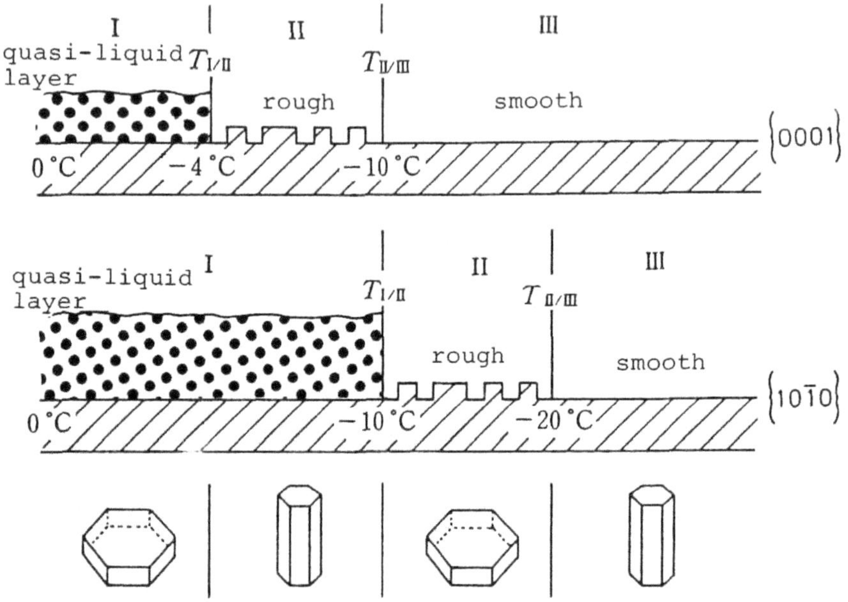

Fig. 18 Variation of structure and growth mechanism of basal {0001} and prism {10$\bar{1}$0} faces with temperature (Kuroda and Lacmann [21]).

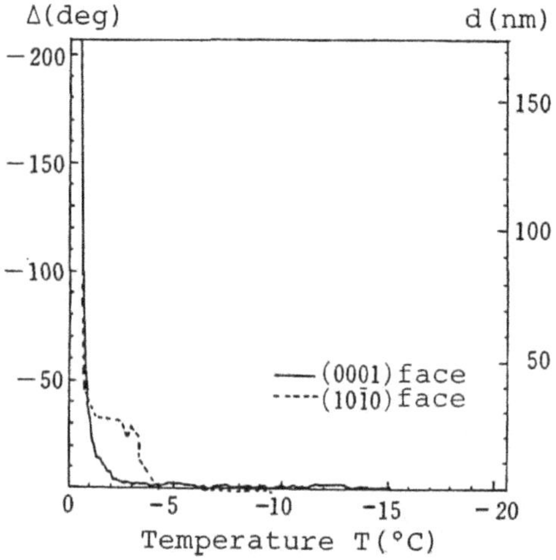

Fig. 19 Temperature dependence of phase shift Δ of the complex relative amplitude attenuation measured for ice surfaces and that of equilibrium thickness d of quasi-liquid layers (Furukawa et al [68]).

It was shown that two transition temperatures $T_{I/II}$ and $T_{II/III}$ were lower for prism $\{10\bar{1}0\}$ faces than for basal $\{0001\}$ faces respectively. If we assign

$$T_{I/II}(0001) = -4^{o}C, \qquad T_{I/II}(10\bar{1}0) = -10^{o}C,$$
$$T_{II/III}(0001) = -10^{o}C, \qquad T_{II/III}(10\bar{1}0) = -20^{o}C,$$

we can divide the temperature range into four parts according to the combination of growth mechanisms of $\{0001\}$ and $\{10\bar{1}0\}$ faces (Fig.18). The complicated habit change may be caused by this change in combination of growth mechanisms with temperature.[21]

After this new interpretation, ellipsometric study of the quasi-liquid layer on ice basal and prism faces has been carried out by Furukawa, Yamamoto and Kuroda.[68] Ellipsometry gives us detailed information on the refractive index and the thickness of a transition layer on a crystal surface. It was found that the transition layers appear at $T_{I/II}(0001)=-2^{o}C$ on the basel face and $T_{I/II}(10\bar{1}0)=-4^{o}C$ on the prism face and the refractive indices of the transition layers on both faces are 1.330 very close to 1.3327 of bulk water at $0^{o}C$. On the other hand, the refractive index of ice is 1.307. It means that the transition layers on ice surfaces can be regarded as quasi-liquid layer. The temperature dependence of the thickness of the quasi-liquid layer showed the difference between basal and prism faces (Fig.19). The results are qualitatively in a good agreement with the theoretical predictions by Kuroda and Lacmann.[21] However, the measured transition temperatures $T_{I/II}$ (0001) and $T_{I/II}$ $(10\bar{1}0)$ are higher than those assigned for interpretation of habit change (Fig.18). Thus, the habit change of snow crystals should be theoretically pursued further.

ACKNOWLEDGMENT

We wish to thank Dr. Y. Furukawa, Institute of Low Temperature Science, Hokkaido University for providing the photographs of snow crystals. This work was partially supported by Grant-in-Aid for Scientific Research by the Ministry of Education, Science and Culture, Japan, to which the authors wish to express their gratitude.

REFERENCES

1) Nakaya,U., "Snow Crystals - Natural and Artificial", Harvard Univ. Press, Cambridge, 1954.
2) Kobayashi,T. and Kuroda,T., "Morphology of Crystals (Part B)", ed. Sunagawa,I., Terra Sci. Pub., Tokyo, 645-743 (1987).
3) Frank,F.C., Forward for a book "Morphology of Crystals", ed. Sunagawa,I., Terra Sci. Pub., Tokyo, 1987.
4) Nakaya,U., "Compendium of Meteorology", ed. Malone,T.F., Am. Meteorol.Soc. 207-220 (1951).
5) Aufm Kampe,H.J., Weickman,H.K. and Kelly,J.J., J.Meteorol. 8, 168 (1951).
6) Mason,B.J., Quat.J.R.Meteorol.Soc. 79, 104(1953).
7) Kobayashi,T., J.Meteorol.Soc.Jpn. 75th Ann. Vol., 38 (1957).
8) Hallett,J. and Mason,B.J., Proc.R.Soc. A247, 440(1958).
9) Kobayashi,T., Philos.Mag. 6, 1363(1961).
10) Toschev,S., "Crystal Growth: An Introduction", ed. Hartman,P., North Holland, Amsterdom, 328-341 (1973).
11) Nittman,J. and Stanley,H.E., Nature 321, 663(1986).
12) Family,F., Platt,E.D. and Vicsek,T., J.Phys. A20, L1177 (1987).
13) Xiao,R., Alexander,J.I.D. and Rosenberger,F., Phys.Rev. A38, 2447 (1988).
14) Yokoyama,E. and Kuroda,T., "Dynamics of Ordering Processes in Condensed Matter", ed. Komura,S. and Furukawa,H., Plenum Pub. Co., New York, 95-100, 1988.
15) Yokoyama,E., Thesis for a doctorate at the Faculty of Science, Hokkaido Univ., 1988.
16) Yokoyama,E. and Kuroda,T., submitted to Phys.Rev.A.
17) Stranski,I.N. and Totomanov,D., Z.Phys.Chem. A163, 399(1933).
18) Dufour,L. and Defay,R., "Thermodynamics of Clouds", Academic Press, New York, 1963.
19) Kikuchi,K. and Hogan,A.W., J.Meteorol.Soc.Jpn. 57, 180(1979).
20) Gonda,T., J.Meteorol.Soc.Jpn, 55, 142(1977).
21) Kuroda,T. and Lacmann,R., J.Cryst.Growth 56, 189(1982).
22) Lee,C.W., J.Meteorol.Soc.Jpn. 50, 171(1972).

23) Uyeda,H. and Kikuchi,K., J.Fac.Sci., Hokkaido Univ. (Ser VII) 5, 21(1976)

24) Kobayashi,T., Furukawa,Y., Kikuchi,K. and Uyeda,H., J.Cryst.Growth 32, 233(1976).

25) Kobayashi,T., Furukawa,Y., Takahashi,T. and Uyeda,H., J.Cryst. Growth 35, 262(1976).

26) Furukawa,Y. and Kobayashi,T., J.Cryst.Growth 45, 57(1978).

27) Takahashi,T., J.Cryst.Growth 59, 441(1982).

28) Takahashi,T., and Kobayashi,T., J.Cryst.Growth 64, 593(1983).

29) Burton,W.K., Cabrera,N. and Frank,F.C., Philos.Trans.R.Soc. A243, 299(1951).

30) Kuroda,T., Irisawa,T. and Ookawa,A., J.Cryst.Growth 42, 41(1977).

31) Seeger,A., Philos.Mag. 44, 1(1953).

32) Berg,W.F., Proc.R.Soc. A164, 79(1938).

33) Bunn,C.W., Discuss.Faraday Soc. 5, 132(1949).

34) Humphreys-Owen,S.P.F., Discuss.Farady Soc. 5, 144(1949).

35) Krueger,G.C. and Miller,C.W., J.Chem.Phys. 21, 2018(1953).

36) Kern,R., Bull.Soc. Fr. Mineral.Crystallogr. 76, 325(1953).

37) Goldsztaub,S. and Kern,R., Acta Crystallogr. 6, 842(1953).

38) Follenius,M., Bull.Soc.Fr.Mineral.Crystallogr. 82, 343(1959).

39) Boscher,J., Goldsztaub,S and Quivy,M., Comp.Rend(Paris) 255, 1133(1962).

40) Goldsztaub,S., Itti,R. and Mussard,F., J.Cryst.Growth 6, 842(1970).

41) Hirano,E. and Ogawa,T., J.Cryst.Growth 51, 113(1981).

42) Chernov,A.A., Sov.Phys.Crystallogr. 8, 401(1964).

43) Chernov,A.A., J.Cryst.Growth 24/25, 11(1974).

44) Cahn,J.W., "Crystal Growth", ed. Peiser,H.S., Pergamon, Oxford, 681-690 (1967).

45) Frank,F.C., J.Cryst.Growth 24/25, 3(1974).

46) Frank,F.C., Contemp.Phys. 23, 3(1982).

47) Wilcox,W.L., J.Cryst.Growth 37, 229(1977); 38, 73(1977).

48) Irisawa,T. Kuroda,T. and Ookawa,A., presented at the 7th Int. Conf. Crystal Growth, Stuttgart, September 1983.

49) Kuroda,T., Irisawa,T. and Ookawa,A., "Morphology of Crystals (Part B)", ed. Sunagawa,I., Terra Sci. Pub., Tokyo, 589-612 (1987).

50) Mullins,W.W. and Sekerka,R.F., J.Appl.Phys. 34, 323(1963).

51) Langer,J.S., Rev.Mod.Phys. 52, 1(1980).

52) O'Hara,S., J.Cryst.Growth 3/4, 555(1980).

53) Gonda,T. and Koike,T., J.Cryst.Growth 65, 36(1983).

54) Nenow,D., Stoyanova,V., and Ganadiev,N., J.Cryst.Growth 66, 36 (1984).

55) Kuroda,T., J.Cryst.Growth 65, 27(1983).

56) Gonda,T., J.Meteorol.Soc.Jpn. 54, 233(1976).

57) Gonda,T., J.Cryst.Growth 49, 173(1980).

58) Nenow,D. and Stoyanova,V., J.Cryst.Growth 41, 73(1977).

59) Keller,V.W., and Hallett,J., J.Cryst.Growth 60, 91(1982).

60) Arakawa,K. and Higuchi,K., J.Fac.Sci.,Hokkaido Univ.(Ser II) IV, 201(1952).

61) Kuroda,T., "Morphology and Growth Unit of Crystals", (Proc. of Int. Oji Semin., Zao Hot Spring, 1985) ed. Sunagawa,I., Terra Sci. Pub., Tokyo, in press.

62) Komabayashi,M., J.Rech.Atmos. 6, 307(1972).

63) Sei,T. and Gonda,T., J.Cryst.Growth 94, 697(1989).

64) Yokoyama,E. and Kuroda,T., J.Meteorol.Soc.Jpn. 66, 927(1988).

65) Kiefer,J. and Hale,B.N., J.Chem.Phys. 7, 3206(1977).

66) Kuroda,T. and Gonda,T., J.Meteorol.Soc.Jpn. 62, 563(1984).

67) Beckmann,W. and Lacmann,R., J.Cryst.Growth 58, 443(1982).

68) Furukawa,Y., Yamamoto,M. and Kuroda,T., J.Cryst.Growth 82, 665(1987).

69) Gonda,T. and Yamazaki,T., J.Cryst.Growth 45, 66(1978).

INSTABILITIES AND PATTERNS IN ELASTIC MATERIALS: GELS AND SOLID SOLUTIONS

Akira ONUKI
Research Institute for Fundamental Physics
Kyoto University, Kyoto 606, Japan

ABSTRACT

We treat problems arising from coupling between phase transitions and elasticity in two typical elastic materials, gels near the volume phase transition and solid solutions near the phase separation. Gels can be highly compressible near the transition and undergo three types of instabilities, macroscopic ones, those in the bulk region, and those on the surface, the latter two resulting in a number of pattern phenomena. Solid solutions are also known to exhibit various patterns or modulated phases after the phase separation. We present a new approach to the elastic effect in solids. It is very instructive to compare these two elastic systems.

1. INTRODUCTION

Polymer gels can swell enormously in solvent[1] and undergo discontinuous volume phase transitions in poor solvent upon a small change of some external parameter such as the temperature.[2] [3] The discontinuity of the volume change can even be diminished to zero, resulting in a critical point, by appropriately changing the degree of ionization of the network.[4] In early theories this phase transition was analyzed within the Flory-Huggins mean-field theory, where gels were assumed to be isotropic and homogeneous at a constant osmotic pressure. Then it looked to belong to the broad class of phase transitions in Ising-type systems. However, such a treatment is very insufficient to cover a rich variety of phenomena in gels such as the phase transition in affinely deformed gels, the phase separation, and inhomogeneous patterns. It also misses a fundamental difference between macroscopic and bulk instabilities. We stress that a number of unique features arise in gels due to the elastic nature of the network and their theoretical study requires a new or more generalized framework.[5] [6] This paper first aims to explain some of such features, thus reporting the present status of the field which is still waiting to be explored.

The second aim of this paper is to present recent preliminary analysis on similar problems in solid solutions. There, the phase separation of a concentration field c is greatly affected by coherently induced elastic fields.[7-9] The study of this effect has a long history in the field of metallurgy. Our new statistical-mechanical approach is to set up a Ginzburg-Landau model,[10] where the strain tensor is a subsidiary variable coupled to c. This model contrasts with the new model of gels where the strain tensor is the order parameter itself. However it will turn out to be very instructive to compare the two systems because a number of common problems arise as combined effects of thermodynamic instability and elasticity in the two systems. Examples are given by (i) elastic interactions among domains in the phase separation, (ii) effects of external stress on the phase transition, (iii) emergence of various patterns where the elasticity is crucial, and so on.

The organization of this paper is as follows. In Section 2 we will present an elastic model for statics and dynamics of gels. Based on its linearized model we will explain three types of instabilities; they are macroscopic ones, those in the bulk region, and those on the surface. Then a cusp-like pattern will be examined, which is formed by periodic folding of an unstable, corrugated surface. In Section 3 my recent calculations on the phase transition in solid solutions will be presented. This part has a very limited scope neglecting many important features because of the present level of my study in the field. However, it will predict a number of new effects especially in the presence of external stress and shed new light on previous results derived from the classical elastic theory of inclusions.

2. PHASE TRANSITION AND PATTERN FORMATION IN GELS.

2.1 Ginzburg-Landau Model for Gels

We are interested in gels swollen by a θ or poor solvent. Let ΔF be the Gibbs free energy change after the mixing of the solvent and an initially pure, unstrained polymer network.[1] When the gel is isotropic and is immersed in a pure solvent, ΔF is a thermodynamic potential dependent on the temperature T, the pressure p inside the gel, and the solvent particle number N_s inside the gel. Namely,

$$d(\Delta F) = -(\Delta S)dT + (\Delta V)dp + (\Delta\mu_s)dN_s \ , \tag{2.1}$$

where ΔS and ΔV are the changes of entropy and volume, respectively, after the mixing and $\Delta\mu_s$ is the difference of the solvent chemical potential $\mu_s(p, T, N_s)$ in the gel and that of pure solvent $\mu_s^0(p, T)$. We fix the pressure p_0 in the surrounding solvent and assume the incompressibility of the system of polymer + solvent. This means that $\Delta V = 0$, $dN_s = a^{-3}dV$, and $\Pi = p - p_0 = -a^{-3}(\Delta\mu_s)$ from $\mu_s(p, T, N_s) = \mu_s^0(p_0, T)$, where Π is the osmotic pressure and a^3 is the volume per solvent particle. Eq.(2.1) now becomes the usual form,

$$d(\Delta F) = -(\Delta S)dT - \Pi dV \ . \tag{2.2}$$

If we express ΔF in terms of the polymer volume fraction $\phi(\propto 1/V)$, we may discuss the phase transition in the usual context (see the appendix).[1-4]

However, our purposes require a free energy for generally inhomogeneous gels and we express it in terms of the deformation tensor $\partial X_i/\partial x_j^0$. Here

$\mathbf{X} = (X_1, X_2, X_3)$ are the Cartesian coordinates of the deformed gel representing the real spatial position, while $\mathbf{x}_0 = (x_1^0, x_2^0, x_3^0)$ are those in some reference relaxed state representing the original position before deformation. We assume the following form,[5) 11)]

$$\Delta F = k_B T \int_V dX[f(\phi, T) + \frac{1}{2}\nu_0 \left(\frac{\phi}{\phi_0}\right) I_1] + \Delta F_{\rm inh} , \tag{2.3}$$

where the integral is within the volume V of the deformed gel, f is a free energy density dependent on ϕ and T, and ϕ is related to the determinant of the deformation tensor by

$$\frac{\phi_0}{\phi} = \det\left(\frac{\partial X_i}{\partial x_j^0}\right) , \tag{2.4}$$

ϕ_0 being the volume fraction in the relaxed state. The ν_0 is the chain number density in the relaxed state. The quantity I_1 is the so-called first invariant in the nonlinear elasticity theory:[12) 13)]

$$I_1 = \sum_{i,j}\left(\frac{\partial}{\partial x_j^0} X_i\right)^2 . \tag{2.5}$$

The last term $\Delta F_{\rm inh}$ is the gradient free energy indispensable in describing short wavelength fluctuations and interfaces after phase separation.[14)] We neglect $\Delta F_{\rm inh}$ in this paper; then, the stress tensor Π_{ij} due to the network deformation is given by[11) 13)]

$$\frac{\Pi_{ij}}{k_B T} = (\phi f' - f)\delta_{ij} - \nu_0\left(\frac{\phi}{\phi_0}\right)\sum_l \left(\frac{\partial X_i}{\partial x_l^0}\right)\left(\frac{\partial X_j}{\partial x_l^0}\right) , \tag{2.6}$$

where $f' = \partial f/\partial\phi$. The expression for Π_{ij} in the presence of $\Delta F_{\rm inh}$ is given in the letter paper of Ref.5.

If we assume that the gel is isotropically expanded as $X_i = (\phi_0/\phi)^{1/3} x_i^0$, Flory's results are recovered:[1)]

$$\Delta F = Vf + \frac{3}{2}\nu_0 V_0 (\phi_0/\phi)^{2/3} , \tag{2.7}$$

where V_0 is the volume of the relaxed gel and

$$V = V_0\phi_0/\phi \ . \tag{2.8}$$

It follows that $\nu_0 V_0$ is the total number of effective chains and $\phi_0 V_0$ is the total monomer number of the network. From (2.2) the osmotic pressure is given by

$$\Pi = -\partial(\Delta F)/\partial V = k_B T[\phi f' - f - \nu_0(\phi/\phi_0)^{1/3}] \ , \tag{2.9}$$

where use has been made of (2.8). Note that Π coincides with the diagonal component of Π_{ij}, (2.6), in the isotropic case. The bulk modulus K is written as

$$K = \phi(\frac{\partial \Pi}{\partial \phi})_T = k_B T[\phi^2 f'' - \frac{1}{3}\nu_0(\phi/\phi_0)^{1/3}] \ , \tag{2.10}$$

where $f'' = \partial^2 f/\partial\phi^2$. The elastic term ($\propto I_1$) in ΔF gives rise to the negative contribution to K (while a positive contribution to $K + \frac{4}{3}\mu$ as will be shown in (2.33) below).

Let us consider small deformations around a homogeneous, isotropic state. That is, we set

$$X_i = \lambda x_i^0 + u_i \ , \tag{2.11}$$

where λ is a constant expansion ratio with respect to the relaxed state and u_i are small. We expand Π_{ij} in powers of u_i to obtain its deviation in the usual form,[15]

$$\delta\Pi_{ij} = -(K - \frac{2}{3}\mu)\sum_i(\frac{\partial}{\partial x_i^s}u_i) - \mu[\frac{\partial}{\partial x_i^s}u_j + \frac{\partial}{\partial x_j^s}u_i] \ , \tag{2.12}$$

where $x_i^s = \lambda x_i^0$ are the Cartesian coordinates fixed to the isotropic state under consideration. The bulk modulus K is equal to (2.10) if we set $\lambda = (\phi_0/\phi)^{1/3}$ and

$$\mu/k_B T = \nu_0\lambda^{-1} = \nu_0(\phi/\phi_0)^{1/3} \ . \tag{2.13}$$

The bilinear part of the free energy ΔF assumes the usual form in terms of K and μ:

$$\Delta F = \int d\mathbf{r}[\frac{1}{2}K(\nabla\cdot\mathbf{u})^2 + \frac{\mu}{4}\sum_{i,j}(\frac{\partial}{\partial x_i}u_j + \frac{\partial}{\partial x_j}u_i - \frac{2}{3}\delta_{ij}\nabla\cdot\mathbf{u})^2] + \cdots \ , \qquad (2,14)$$

where x_i^s are rewritten as x_i to avoid the heavy notation.

The dynamics is as follows:[16)5)] The relative motion between the network and the velocity field $\mathbf{v}_s$ gives rise to a frictional force which is balanced with the force due to the network deformation,

$$\gamma(\frac{\partial}{\partial t}X_i - v_{si}) = -\sum_j \frac{\partial}{\partial x_j}\Pi_{ij} = -(\frac{\phi}{\phi_0})\frac{\delta}{\delta X_i}(\Delta F) . \tag{2.15}$$

Here γ is the friction coefficient (equal to f in Tanaka's notation)[16)] between the network and solvent, and $\gamma \propto \phi^2$ from the scaling theory for θ solvent.[17)] The $\mathbf{v}_s$ on the left hand side is taken to be the value at the position $\mathbf{X}$. The relation $\nabla \cdot \Pi = (\phi/\phi_0)\delta(\Delta F)/\delta\mathbf{X}$ is derivable from the definition of Π_{ij}, (2.6).[11)] The velocity field obeys

$$\rho_0\frac{\partial}{\partial t}v_{si} = -\frac{\partial p}{\partial x_i} - \sum \frac{\partial}{\partial x_j}\Pi_{ij} + \eta_o\nabla^2 v_{si} , \tag{2.16}$$

where ρ_0 is the mass density of the solvent and is nearly equal to the constant, total mass density, the role of the pressure p is to ensure the incompressibility condition,

$$\nabla \cdot \mathbf{v}_s = 0 , \tag{2.17}$$

and η_0 is the solvent viscosity. The linearized versions of (2.15) and (2.16) around an isotropic state become

$$\frac{\partial}{\partial t}\mathbf{u} = \mathbf{v}_s + \frac{1}{\gamma}[(K + \frac{1}{3}\mu)\nabla(\nabla \cdot \mathbf{u}) + \mu\nabla^2\mathbf{u}] , \tag{2.18}$$

$$\rho_0\frac{\partial}{\partial t}\mathbf{v}_s = \mu[\nabla^2\mathbf{u} - \nabla(\nabla \cdot \mathbf{u})] + \eta_0\mathbf{v}_s . \tag{2.19}$$

In the linear regime the longitudinal part of $\mathbf{u}$ is decoupled from $\mathbf{v}_s$, while the transverse part couples with $\mathbf{v}_s$ giving rise to transverse sound modes[18)] at long wavelengths, $k \lesssim 2(\rho_0\mu)^{1/2}/\eta_0$.

The critical dynamics near the spinodal point is also of great interest. Tanaka[19] performed dynamic light scattering from isotropic gles near the spinodal point $K + \frac{4}{3}\mu = 0$ and claimed that the diffusion constant deduced from the decay rate could be fitted to the Kawasaki formula $D = k_B T/6\pi\eta_0\xi$ as in critical fluids. Here $\xi \propto |\Delta T|^{-1/2}$ is the correlation length for small $\Delta T = T - T_s$, T_s being the spinodal temperature. However, his claim holds only in some limited cases if the shear modulus μ remains finite at the spinodal point. In gels the network structure does prohibit translational motion of the fluctuations over some characteristic distance. This makes the dynamics somewhat complicated in gels with small μ. A preliminary calculation was given in Ref.20.

2.2 Macroscopic and Bulk Instabilities

Gels at a constant osmotic pressure first become unstable macroscopically while the finite shear modulus still suppress fluctuations smaller than the system size. As a result the spinodal line, on which light scattering amplitudes diverge, is separated from the critical point and is pushed into an unstable region. The same situation can also be found in solid solutions at the valence instability such as $Sm_{1-x}Y_xS$ or $Ce_{1-x}Th_x$.[21] This is a fundamental aspect of the transition unnoticed so far.

Due to the coupling between the transverse part of the displacement and the velocity field, deformations without volume changes can take place relatively quickly. That is, if the network and the solvent move together, the mutual friction becomes small[22] (However, this point was overlooked in Refs.11 and 23.) Because the effect of this coupling on kinetics has not yet been fully analyzed, let us examine a simple case of purely longitudinal motion;[24][23] a spherical gel is immersed in solvent at zero osmotic pressure and a small deviation of the displacement is assumed to be of the form,

$$u_i(\mathbf{r}, t) = u(r, t)x_i/r \ . \tag{2.20}$$

Then we find rot $\mathbf{u} = 0$ and hence $\mathbf{v}_s = 0$ from (2.19). From (2.18) we have the following eigenvalue problem,[23]

$$\mu\nabla^2 u_i + (K + \frac{1}{3}\mu)\frac{\partial}{\partial x_i}(\nabla \cdot \mathbf{u}) = -\gamma\Omega\mathbf{u} \ , \tag{2.21}$$

Ω being the relaxation rate. At the boundary $r = R$ the normal unit vector is x_i/r and

$$\sum_j (\delta\Pi_{ij})x_j/r = 0 \quad \text{at} \quad r = R\ , \tag{2.22}$$

which is rewritten as

$$2(1-\epsilon)u(R) = (1+\epsilon)Ru'(R)\ . \tag{2.23}$$

Hereafter ϵ is defined by

$$\epsilon = K/\mu + 1/3\ . \tag{2.24}$$

After solving (2.21) we obtain

$$u(r) = r^{-1/2}J_{3/2}(\zeta r/R) = \text{const.}(\sin z/z^2 - \cos z/z)\ , \tag{2.25}$$

where ζ is a constant to be determined below, $z = \zeta r/R$, $J_\nu(z)$ is the Bessel function of order ν, and (2.23) leads to

$$G(\zeta) \equiv \frac{1}{\zeta^2} - \frac{1}{\zeta \tan\zeta} = \frac{1}{4}(1+\epsilon)\ . \tag{2.26}$$

The relaxation rate Ω is related to ζ by

$$\Omega = \mu(1+\epsilon)\zeta^2/(\gamma R^2)\ . \tag{2.27}$$

The local volume change is known from $\nabla\cdot\mathbf{u}$, which is proportional to $\sin z/z$ with $z = \zeta r/R$.

It is easy to find a series, $\zeta_0^2 < \zeta_1^2 < \zeta_2^2 < \cdots$, satisfying (2.26). In particular, the first mode becomes marginal as $K \to 0$. In fact, from $G \cong \frac{1}{3} + \frac{1}{45}\zeta^2 + \cdots$ for $|\zeta| \ll 1$, ζ_0 is determined by

$$\zeta_0^2 \cong (45/4)K/\mu \quad \text{for} \quad |K/\mu| \ll 1\ . \tag{2.28}$$

For $K < 0$, ζ_0 is purely imaginary. In Fig.1 we plot the corresponding series of the eigenvalues as functions of $K/\mu = \epsilon - 1/3$,

$$\Omega_n = (\pi^2\mu/\gamma R^2)\Omega_n^\star \tag{2.29}$$

where $\Omega_n^\star = (\epsilon+1)\zeta_n^2/\pi^2$. The first mode $n = 0$ is unstable in the region $-1 < \epsilon < \frac{1}{3}$ and is nonexistent for $\epsilon < -1$. It limiting behavior is

$$\Omega_0^\star \cong \begin{cases} K/\mu + 4/3 - 8/\pi^2 & \text{for } K/\mu \gtrsim 1 \\ (15/\pi^2)K/\mu & \text{for } |K/\mu| \lesssim 1 \\ -(16/\pi^2)(K/\mu + 4/3)^{-1} & \text{as } K/\mu \to -4/3\,. \end{cases} \tag{2.30}$$

The growth rate of this mode increases as $K/\mu \to -4/3$. The other eigenvalues become negative for $K/\mu < -4/3$, implying onset of spinodal decomposition in the bulk region.

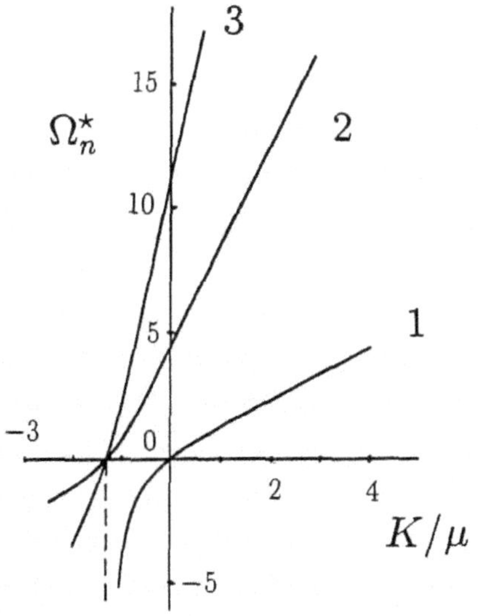

Fig.1 The dimensionless decay or growth rates $\Omega_n^\star$ defined by (2.29) for $n = 1, 2$, and 3.

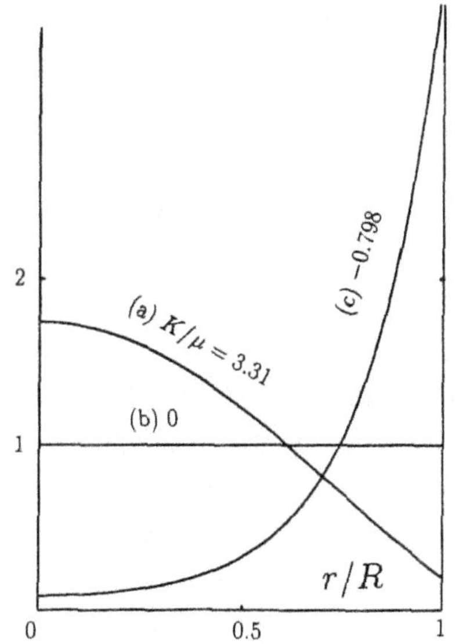

Fig. 2 The profiles of the first mode $C(\epsilon)\sin(\zeta_0 r/R)/(\zeta_0 r/R)$ defined by (2.31): (a) $K/\mu = 3.31$, $\zeta_0/\pi = 0.894$, (b) $K/\mu = \zeta_0 = 0$, and (c) $K/\mu = -0.798$, $\zeta_0/\pi = 2i$.

A small change of an external parameter will slightly break the zero osmotic pressure condition, leading to a change of the radius. The subsequent time-development is dominated by the first linear mode $n = 0$ after a short transient time if the change is not of first-order in the linearly stable region $K > 0$ or in an early stage in the unstable region $-\frac{4}{3}\mu < K < 0$. Retaining only the first mode we may express $\nabla \cdot \mathbf{u}$ as

$$\nabla \cdot \mathbf{u}(\mathbf{r}, t) \cong 3a_0 \exp[-\Omega_0 t] C(\epsilon) \sin(\zeta_0 r/R)/(\zeta_0 r/R) \ , \tag{2.31}$$

where we have assumed $\mathbf{u}(\mathbf{r}, 0) = a_0 \mathbf{r}$ as the initial condition and the coefficient $C(\epsilon)$ increases from 0 to 2 as $\epsilon = K/\mu + 1/3$ increases from -1 to ∞. In particular $C(1/3) = 1$. The quantity $\nabla \cdot \mathbf{u}/[3a_0 \exp(-\Omega_0 t)] = C(\epsilon) \sin(\zeta_0 r/R)/(\zeta_0 r/R)$ is plotted in Fig.2. For $\epsilon > 1/3$ the boundary region relaxes more rapidly than the center region. For $-1 < \epsilon < 1/3$ the growth of the variation is largest at the boundary. In both the cases the temporal change is slowest at the center as a natural result. At $\epsilon = 1/3$ (or $K = 0$) the mode $n = 0$ represents homogeneous expansion or shrinkage.

Thus the gel first undergoes a macroscopic instability, as K is lowered to negative, tending to a new swollen or shrunken state. In the region $-4/3 < K/\mu < 0$ small scale fluctuations decay thermally and there will not emerge small-scale inhomogeneities such as interfaces separating two phases. To check the last point we may examine (i) a spherical droplet-solution to $\nabla \cdot \Pi = (\phi/\phi_0) \delta F / \delta \mathbf{X} = 0$ and (ii) a thin uniaxially deformed layer at the boundary separated from the interior region by a sharp interface. The former is expected to emerge in the bulk region and the latter near the boundary after quenching. In both the two cases, due to the elastic effect, a sharp interface can appear only when $|K/\mu|$ is greater than some critical values of order 1. Thus the two-phase coexistence is strongly affected by the presence of the shear-deformation free energy for relatively shallow quenching. Details will be reported in another paper. We should mention an observation of coexistence of two isotropic phases in a rod with a thick, diffuse transition region.[25]

The scattering intensity diverges only in the limit $K/\mu \to -4/3$ and the point $K/\mu = -4/3$ is the cloud point or the spinodal point. In practice we may easily

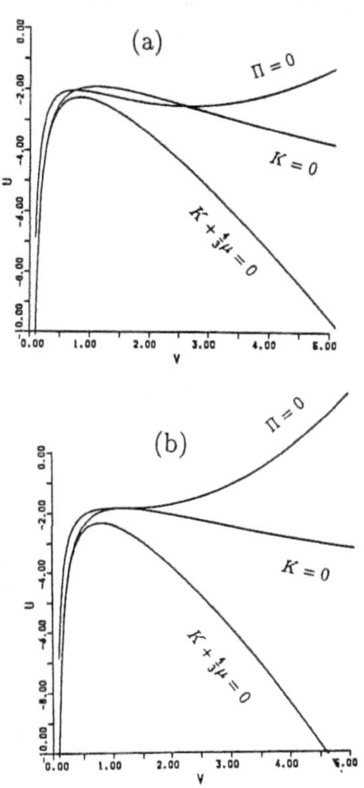

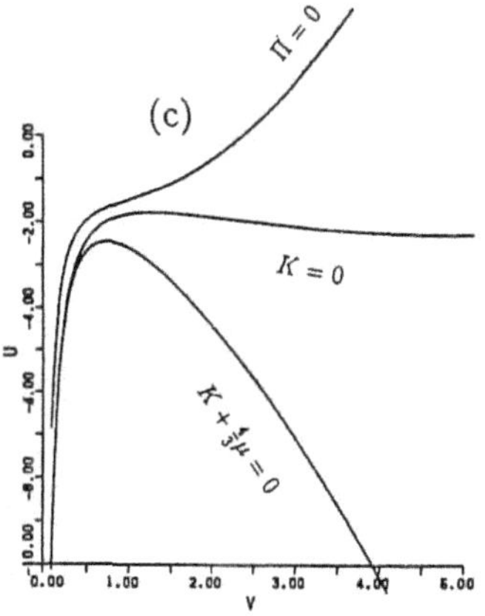

Fig.3 The three curves $\Pi = 0, K = 0$, and $K + \frac{4}{3}\mu = 0$ from the model in the appendix. The abscissa denotes $V = v/\gamma^{1/2} = 1/(\gamma^{1/2}\phi)$ and the ordinates $U = (1 - 2\chi)/\gamma^{1/2}$, where γ is defined in (A.4). In (a) the volume changes discontinuously along the curve $\Pi = 0$. In (b) the critical point is realized at the intersection point of $\Pi = 0$ and $K = 0$. In (c) there is no first-order change.

approach to the point $K + \frac{4}{3}\mu = 0$ in the interior region of large gels because of the slow growth rate ($\propto 1/R^2$) for $K < 0$ (although the boundary region must be uniaxially deformed in such experiments). Notice that, in calculating the scattering amplitude $I_k =< |\phi_{\mathbf{k}}|^2 >$, the boundary conditions are irrelevant and the system size may be pushed to infinity because the scattering wave number k is much larger than the inverse system size $1/R$. From (2.14) it readily follows

$$I_k \propto 1/(K + \frac{4}{3}\mu) \tag{2.32}$$

where $1/R \ll k \ll 1/\xi$, ξ being than the thermal correlation length. Obviously, we may expect small-scale fluctuation enhancement or spinodal decomposition only when $K/\mu < -4/3$. However, a number of papers have claimed that spinodal decomposition occurs for $K < 0$.[14)26)27)] In Fig.3 we plot the three curves of

$\Pi = 0$, $K = 0$, and $K + \frac{4}{3}\mu = 0$ on the basis of the simple Flory-Huggins model (see the appendix). Here $\Pi > 0$, $K > 0$, and $K + \frac{4}{3}\mu > 0$ above the corresponding curves. They are obtained from (2.9), (2.10), and

$$K + \frac{4}{3}\mu = k_B T[\phi^2 f'' + \nu_0(\phi/\phi_0)^{1/3}] , \tag{2.33}$$

where the elastic contribution is positive in contrast to the case of K. Note that the elastic contribution is increasingly important at small ϕ. As a result the condition $K = 0$ may be realized even for good solvent in some cases,[14] whereas the condition $K + \frac{4}{3}\mu = 0$ is satisfied only for poor solvent. See Eqs.(A.7)~(A.9).

In the above arguments we have assumed that the shear modulus μ is nearly constant in the region $|K| \lesssim \mu$. As its mean-field expression (2.13) suggests, its critical singularity is at most very weak from our model. In fact Hirotsu has recently found that μ remains finite at the critical point (private communication). Note that the width of the temperature region $|K|/\mu < 1$ should have been of order 1 deg in gels with small μ in early Tanaka's work.[4) 16) 19)] However, the difference of the instabilities at the two points, $K = 0$ and $K + \frac{4}{3}\mu = 0$, has not yet been reported in the literature.

2.3 Uniaxial Gels

When gels are uniaxially deformed,[5) 28)] we encounter three types of instabilities depending on the boundary conditions and the values of parameters involved. They are (i) macroscopic instabilities, (ii) bulk instabilities with small-scale domains, and (iii) surface instabilities, where corrugations on the surface grow leading to fine patterns on the surface.

To illustrate these instabilities let us start with a homogeneous, uniaxial state. The displaced spatial position (X, Y, Z) is composed of the average and a displacement $\mathbf{u}$ defined by

$$X = \alpha_{\|}x^0 + u_x , Y = \alpha_{\perp}y^0 + u_y , Z = \alpha_{\perp}z^0 + u_z . \tag{2.34}$$

The elongation ratios $\alpha_{\|}$ and $\alpha_{\perp}$ are generally different and we define

$$\hat{\delta} = \alpha_{\perp}/\alpha_{\|} . \tag{2.35}$$

For the mathematical convenience we hereafter write

$$x = \alpha_{||} x^0 \ , y = \alpha_{\perp} y^0 \ , z = \alpha_{\perp} z^0 \ , \tag{2.36}$$

or $x_i = \alpha_i x_i^0$ with $i = x, y, z$. The first invariant I_1, (2.5), is of the following form,

$$I_1 = \sum_i \alpha_i^2 + 2 \sum_i \alpha_i^2 (\frac{\partial u_i}{\partial x_i}) + \sum_{i,j} \alpha_i^2 (\frac{\partial u_j}{\partial x_i})^2 \ . \tag{2.37}$$

Picking up terms bilinear in the fluctuations in ΔF, (2.3), we find the density correlation function in the Gaussian approximation,

$$< |\delta \phi_{\mathbf{q}}|^2 > \cong \phi^2 / [\phi^2 f'' + \nu_0 \frac{\phi}{\phi_0} (\alpha_{||}^2 \frac{q_x^2}{q^2} + \alpha_{\perp}^2 \frac{q_y^2 + q_z^2}{q^2}) + C_{||} q_x^2 + C_{\perp} (q_y^2 + q_z^2)] \ , \tag{2.38}$$

where $\phi = \phi_0 / (\alpha_{||} \alpha_{\perp}^2)$ and the scattering vector $\mathbf{q}$ must be much larger than the inverse of the system dimension. The last term of (2.37) gives rise to the angle-dependent terms ($\propto \alpha_i^2 q_i^2 / q^2$) in the denomination of (2.38). The last two terms ($\propto q_i^2$) arise from the gradient term ΔF_{inh} in (2.3). Recently a light scattering experiment from a uniaxial gel was reported qualitatively in accord with (2.38).[29] The solvent quality there was good, however, and it is then of great interest to perform such experiments in poor solvent near the spinodal curve where the anisotropy is more enhanced and we expect a butterfly-like scattering pattern. See Fig.4.

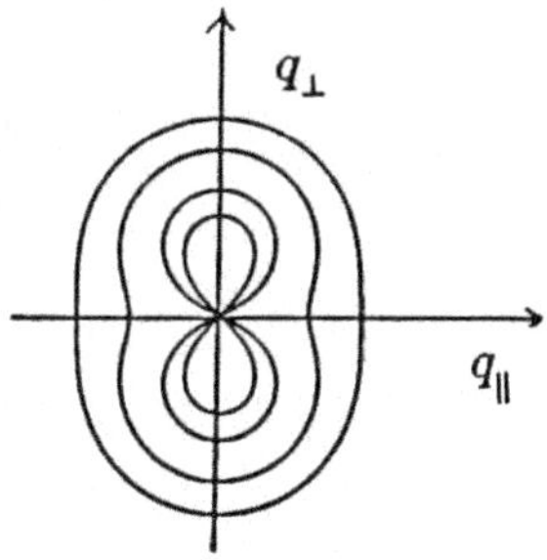

Fig.4 The scattering intensity (2.38) close to the instability point for $\hat{\delta} < 1$. Isointensity curves are written.

The dynamic equations (2.15) and (2.16) will be linearized around the uniaxial state:

$$\frac{\partial}{\partial t}\mathbf{u} = \mathbf{v}_s + D[\epsilon \frac{\partial}{\partial \mathbf{r}}(\nabla \cdot \mathbf{u}) + \{\frac{\partial^2}{\partial x^2} + \hat{\delta}^2(\frac{\partial^2}{\partial y^2} + \frac{\partial^2}{\partial z^2})\}\mathbf{u}] \, , \tag{2.39}$$

$$\rho_0 \frac{\partial}{\partial t}\mathbf{v}_s = -\nabla p + \mu[\frac{\partial^2}{\partial x^2} + \hat{\delta}^2(\frac{\partial^2}{\partial y^2} + \frac{\partial^2}{\partial z^2})]\mathbf{u} + \eta_0 \nabla^2 \mathbf{v}_s \, , \tag{2.40}$$

where p ensures $\nabla \cdot \mathbf{v}_s = 0$. The ϵ is redefined as

$$\epsilon = \phi_0^2 f''/(\nu_0 \alpha_{\parallel}^3 \alpha_{\perp}^2) \tag{2.41}$$

and

$$D = k_B T \nu_0/(\gamma \hat{\delta} \alpha_{\perp}) \, , \tag{2.42}$$

f'' being the value at $\phi = \phi_0/(\alpha_{\parallel}\alpha_{\perp}^2)$. The parameter ϵ represents the degree of incompressibility and its concrete expression in terms of the Flory-Huggins parameters is shown in Ref.11. The density deviation $\delta\phi = -\phi(\nabla \cdot \mathbf{u})$ is decoupled from $\mathbf{v}_s$ in the linear regime as

$$\frac{\partial}{\partial t}\delta\phi = D[(1+\epsilon)\frac{\partial^2}{\partial x^2} + (\hat{\delta}^2 + \epsilon)(\frac{\partial^2}{\partial y^2} + \frac{\partial^2}{\partial z^2})]\delta\phi \, . \tag{2.43}$$

With lowering of ϵ a bulk instability occurs with a spinodal point given by

$$\epsilon = -1 \quad \text{for} \quad \hat{\delta} > 1 \, , \tag{2.44a}$$

$$\epsilon = -\hat{\delta}^2 \quad \text{for} \quad \hat{\delta} < 1 \, . \tag{2.44b}$$

The resultant spinodal decomposition is triggered by fluctuations with wave vectors in particular directions, which are those varying along the x axis for the uniaxially compressed case $\hat{\delta} > 1$, and those varying in the yz plane for the uniaxially stretched case $\hat{\delta} < 1$. Thus we expect emergence of lamellar domains for $\hat{\delta} > 1$ and cylindrical domains for $\hat{\delta} < 1$.[5)]

Recently a numerical simulation was performed on a quenched, two-dimensional network model,[30)] which shows intriguing morphology of domains in late stages of the spinodal decomposition. However, there is still no theory on the late-stage spinodal decomposition in gels.[14)] Light scattering experiments should be very informative on such domain structures.[31)] To get unambiguous data, however, it seems to be necessary to clamp all the boundaries of gels to avoid the surface instability.[31)]

2.4 Pattern Formation in a Clamped Gel

The macroscopic and surface instabilities strongly depend on the boundary conditions. As a striking observation Tanaka found that transient patterns consisting of numerous line segments of cusps emerge on the surface in the course of swelling when the volume change is large.[3)] Furthermore, he showed that the same patterns can be formed permanently in equilibrium gels whose lower surfaces are fixed to a substrate and whose upper surfaces are in contact with a solvent. In these cases, the gels are uniaxially stretched in the direction normal to the surface and compressed in the lateral directions. Note that in shrinking gels another type of patterns was observed.[3)]

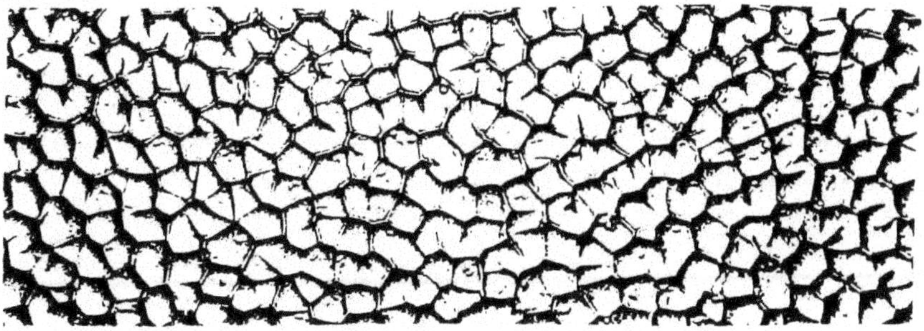

Fig.5 A picture of the cusp-like pattern on a clamped gel taken by Tanaka's group in 1986. Other interesting pictures can be found in Ref.3.

(a) Linear Stability Analysis

Let us hence analyze the linear stability of a gel slab clamped at $x = 0$ and in contact with solvent at $x = L_{\|}$.[32)11)] For simplicity we consider only two dimensional deformations in which $u_z = 0$ and there is no z-dependence. Then the bilinear part of the free energy is written as

$$\Delta F/k_BT = \nu_0(\phi/\phi_0)\alpha_{\|}^2 L_z \int dx \int dy[\frac{1}{2}\epsilon(\frac{\partial}{\partial x}u_x + \frac{\partial}{\partial y}u_y)^2 + \frac{1}{2}(\frac{\partial}{\partial x}\mathbf{u})^2 + \frac{1}{2}\hat{\delta}^2(\frac{\partial}{\partial y}\mathbf{u})^2] , \tag{2.45}$$

where L_z is the length of the gel in the z axis. As the boundary conditions we require

$$u_x = u_y = 0 \quad \text{at} \quad x = 0 \tag{2.46}$$

$$(1+\epsilon)\frac{\partial}{\partial x}u_x - (1-\epsilon)\frac{\partial}{\partial y}u_y = 0 \, , \, \frac{\partial}{\partial x}u_y + \frac{\partial}{\partial y}u_x = 0 \quad \text{at} \quad x = L_{\|} \, . \tag{2.47}$$

The boundary conditions at the upper surface (2.47) have been derived from $\sum_j \Pi_{ij} n_j = 0$, n_j being the normal unit vector to the surface.

We are led to the following eigenvalue problem,

$$\epsilon\frac{\partial}{\partial \mathbf{r}}(\nabla \cdot \mathbf{u}) + (\frac{\partial^2}{\partial x^2} + \hat{\delta}^2\frac{\partial^2}{\partial y^2})\mathbf{u} = -\lambda \mathbf{u} \tag{2.48}$$

with the boundary conditions (2.46) and (2.47). We can generally prove that the eigenvalue λ is real and there is an orthogonal complete set of eigenvectors $\{\mathbf{u}_p\}$ with eigenvalue λ_p where p is an appropriate subscript. If we set $\mathbf{u} = \sum_p A_p \mathbf{u}_p$, the free energy ΔF, (2.45), can be expressed as

$$\Delta F = \sum_p \lambda_p C_p A_p^2 \, , \tag{2.49}$$

where the $Cp = \int\int dxdy|\mathbf{u}_p|^2$ are the normalization coefficients. If the smallest eigenvalue λ tends to zero, we have an instability. If the velocity field $\mathbf{v}_s$ is neglected, (2.39) shows that the relaxation rate is expressed by $\Omega = D\lambda$. However, the eigenvalue problem in the presence of $\mathbf{v}_s$ has not yet been solved. Here we can only claim that, if some λ becomes negative, some corresponding Ω also becomes negative.

The criterion of instability turns out to be very simple:[11] A deformation localized near the surface becomes unstable if

$$\hat{\delta} < \hat{\delta}_c \,, \tag{2.50}$$

where $\hat{\delta}_c$ is a function of ϵ and is plotted in Fig.6. When $\hat{\delta} - \hat{\delta}_c$ is small, λ can be written as

$$\lambda \simeq -(\hat{\delta}_c^2 - \hat{\delta}^2)\{1 - \exp[-2\hat{\delta}(k - k_c)L_{||}]\}k^2 \,, \tag{2.51}$$

where k is the wave number in the y axis and

$$k_c \simeq \frac{1}{2\hat{\delta}L_{||}} \ln[A_c/(\hat{\delta}_c^2 - \hat{\delta}^2)] \propto \frac{1}{L_{||}} \,, \tag{2.52}$$

A_c being a constant. Thus λ is negative for $\hat{\delta} < \hat{\delta}_c$ and $k > k_c$. If k considerably exceeds k_c, we have $\lambda \cong -(\hat{\delta}_c^2 - \hat{\delta}^2)k^2$. The lower bound k_c arises from the fact that the lower surface of the gel is fixed. On the other hand, we should recall that we have neglected the gradient term ΔF_{inh} in the free energy.[14] It will provide the upper bound of the instability and λ should be modified as

$$\lambda \simeq -(\hat{\delta}_c^2 - \hat{\delta}^2)k^2 + bk^4 \,. \tag{2.53}$$

Here $k \gg k_c$ and $b \sim Na^2 \sim R_G^2$, where $N = \phi_0/(\nu_0 a^3)$ is the effective polymerization index and R_G is the gyration radius of one effective chain. This suggests that very fine surface perturbations should grow as a spinodal decomposition process, with the initial peak wave number given by

$$k_m = [(\hat{\delta}_c^2 - \hat{\delta}^2)/2b]^{1/2} \sim (\hat{\delta}_c^2 - \hat{\delta}^2)^{1/2}/R_G \,. \tag{2.54}$$

As the fluctuations on the surface grow, some nonlinear mechanism should come into play serving to suppress the growth and thus leading to some final, stationary pattern. We expect the following: If the gel is sufficiently incompressible, nonlinear terms in ΔF can stop the growth at a relatively small saturation level. (ii) However, if the gel is sufficiently compressible (or for small ϵ), such nonlinear terms are not effective enough, allowing the corrugated surfaces to fold periodically.

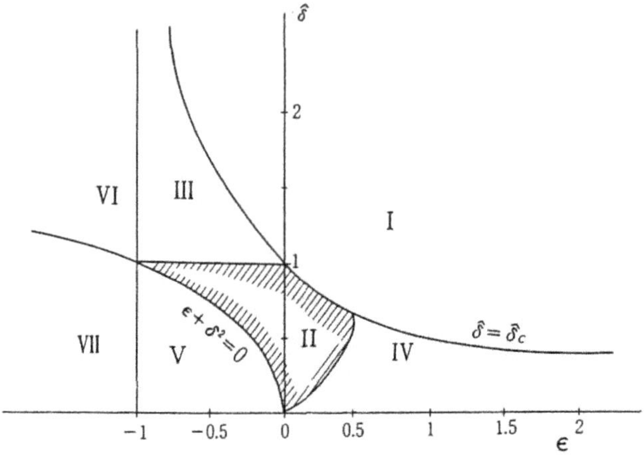

Fig.6 The phase diagram from the linear theory.[11] The system is stable in the region I, while in II~IV it is unstable against surface perturbations but still stable with respect to bulk instabilities. In II we expect folding of the surface. In III the following is unstable in the sense that $\Pi_{yy} < 0$ at the contact for $\hat{\delta} > 1$. In IV there is no meaningful solutions in the linear scheme representing periodic cusps. In V we expect anisotropic spinodal decomposition in the bulk with cylindrical domains, while in VI that with lamellar domains. In VII the gel is unstable against plane wave perturbations of any directions.

Finally we mention analysis of another situation:[11] When both the upper and lower surfaces are freely exposed to solvent, we find a very different instability. That is, if $\epsilon > 0$, the gel plate bends macroscopically (without small-scale structures) for $\hat{\delta} < 1$. This case corresponds to the usual elastic instability of plates or rods subject to compressional forces (Euler's problem).[15] However, a surface instability occurs for $\epsilon < 0$.

(b) Surface Folding

The present author has calculated inhomogeneous solutions to $\nabla \cdot \Pi = 0$ which represent periodically folded patterns on the surface.[11] We have to linearize the equations around the homogeneous uniaxial state to get analytic results, but we fully take into account of a drastic change of the boundary conditions at the folded part where surface parts press each other. Nevertheless they lower the free energy below its value for the homogeneous state. Subsequent numerical simulations on simple elastic networks by Hwa and Karder[33] and by Sekimoto and Kawasaki[34] have confirmed that the surface tends to touch and fold as surface corrugations grow.

Interestingly, the mathematics for the pattern has turned out to be very similar to that for fractures in elastic bodies. In both the cases, fundamental elements of the phenomena are surface singularities, folded parts or cracks, accompanied by large elastic strains. To give some idea of the mathematics involved, let us take the simplest case: We assume a single folded part, $-d < y < d$, at the upper surface $x = L_{\parallel}$ for $\epsilon = 0$. Here we have $Y = y + u_y =$ const. for $|y| < d$ and $x = L_{\parallel}$ because the two surface parts, $-d < y < 0$, and $0 < y < d$, are in contact with each other. Further imposing the zero shear stress condition $\Pi_{xy} = 0$ at the folded part, we have

$$(\partial u_x/\partial x)_b = (\partial u_y/\partial y)_b = -1 \tag{2.55}$$

for $|y| < d$ and $x = L_{\parallel}$. The notation $(\cdots)_b$ denotes values at $x = L_{\parallel}$. The boundary condition (2.47) is supposed for $|y| > d$ and $x = L_{\parallel}$. Then the strain $U(y) \equiv (\partial u_x/\partial y + \partial u_y/\partial x)_b$ satisfies on the upper surface

$$(\partial u_y/\partial y)_b = -\frac{\hat{\delta}}{1-\hat{\delta}^2}(\hat{H}U) , \tag{2.56}$$

where $\hat{H}$ is the Hilbert transformation defined by $\hat{H}f = \pi^{-1}\int dy' f(y')/(y'-y)$ (the Cauthy principal value being taken). Use of (2.55) leads to

$$U(y) = [(1-\hat{\delta}^2)/\hat{\delta}]y\theta(d-|y|)/\sqrt{d^2-y^2} , \tag{2.57}$$

$$(\partial u_y/\partial y)_b = -1 + |y|\theta(|y|-d)/\sqrt{y^2-d^2} , \tag{2.58}$$

$\theta(p)$ being the step function. It is straightforward to calculate the behavior of u_x and u_y in the interior region.

Generally we may analytically calculate periodic, roll patterns in the region II of Fig.6 in the linear scheme using a conformal transformation.[11)] The solutions exist for arbitrary period $L_{\perp}$ in the y axis and for not too large depth d of the folding. In Fig.7 we show theoretical cusp patterns for the thickness $L_{\parallel} = 1$ and the period $L_{\perp} = 0.8$. They show the following: (i) The gel is strongly compressed just below

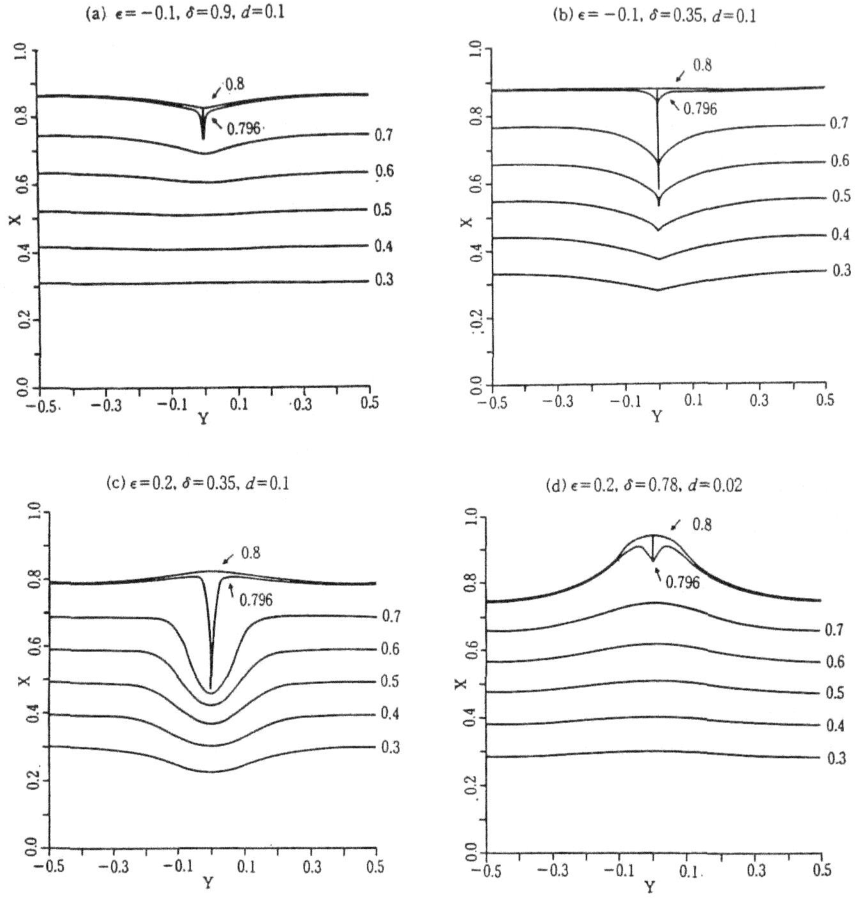

Fig.7 Theoretical cusp profiles in one period for $L_\perp = 1$ and $L_{||} = 0.8$. Here, $\epsilon = -0.1, \delta = 0.9$, and $d = 0.1$ in (a); $\epsilon = -0.1, \delta = 0.35$, and $d = 0.1$ in (b); $\epsilon = 0.2, \delta = 0.35$, and $d = 0.1$ in (c); $\epsilon = 0.2, \delta = 0,78$, and $d = 0.02$ in (d). The numbers attached to each curve are the initial height x in the homogeneous state before the cusp formation. In (d) δ is close to the critical value $\delta_c = 0.82$ at $\epsilon = 0.2$ and deformations are large even for $d = 0.02$.

the apex of the cusp. The surface $x = 0.8$ and the curve of $x = 0.796$ are noticeably separated at $y \cong \pm d$ indicating enormous and unphysical expansion here. In these two regions the linear scheme should be inadequate. (ii) Just below the critical line $\hat{\delta} = \hat{\delta}_c$ the cusp solution is meaningful only for small depth d of the folding,

$$|\epsilon| d / L_\perp \lesssim \hat{\delta}_c^2 - \hat{\delta}^2 \quad \text{as} \quad \hat{\delta} \to \hat{\delta}_c \, . \tag{2.59}$$

(iii) For $\epsilon \gtrsim 1$ (in the region IV of Fig.6) the solutions exhibit unphysical overlapping and no meaningful solutions exist in our linear approximation.

Final comments in this subsection are as follows: (i) Our solutions to the cusp patterns give extremum to the free energy as a functional of the displacement. However, they do not give the minimum. Hence there should be some coarsening of the pattern and gradual increase of d, until $L_\perp$ and d minimize the free energy. This process has indeed been observed, and a coarsening law, $L_\perp(t) \propto t^\alpha$, has been found.[3)31)] To calculate the final pattern we should take into account of the nonlinear elastic effect and the effect of finite thickness. (ii) In the experiment[3)] hexagonal patterns were observed, although we have analyzed only roll patterns. (iii) The cusp formation for large ϵ is very improbable. For large ϵ the nonlinear elasticity will suppress the surface corrugation before folding.[11)] On the other hand, in the region III we can calculate cusp solutions, but the folded parts turn out to be not pressing each other ($\Pi_{yy} < 0$). We stress that nothing is known about the pattern formation in the regions III and IV. Observations of the pattern, with changing external parameters such as the temperature, is strongly needed. (iv) If the system is situated very close to the curve $\hat{\delta} = \hat{\delta}_c(\epsilon)$ in Fig.6, the amplitude of the surface inhomogeneities will be small and cubic, and quartic terms in the free energy will suffice to describe the patterns. Here, the transition can well be of first-order due to the general presence of the cubic terms. To see such a discontinuity, we propose experiments on a clamped gel in the presence of a temperature gradient in the lateral direction. (v) The dynamics of the pattern development has not yet been studied.

2.5 Some Thermodynamic Relations

We finally mention some thermodynamic relations valid for isotropic gels immersed in solvent at a constant osmotic pressure with the end of removing confusions in previous experiments. As (2.2) indicates, the fluctuation variance of the total volume is given by the following thermodynamic derivative,

$$< (\delta V)^2 >= -(\frac{\partial V}{\partial \Pi})_T = V/K \ , \tag{2.60}$$

where K is the bulk modulus given by (2.10). Surely the volume (= the order parameter) undergoes fluctuation-enhancement at the critical point. In accord with this, the first mode $n = 0$ obtained in this subsection gives rise to the following contribution to the density-correlation function,[23)]

$$< \delta\phi(\mathbf{r}_1)\delta\phi(\mathbf{r}_2) > \cong \frac{1}{VK} + \cdots \ . \tag{2.61}$$

Here the first term on the right hand side becomes (2.60) if integrated order $\mathbf{r}_1$ and $\mathbf{r}_2$ within the gel, whereas the remaining terms, which are not explicitly written, are well-defined in the region $K + \frac{4}{3}\mu > 0$ and tend to the usual Ornstein-Zernike correlation function in the limit $V \to 0$ (the gradient free energy being taken into account). In practice, if a large number of samples of gels are prepared, their volumes fluctuate thermally according to (2.60) while being kept transparent even for $K \cong 0$.

It is also worth mentioning another thermodynamic relation for the specific heat C_π at constant osmotic pressure,

$$C_\pi = C_V + V(\partial \Pi/\partial T)_V^2/K \tag{2.62}$$

where C_V is the specific heat at constant volume. We expect that C_V and $(\partial \Pi/\partial T)_V$ should remain finite at $K = 0$, so that C_π should diverge as $1/K$ at the critical point. The thermal expansion coefficient $\beta_\pi = V^{-1}(\partial V/\partial T)_\pi = (\partial \Pi/\partial T)_V K^{-1}$ should also diverge as $1/K$. We believe that the asymptotic critial statics at $K = 0$ is exactly described by the mean field theory. Recently Li and Tanaka measured C_V near $K = 0$ where the thermal relaxation time was much shorter than the swelling time.[35] They claimed that the critical statics of gels is the same as that of Ising systems. We definitely disagree with their interpretation. In gels usually used in experiments, the two points, $K = 0$ and $K + \frac{4}{3}\mu = 0$, are rather close as a complicating fact. Consequently, there should be considerable enhancement of small-scale density fluctuations at $K \cong 0$, which will make C_V somewhat peaked at $K \cong 0$. (Furthermore, if there is some swelling of the sample, there should be an extra absorption of heat.) In support of our conclusion the critical behavior of solid solutions of $Ce_{1-x}Th_x$ at the valance instability was found to be described by the mean field theory.[36] Such solid solutions and gels belong to the same universality class as regards the statics;[21] the difference is that the two points, $K = 0$ and $K + \frac{4}{3}\mu = 0$, are much separated in the former and slightly separated in the latter.

3. GINZBURG-LANDAU MODEL FOR SOLID SOLUTIONS

3.1 General Relations for the Cubic Case

In solid solutions elastic fields are induced in the phase separation process due to the lattice misfit between the emerging two phases and they radically influence the domain morphology.[7] Since the pioneering work by Cahn,[8] the effect has been extensively studied, but there still remain many aspects to be investigated due to the complicated and varied nature of the problem.[9] The traditional approach here has been to calculate elastic energies in the presence of precipitates assumed to have some given elastic properties and shapes. The calculations are in most cases very complicated. The present author has recently proposed a Ginzburg-Landau approach to analyze the problem more systematically and also to treat other situations such as the case of spinodal decomposition.[10]

In our scheme the elastic strain $\partial u_i/\partial x_j$ is a subsidiary tensor variable coupled to a conserved order parameter c in the free energy. After the phase separation we expect the so-called coherent case in which the lattice constants of the two phases are different but the lattice planes are continuous through the two phase regions without dislocations at the interfaces. For simplicity the solid is assumed to have the cubic symmetry before and after the phase separation.[37] The total free energy

F is of the form,

$$F = \int d\mathbf{r}[f(c) + \frac{1}{2}(\nabla c)^2 + \alpha c(\nabla \cdot \mathbf{u}) + f_{el}] \tag{3.1}$$

where $f(c)$ is the Ginzburg-Landau free energy density for c, c is measured from some reference value, and α is the coupling constant between c and $\nabla \cdot \mathbf{u}$. The f_{el} is the usual elastic free energy in the reference state in which $c = 0$,

$$f_{el} = \frac{1}{2}(C_{11} - C_{12}) \sum_i (\frac{\partial}{\partial x_i} u_i)^2 + \frac{1}{2} C_{12} (\nabla \cdot \mathbf{u})^2 + \frac{1}{2} C_{44} (u_{xy}^2 + u_{yz}^2 + u_{zx}^2) \, , \tag{3.2}$$

where $u_{ij} \equiv \partial u_i / \partial x_j + \partial u_j / \partial x_i$ and C_{ij} are the stiffness constants. The usual isotropic case can be recovered under the condition $C_{11} - C_{12} - 2C_{44} = 0$, where the bulk and shear moduli are expressed as

$$K = (C_{11} + 2C_{12})/3 \; , \mu = C_{44} \; . \tag{3.3}$$

Then the usual form for isotropic elastic bodies follows:[15)]

$$f_{el} = \frac{1}{2} K (\nabla \cdot \mathbf{u})^2 + \frac{1}{4} \mu \sum_{i,j} (\frac{\partial}{\partial x_i} u_j + \frac{\partial}{\partial x_j} u_i - \frac{2}{3} \delta_{ij} \nabla \cdot \mathbf{u})^2 \, . \tag{3.4}$$

We shall see that the domain morphology can be influenced by external stresses only when the two phases have different stiffness constants as long as f_{el} is composed of terms bilinear in $\mathbf{u}$ (or within the linear elastic theory). To take into account such an effect we allow C_{ij} to depend linearly on c as

$$C_{ij} = C_{ij}^0 + c C_{ij}^1 \; . \tag{3.5}$$

In homogeneous, disordered states without external stress we have an isotropic homogeneous deformation, $\partial u_i / \partial x_j = \delta_{ij} \zeta$, by neglecting inhomogeneous fluctuations of c. Here (3.1) becomes

$$F = V[f(c) + 3\alpha c\zeta + \frac{9}{2}K\zeta^2] \tag{3.6}$$

where $K = (C_{11}+2C_{12})/3$ and V is the total volume. Minimization of F determines ζ in the form,

$$\partial u_i/\partial x_j = \zeta\delta_{ij} = -(\alpha c/3K)\delta_{ij} \ . \tag{3.7}$$

F may be written as $F = Vf'$, where

$$f' = f - \frac{1}{2}(\alpha^2/K)c^2 \ . \tag{3.8}$$

Here f' is the free energy density obtained if f_{el} is measured from the c-dependent homogeneous state (3.7). It coincides with the definition of the free energy density in Cahn's theory.[8] The lattice constant a in the disordered phase is roughly a linear function of c (Vegard's law).[7] In our scheme, further taking into account the c-dependence of K, we find the concentration expansion coefficient η in the form

$$\eta \equiv \frac{\partial}{\partial c}\ln a = \frac{\partial}{\partial c}\zeta = -\hat{\alpha}/3K \ , \tag{3.9}$$

where

$$\hat{\alpha} = \alpha - cK_1/K = \alpha + K_1(\nabla\cdot\mathbf{u}) \ , \tag{3.10}$$

with

$$K_1 = \partial K/\partial c = (C_{11}^1 + 2C_{12}^1)/3 \ . \tag{3.11}$$

Next we calculate the elastic stress tensor σ_{ij}. To find it we make a small incremental displacement $u_i \to u_i + \delta u_i$ with c held fixed. Then the resultant change of F should be given by

$$\delta F = \int d\mathbf{r} \sum_{i,j} \sigma_{ij}(\frac{\partial}{\partial x_i}\delta u_j) \ . \tag{3.12}$$

It readily follows that

$$\sigma_{ii} = \alpha c + C_{11}\frac{\partial}{\partial x_i}u_i + C_{12}(\nabla \cdot \mathbf{u} - \frac{\partial}{\partial x_i}u_i) \ , \tag{3.13}$$

$$\sigma_{ij} = C_{44}(\frac{\partial}{\partial x_i}u_j + \frac{\partial}{\partial x_j}u_i) \quad for \quad i \neq j \ . \tag{3.14}$$

Since we have assumed that c is a conserved variable, the relaxation time of c is much slower than that of $\mathbf{u}$ at long wavelengths and we may assume that the elastic field instantaneously relaxes to adjust to a given concentration field. This is the condition of mechanical equilibrium,

$$\delta F/\delta u_i = -\sum_j \frac{\partial}{\partial x_j}\sigma_{ij} = 0 \ . \tag{3.15}$$

We also notice that F can be transformed after some manipulations into

$$F = \int d\mathbf{r}[f + \frac{1}{2}(\nabla c)^2 + \frac{1}{2}\alpha c\nabla \cdot \mathbf{u} + \frac{1}{2}\sum_{i,j}\sigma_{ij}\frac{\partial u_i}{\partial x_j}] \ . \tag{3.16}$$

In this paper we suppose the presence of an external stress $\vec{\sigma}_{ex}$ imposed on the solid at the boundary, which then gives rise to an affine deformation of the solid,

$$A_{ij} = <\frac{\partial}{\partial x_j}u_i> \ , \tag{3.17}$$

where $<\cdots>$ is the spatial average. Writing $\delta u_i = u_i - <u_i>$ and using (3.15) we may rewrite (3.16) as

$$F = \int d\mathbf{r}[f + \frac{1}{2}(\nabla c)^2 + \frac{1}{2}\alpha c(\nabla \cdot \mathbf{u}) + \frac{1}{2}Tr(\vec{\sigma} \cdot \vec{A}) + \frac{1}{2}\sum_{i,j}(\sigma_{ex})_{ij}(\frac{\partial}{\partial x_j}\delta u_i)] \ . \tag{3.18}$$

The last term in the brackets is negligible in most cases for macroscopic crystal sizes. An effective free energy of c only will follow if (3.15) is solved and $\mathbf{u}$ is eliminated in F as will be done in Section 3.5 below.

The dynamic equation for c is assumed to be of the diffusion type,

$$\frac{\partial}{\partial t}c = \lambda\nabla^2(\delta F/\delta c) = \lambda\nabla^2[f' - \nabla^2 c + \mu_{el}] \,, \tag{3.19}$$

where λ is the kinetic coefficient, $f' = \partial f/\partial c$, and μ_{el} is the elastic part of the chemical potential,

$$\mu_{el} = \alpha(\nabla\cdot\mathbf{u}) + \frac{1}{2}(C_{11}^1 - C_{12}^1)\sum_i(\frac{\partial}{\partial x_i}u_i)^2 + \frac{1}{2}C_{12}^1(\nabla\cdot\mathbf{u})^2 + \frac{1}{2}C_{44}^1(u_{xy}^2 + u_{yz}^2 + u_{zx}^2) \,. \tag{3.20}$$

Notice that the last three terms arise from the $\dot{c}$-dependence of C_{ij} and can depend on the externally given deformation A_{ij}, (3.17), because $\{\partial u_i/\partial x_j\}$ consists of the average and the deviation.

3.2 Linear Approximation

Expecting only small fluctuations we assume that c is composed of a homogeneous average and a small sinusoidal deviation,

$$c = \bar{c} + c_1\exp(i\mathbf{k}\cdot\mathbf{r}) \,. \tag{3.21}$$

The strain is then of the form,

$$\partial u_i/\partial x_j = A_{ij} + ik_j u_{1i}\exp(i\mathbf{k}\cdot\mathbf{r}) \,. \tag{3.22}$$

Here it is convenient to introduce the following traceless tensor,

$$S_{ij} = A_{ij} + A_{ji} - \frac{2}{3}\delta_{ij}(Tr\overleftrightarrow{A}) \,. \tag{3.23}$$

The linearized version of (3.15) reads

$$[(C_{11} - C_{12} - 2C_{44})k_j^2 + C_{44}k^2]u_{1j} - i(C_{12} + C_{44})k_j g_{\mathbf{k}} = i\sum_m(\hat{\alpha}\delta_{jm} + W_{jm})k_m c_1 \,, \tag{3.24}$$

where the averages $< C_{ij} >= C_{ij}^0 + \bar{c}C_{ij}^1$ are simply written as C_{ij}, $g_{\mathbf{k}} = i\mathbf{k}\cdot\mathbf{u}_1$, $\hat{\alpha}$

is defined by (3.10), and

$$W_{ij} = C^1_{44} S_{ij} + \frac{1}{2}(C^1_{11} - C^1_{12} - 2C^1_{44}) S_{ii}\delta_{ij} \ . \tag{3.25}$$

We further introduce $\chi(\hat{\mathbf{k}})$, $\psi_1(\hat{\mathbf{k}})$, and $\psi_2(\hat{\mathbf{k}})$ by

$$\chi(\hat{\mathbf{k}}) = \sum_j \hat{k}_j^2/[C_{44} + (C_{11} - C_{12} - 2C_{44})\hat{k}_j^2] \ , \tag{3.26}$$

$$\psi_n(\hat{\mathbf{k}}) = \sum_{ij} \hat{k}_i (W^n)_{ij} \hat{k}_j/[C_{44} + (C_{11} - C_{12} - 2C_{44})\hat{k}_j^2] \ , \tag{3.27}$$

where $\hat{\mathbf{k}} = k^{-1}\mathbf{k}$ denotes the direction of $\mathbf{k}$. In terms of these quantities (3.24) is explicitly solved to give

$$\begin{aligned}[C_{44}k^2 + (C_{11} - C_{12} - 2C_{44})k_j^2]u_{1j} = &-iq_j \frac{C_{12} + C_{44}}{1 + (C_{12} + C_{44})\chi(\hat{\mathbf{k}})}[\hat{\alpha}\chi(\mathbf{k}) + \psi_1(\hat{k})]c_1 \\ &+ i\sum_m q_m(\hat{\alpha}\delta_{jm} + W_{jm})c_1 \ .\end{aligned} \tag{3.28}$$

Our desired quantity is the coefficient $\tau_{el}(\hat{\mathbf{k}}) \equiv \delta\mu_{el}/c_1$, where $\delta\mu_{el}$ is the deviation of μ_{el}, (3.20). It can be readily calculated if we notice

$$\delta\mu_{el} = \hat{\alpha} g_{\mathbf{k}} + \sum_{ij} W_{ij}(\partial u_{1i}/\partial x_j) \ . \tag{3.29}$$

Substitution of (3.28) yields

$$\tau_{el}(\hat{\mathbf{k}}) = -\frac{\chi(\hat{\mathbf{k}})}{[1 + (C_{12} + C_{44})\chi(\hat{\mathbf{k}})]}[\hat{\alpha} + \frac{1}{\chi(\hat{\mathbf{k}})}\psi_1(\hat{\mathbf{k}})]^2 + [\frac{1}{\chi(\hat{\mathbf{k}})}\psi_1(\hat{\mathbf{k}})^2 - \psi_2(\hat{\mathbf{k}})] \ . \tag{3.30}$$

The linear dynamic equation is now written from (3.19) in the form,

$$\frac{\partial}{\partial t}c_1 = -\lambda k^2[f'' + k^2 + \tau_{el}(\hat{\mathbf{k}})]c_1 \tag{3.31}$$

where $f'' = \partial^2 f/\partial c^2$. It should also be noted that the bilinear part of the free energy is written in the Fourier space as

$$F = \frac{1}{2}\int_k \left[[f'' + k^2 + \tau_{el}(\hat{\mathbf{k}})]|c_{\mathbf{k}}|^2\right] \cdots , \tag{3.32}$$

where $\int_{\mathbf{k}}(\cdots) \equiv (2\pi)^{-3}\int d\mathbf{k}(\cdots)$ and $c_{\mathbf{k}}$ is the Fourier transform of $c(\mathbf{r})$. As a result we find the mean field expression for the structure factor of c in one-phase states,

$$< |c_{\mathbf{k}}|^2 >= 1/[f'' + \tau_{el}(\hat{\mathbf{k}}) + k^2] \, . \tag{3.33}$$

A similar form for the structure factor was also found for affinely deformed gels[5] (see (2.38)).

In the literature the coefficient $\tau_{el}(\hat{\mathbf{k}})$ was calculated for cubic crystals in the simplest case of no external stress and no c-dependence of the moduli.[7)37)38] Remarkably such previous results can be reproduced from our general expression (3.30), although the previous derivations were rather different from ours. To show the coincidence we rewrite the bilinear part of F, (3.32), as

$$F = \frac{1}{2}\int_k [(\frac{\partial^2}{\partial c^2}f^{\prime}) + k^2 + B(\hat{\mathbf{k}})]|c_{\mathbf{k}}|^2 + \cdots \tag{3.34}$$

where $f^{\prime}$ is defined by (3.8). By setting $S_{ij} = 0$ and $C^1_{ij} = 0$ we have

$$B(\hat{\mathbf{k}}) = \alpha^2/K + \tau_{el}(\hat{\mathbf{k}}) = \alpha^2/K - \alpha^2\chi(\hat{\mathbf{k}})/[1 + (C_{12} + C_{44})\chi(\hat{\mathbf{k}})] \, . \tag{3.35}$$

Some manipulations lead to the expression in Ref.7,

$$B(\hat{\mathbf{k}}) = \eta^2(C_{11} + 2C_{12})[3 - (C_{11} + 2C_{12}) \times \frac{1 + 2\xi_a(n_1^2n_2^2 + n_1^2n_3^2 + n_2^2n_3^2) + 3\xi_a^2n_1^2n_2^2n_3^2}{C_{11} + \xi_a(C_{11} + C_{12})(n_1^2n_2^2 + n_1^2n_3^2 + n_2^2n_3^2) + \xi_a^2(C_{11} + 2C_{12} + C_{44})n_1^2n_2^2n_3^2}] \tag{3.36}$$

where η is defined by (3.9), $\xi_a \equiv (C_{11} - C_{12} - 2C_{44})/C_{44}$ is the elastic anisotropy, and $n_i \equiv \hat{k}_i$. Further transformations also result in Hilliard's expression.[38] It goes without saying that $B(\hat{\mathbf{k}})$ simply reduces to Cahn's original expression $B = 2\eta^2 E/(1-\nu)$ in the isotropic case (3.3),[8] where E is the Young modulus and ν is the Poisson ratio.

It is well-known that, if $\xi_a < 0$ (which is the case for most cubic crystals), $B(\hat{\mathbf{k}})$ is minimized in the [100], [010], and [001] directions.[7)37)38)] On the other hands, if $\xi_a > 0$, the minimum is attained in the [111] direction and the equivalent other three directions $[\bar{1}11], [1\bar{1}1]$, and $[11\bar{1}]$. When $B(\hat{\mathbf{k}})$ is made negative in these directions, the resultant domains tend to be elongated in the so-called habit planes, which are those perpendicular to these directions.

So far the above calculations have been conveniently performed in the Fourier space. In the real space the $\hat{\mathbf{k}}$-dependence of $\tau_{el}(\hat{\mathbf{k}})$ gives rise to a long-ranged, angle-dependent interaction. The elastic part of F in the bilinear order is rewritten from (3.32) as[10)39)]

$$\begin{aligned}\Delta F_{el} &= \frac{1}{2}\int_{\mathbf{k}} \tau_{el}(\hat{\mathbf{k}})|c_{\mathbf{k}}|^2 \\ &= \frac{1}{2}\int d\mathbf{r}\frac{1}{2}\bar{\tau}_{el}c(\mathbf{r})^2 + \frac{1}{2}\int\int d\mathbf{r}d\mathbf{r}'\frac{1}{|\mathbf{r}-\mathbf{r}'|^3}\phi(\frac{\mathbf{r}-\mathbf{r}'}{|\mathbf{r}-\mathbf{r}'|})c(\mathbf{r})c(\mathbf{r}')\ .\end{aligned} \tag{3.37}$$

Here $\bar{\tau}_{el}$ is the angle average of $\tau_{el}(\hat{\mathbf{k}})$, and $\phi(\hat{\mathbf{r}})$ depends only on the direction $\hat{\mathbf{r}} = r^{-1}\mathbf{r}$ and

$$\phi(\hat{\mathbf{r}}) = (2\pi)^{-3}r^3\int d\mathbf{k}\bar{e}^{i\mathbf{k}\cdot\mathbf{r}}[\tau_{el}(\hat{\mathbf{k}}) - \bar{\tau}_{el}]\ . \tag{3.38}$$

By definition the angle average of $\phi(\hat{\mathbf{r}})$ vanishes and hence the long range interaction is attractive or repulsive depending on the direction $\hat{\mathbf{r}}$. The second term of (3.37) is present as long as $\xi_a \neq 0$ even if C_{ij} are independent of c and there is no applied stress. In real crystals this term should be crucial in determining the domain morphology as was discussed by Miyazaki et al.[40)] For example, if $|\xi_a| << 1$, $B(\hat{\mathbf{k}})$ in (3.36) may be expanded in powers of ξ_a. Thus, when $S_{ij} = 0$, we obtain to leading order in ξ_a,

$$\tau_{el}(\hat{\mathbf{k}}) - \bar{\tau}_{el} = -\alpha_1[(\hat{k}_x\hat{k}_y)^2 + (\hat{k}_y\hat{k}_z)^2 + (\hat{k}_z\hat{k}_x)^2 - \frac{1}{5}] + \cdots \tag{3.39}$$

where

$$\alpha_1 = \eta^2(1 + 2C_{12}/C_{11})^2(C_{11} - C_{12})\xi_a\ . \tag{3.40}$$

The function $\phi(\hat{\mathbf{r}})$ may be explicitly calculated in the form,

$$\phi(\hat{\mathbf{r}}) = -(\frac{15}{8\pi})\alpha_1[(\hat{x}\hat{y})^2 + (\hat{y}\hat{z})^2 + (\hat{z}\hat{x})^2 - \frac{1}{5}] + \cdots . \tag{3.41}$$

If $\xi_a < 0$, $\phi(\hat{\mathbf{r}})$ is surely minimum (and negative) in the [100] direction and is maximum (and positive) in the [111] direction. The reverse results follow for $\xi_a > 0$.

3.3 Spinodal Curve and Habit Planes in the Cubic Case

In calculating the structure factor (3.33) we have assumed that $f'' + \tau_{el}(\hat{\mathbf{k}})$ is nonnegative. The dynamic equation (3.31) indicates that a linear instability is triggered when $f'' + \tau_{el}(\hat{\mathbf{k}})$ becomes negative for some $\hat{\mathbf{k}}$. Let $\mathbf{n}_0$ be the direction in which $\tau_{el}(\hat{\mathbf{k}})$ is minimum. Then the fluctuations varying along $\mathbf{n}_0$ first become unstable with lowering of the temperature and the spinodal curve is determined by

$$f'' + \tau_{el}(\mathbf{n}_0) = 0 . \tag{3.42}$$

However, the minimization procedure of $\tau_{el}(\hat{\mathbf{k}})$ is rather complicated in the presence of general stress. Here, as an illustrative example, we examine how a small uniaxial stress along the [100] direction can change the cubic symmetry of the solid. Taking the Cartesian coordinates along the crystal axes we have

$$\{A_{ij}\} = \begin{pmatrix} \lambda_{\parallel} & 0 & 0 \\ 0 & \lambda_{\perp} & 0 \\ 0 & 0 & \lambda_{\perp} \end{pmatrix} . \tag{3.43}$$

The difference between the external stress $\sigma_{\parallel}$ along the x axis and that $\sigma_{\perp}$ in the lateral directions is given by

$$\sigma_A \equiv \sigma_{\parallel} - \sigma_{\perp} = 2C_{44}(\lambda_{\parallel} - \lambda_{\perp}) . \tag{3.44}$$

We find $S_{xx} = -2S_{yy} = -2S_{zz} = \frac{4}{3}(\lambda_{\parallel} - \lambda_{\perp})$ and may write $\psi_1(\hat{\mathbf{k}})$, defined by

(3.27), in the form

$$\psi_1(\hat{\mathbf{k}}) = \frac{1}{3C_{44}}(C_{11}^1 - C_{12}^1)(\lambda_\parallel - \lambda_\perp)\Psi(\hat{\mathbf{k}}) \tag{3.45}$$

with

$$\Psi(\hat{\mathbf{k}}) = \frac{2\hat{k}_x^2}{1+\xi_a \hat{k}_x^2} - \frac{\hat{k}_y^2}{1+\xi_a \hat{k}_y^2} - \frac{\hat{k}_z^2}{1+\xi_a \hat{k}_z^2} , \tag{3.46}$$

where $\xi_a = (C_{11} - C_{12} - 2C_{44})/C_{44}$. It is convenient to introduce the following dimensionless number representing the strength of the anisotropic stress σ_A,

$$A_u = -(C_{11}^1 - C_{12}^1)(\lambda_\parallel - \lambda_\perp)/\hat{\alpha} = -(C_{11}^1 - C_{12}^1)\sigma_A/(2C_{44}\hat{\alpha}) . \tag{3.47}$$

Here we limit ourselves to the case $|A_u| << 1$, which simply allows us to neglect the terms bilinear in $\overleftrightarrow{S}$ in $\tau_{el}(\hat{\mathbf{k}})$, (3.30), leading to

$$\tau_{el}(\hat{\mathbf{k}}) \cong -\frac{\hat{\alpha}^2\chi(\hat{\mathbf{k}})}{1+(C_{12}+C_{44})\chi(\hat{\mathbf{k}})}[1 - \frac{2}{3C_{44}\chi(\hat{\mathbf{k}})}A_u\Psi(\hat{\mathbf{k}})] . \tag{3.48}$$

With the above relation we notice the following: (i) If $\xi_a < 0$ and $A_u < 0$, the [100] direction minimizes $\tau_{el}(\hat{k})$, favoring emergence of lamellar domains in the phase separation. (ii) If $\xi_a < 0$ and $A_u > 0$, the minimum is attained for the two lateral directions [010] and [001]. (iii) If $\xi_a > 0$, the degeneracy in four directions is not removed. Assuming $|A_u/\xi_a| << 1$, we find that the habit plane (111) is slightly shifted as

$$\hat{k}_x^2 - \frac{1}{3} \cong -(\frac{2}{3\xi_a})[1 + (1 + \frac{C_{12}}{C_{44}})\frac{1}{1+\frac{1}{3}\xi_a}]A_u , \tag{3.49}$$

where $\hat{k}_y$ and $\hat{k}_z$ are determined from $\hat{k}_y^2 = \hat{k}_z^2 = (1 - \hat{k}_x^2)/2$.

For large external stresses we encounter a very interesting problem on the competition between the inherent cubic anisotropy and the externally induced anisotropy. Effects of shear deformations should be unique and worth examining.

3.4 Spinodal Curve in the Isotropic Case

Under the isotropic condition $C_{11} - C_{12} - 2C_{44} = 0$ the minimization procedure of $\tau_{el}(\hat{\mathbf{k}})$ is much simplified even for large external stresses. The moduli K and μ depend on c as

$$K = K_0 + K_1 c \ , \mu = \mu_0 + \mu_1 c \ . \tag{3.50}$$

Writing $K_L = K + \frac{4}{3}\mu$ we find

$$\tau_{el}(\hat{\mathbf{k}}) = -\frac{1}{K_L}[\hat{\alpha} + \mu_1(\hat{\mathbf{k}} \cdot \overleftrightarrow{S} \cdot \hat{\mathbf{k}})]^2 - (\frac{\mu_1^2}{\mu})[(\hat{\mathbf{k}} \cdot \overleftrightarrow{S}^2 \cdot \hat{\mathbf{k}}) - (\hat{\mathbf{k}} \cdot \overleftrightarrow{S} \cdot \hat{\mathbf{k}})^2] \ . \tag{3.51}$$

Further in the case of uniaxial deformation (3.43) we have

$$\tau_{el}(\hat{\mathbf{k}}) = -\frac{1}{K_L}[\hat{\alpha} + 2\mu_1(\lambda_{\parallel} - \lambda_{\perp})(\hat{k}_x^2 - \frac{1}{3})]^2 - 4\frac{\mu_1^2}{\mu}(\lambda_{\parallel} - \lambda_{\perp})^2 \hat{k}_x^2(1 - \hat{k}_x^2) \ . \tag{3.52}$$

We seek $\mathbf{n}_0$, which minimizes $\tau_{el}(\hat{\mathbf{k}})$, in the uniaxial case. There are three cases depending on the following dimensionless parameter,

$$A_e = (\lambda_{\parallel} - \lambda_{\perp})(\partial ln\mu/\partial c)/(\partial ln a/\partial c) = -(\frac{3}{2})K\mu_1\sigma_A/(\mu^2\hat{\alpha}) \ , \tag{3.53}$$

where $(\partial ln\mu/\partial c) = \mu_1/\mu$ and $(\partial ln a/\partial c)$ is given in (3.9). We draw the following conclusions, the first two being consistent with a recent simulation:[41] (i) For $-3 \leq A_e < 0$, $\mathbf{n}_0$ is parallel to the x-axis. Emerging domains below the spinodal curve should be lamellar. (ii) For $0 < A_e \leq 3K/(K + \frac{2}{3}\mu)$ the minimum of $\tau_{el}(\hat{\mathbf{k}})$ is attained for $\mathbf{n}_0$ in the yz plane. The domain structure is homogeneous in the x axis and is two-dimensional. (iii) When $A_e > 3K/(K + \frac{2}{3}\mu)$ or $A_e < -3$, $\mathbf{n}_0$ is oblique to the x axis. The angle θ_0 with respect to the x axis is determined by

$$\cos^2\theta_0 = \hat{k}_x^2 = \frac{K}{2(K + \frac{1}{3}\mu)}[-\frac{3}{A_e} + 1 + \frac{2\mu}{3K}] \ . \tag{3.54}$$

The domain structure is three-dimensional. The ratio of the period along the x axis and that in the yz plane is given by $\tan\theta_0$.

3.5 Long-Range Interactions in the Isotropic Case

(a) Effective Free Energy

Here we examine long-range interactions arising from the coherently induced elastic field in some detail in the isotropic case. Let the expansion coefficients K_1 and μ_1 in (3.50) be very small. The condition of mechanical equilibrium (3.15) is written as

$$\frac{\partial}{\partial x_i}\Big[\alpha c + (K + \frac{1}{3}\mu_0)g\Big] + \mu_0 \nabla^2 u_i + \mu_1 \sum_j \frac{\partial}{\partial x_j} c\Big[\frac{\partial}{\partial x_i}u_j + \frac{\partial}{\partial x_j}u_i - \frac{2}{3}\delta_{ij}\nabla\cdot\mathbf{u}\Big] = 0 \ , \tag{3.55}$$

where $g = \nabla \cdot \mathbf{u}$. We will calculate the effective interactions up to first order in K_1 and μ_1. To solve (3.55) under the simplest boundary conditions we assume that c is inhomogeneous only in a restricted region far from the boundary. It is convenient to measure c from the value in the undisturbed region. Then $c \to 0$, $K \to K_0$, and $\mu \to \mu_0$ far from the disturbed region. We also apply an external stress at the boundary, which gives rise to a homogeneous affine deformation, $\langle \partial u_i/\partial x_j \rangle = A_{ij} =$ const.

If the last term of (3.55) is neglected, we find up to first order in K_1,

$$g = g_0 - (K_{L0} + K_1 c)^{-1}\hat{\alpha} c + \cdots \ , \tag{3.56}$$

where $g_0 = Tr\overleftrightarrow{A}$ is a constant, $K_{L0} = K_0 + \frac{4}{3}\mu_0$, and $\hat{\alpha} = \alpha + K_1 g_0$. As regards the strain tensor we need only the zero-th order solution,

$$\partial u_i/\partial x_j = A_{ij} - \frac{\alpha}{K_{L0}}\frac{\partial}{\partial x_j}d_i + \cdots \tag{3.57}$$

where the vector d_i is defined as the solution of the equation,

$$\nabla^2 d_i = \frac{\partial}{\partial x_i}c \quad \text{or} \quad d_i = (\frac{1}{\nabla^2}\frac{\partial}{\partial x_i})c \ . \tag{3.58}$$

Hereafter $\frac{1}{\nabla^2}$ denotes the inverse operator of ∇^2. We notice that $\nabla\cdot\mathbf{d} = c$, $\partial d_i/\partial x_j = \partial d_j/\partial x_i$, and d_i decays as r^{-2} far from the disturbed region (since the *charge* $\partial c/\partial x_i$

vanishes if integrated over space). To first order in μ_1, (3.57) may be substituted into the last term of (3.55) and the resultant change of g, denoted by δg, is determined by

$$K_{L0}\nabla^2 \delta g = -\mu_1 \sum_{ij} S_{ij}\frac{\partial^2}{\partial x_i \partial x_j}c + 2(\alpha\mu_1/K_{L0})\sum_{ij}\frac{\partial^2}{\partial x_i \partial x_j}c(\frac{\partial}{\partial x_i}d_j - \frac{1}{3}c\delta_{ij}) \ , \tag{3.59}$$

where S_{ij} is defined in (3.23).

Now we may calculate the effective free energy F_{eff} for c only by substituting the above results for $\mathbf{u}$ into (3.18). Up to first order in K_1 and μ_1 we obtain

$$F_{eff} = \int d\mathbf{r}\left[f_{eff} + f_d + f_E + \frac{1}{2}|\nabla c|^2\right] \ . \tag{3.60}$$

The f_{eff} is the local interaction term,[8)]

$$f_{eff} \simeq f - \frac{1}{2}\hat{\alpha}^2 c/(K + \frac{4}{3}\mu_0) + \cdots \ . \tag{3.61}$$

The f_d arises from the anisotropic deformation and turns out to be a long-ranged, dipolar interaction,

$$\begin{aligned} f_d &= (\mu_1\alpha/K_{L0})\sum_{ij} S_{ij}(\frac{\partial c}{\partial x_i})\frac{1}{\nabla^2}(\frac{\partial c}{\partial x_j}) \\ &= -(\mu_1\alpha/4\pi K_{L0})\sum_{i,j}\int d\mathbf{r}' S_{ij}[\frac{\partial}{\partial x_i}c(\mathbf{r})]\frac{1}{|\mathbf{r}-\mathbf{r}'|}[\frac{\partial}{\partial x_j'}c(\mathbf{r}')] \ . \end{aligned} \tag{3.62}$$

The third term f_E has a cubic dependence on c and corresponds to an interaction first calculated by Eshalby,[42)]

$$f_E = \mu_1(\alpha/K_{L0})^2 c\sum_{ij}(\frac{\partial}{\partial x_i}d_j - \frac{1}{3}c\delta_{ij})^2 \ . \tag{3.63}$$

(b) Isolated Spheroidal Inclusion

To examine the meaning of f_d and f_E we treat the uniaxial case (3.43) and first consider an isolated spheroidal precipitate represented by $x^2/R_{\|}^2+(y^2+z^2)/R_{\perp}^2 = 1$. The two lengths $R_{\|}$ and $R_{\perp}$ are assumed to be much longer than the interfacial width and then we may set $c = \Delta c =$ const. inside the spheroid. Using the well-known results of electrostatics[43] we may calculate d_i, (3.58), for a general ellipsoid represented by $\sum_j x_j^2/R_j^2 = 1$. That is,

$$d_i = (\Delta c)N_i x_i \quad \text{inside the ellipsoid} \ , \tag{3.64}$$

where N_i with $i = x, y, z$ are the depolarization factors. (The quantity $\phi \equiv d_i - \epsilon_i x_i$ is the electrostatic potential for an uncharged conducting ellipsoid in a uniform external field $\epsilon_i = (\Delta c)N_i$.) For the spheroidal case, using a partial integration, the spatial integral of f_d is calculated as

$$F_d = \int d\mathbf{r} f_d = 2(\Delta\mu)\hat{\delta}(\lambda_{\|} - \lambda_{\perp})(N_x - \frac{1}{3})V_s \ , \tag{3.65}$$

where use has been made of $N_y = N_z = \frac{1}{2}(1 - N_x)$. Here $\Delta\mu = \mu_1(\Delta c)$ is the difference of the shear moduli between the precipitate and the matrix, $V_s = (4\pi/3)R_{\|}R_{\perp}^2$ is the volume of the spheroid, and

$$\hat{\delta} = -\alpha(\Delta c)/K_{L0} \simeq (\Delta a/a)/(1 + \frac{4}{3}\mu_0/K_0) \ , \tag{3.66}$$

$\Delta a = a_p - a_m$ being the difference of the lattice constants of the precipitate and the matrix. In the literature[41)44)] the misfit parameter δ is defined by $\delta = \Delta a/a = \hat{\delta}(1 + \frac{4}{3}\mu_0/K_0)$, while the applied stress σ_A is equal to $2(\lambda_{\|} - \lambda_{\perp})\mu_0$. The N_x increases from 0 to 1 as $R_{\perp}/R_{\|}$ increases form 0 to ∞.[43)] In particular, $N_x - \frac{1}{3} \simeq (2/15)(R_{\perp}^2/R_{\|}^2 - 1)$ for $R_{\perp} \simeq R_{\|}$. We notice that, if $(\Delta\mu)\hat{\delta}(\lambda_{\|} - \lambda_{\perp}) < 0$ or $A_e < 0$ (see (3.53)), F_d decreases with increasing $R_{\perp}/R_{\|}$ favoring oblate spheroids. On the other hand, if $(\Delta\mu)\hat{\delta}(\lambda_{\|} - \lambda_{\perp}) > 0$ or $A_e > 0$, F_d is smaller for prolate spheroids. These results are generally consistent with an elastic theory of Pineau,[44)] a computer simulation of Gayda and Srolovitz,[41)] and the results in Section 3.4. However, there seem to be discrepancies with our results and those of Refs. 44 and 41 in the case,

$(\Delta\mu)\hat{\delta}(\lambda_{\parallel}-\lambda_{\perp}) > 0$ and $|\sigma_A/E_m\delta| \gtrsim 10$, where $\sigma_A/E_m\delta = (\lambda_{\parallel}-\lambda_{\perp})/[3(1-\nu_0)\hat{\delta}]$ in our notation, ν_0 being the Poisson ratio of the matrix. In this case the solid is very strongly compressed or stretched by the applied load. The origin of the discrepancies are not clear at present.

Similarly f_E is nonvanishing only inside the spheroid due to the presence of the factor c and

$$F_E = \int dr f_E = \frac{3}{2}(\Delta\mu)\hat{\delta}^2(N_x - \frac{1}{3})^2 V_s \ . \tag{3.67}$$

This term vanishes for spheres and changes its sign depending on the sign of $\Delta\mu = \mu_1(\Delta c)$. We discuss the two ceases, $\Delta\mu < 0$ and $\Delta\mu > 0$, separately when $S_{ij} = 0$ for simplicity: (i) If $\Delta\mu < 0$, the precipitate is softer than the matrix. Remarkably F_E decreases as the shape deviates from spheres. This will lead to a shape-charge transition as the precipitate size increases.[45)] Recall that the interface free energy contribution stabilizes spheres and dominants over F_E for small sizes. Let us compare the two contributions for a nearly spherical spheroid at a fixed volume V_s. We assume deep quenching where the interfacial free energy σ (per unit area) is nearly independent of the direction of the normal to the interface. Noting that the surface area A is expanded as $A = 4\pi\bar{R}^2[1+(8/15)(R_{\perp}/R_{\parallel}-1)^2+\cdots]$ with $\bar{R} \equiv (3V_s/4\pi)^{1/3}$, we find a crossover radius R_E given by

$$R_E = 5\sigma/(|\Delta\mu|\hat{\delta}^2) \ . \tag{3.68}$$

For $\bar{R} > R_E$ the spherical shape is unstable against deformations (even without external loads), leading to anisotropic shapes with larger surface areas. (ii) On the other hand, if $\Delta\mu > 0$, the precipitate is harder than the matrix and F_E serves to stabilize spherical shapes.

In the presence of external loads the shape development of precipitates is very complicated even if they are isolated. Here we only remark the following: (i) For small sizes, $\bar{R} < R_E$, the interface free energy and F_d are balanced resulting in some characteristic shape. (ii) For large sizes, $\bar{R} > R_E$, F_E will become important. (iii) However, if the shape greatly deviates from spheres, the interface free energy should again come into play even for large domains. If $\Delta\mu < 0$, the external load gives a preferred direction or plane to the shape instability and the growth of the deformations will be eventually saturated with increase of the surface area. If $\Delta\mu > 0$ and $\bar{R} \gg R_E$, the minimization of the sum $F_d + F_E$ yields the following optimal value of N_x,

$$N_x = \frac{1}{3} - \frac{2}{3}(\lambda_{\parallel} - \lambda_{\perp})/\hat{\delta} \ . \tag{3.69}$$

Because $0 < N_x < 1$ the above result is meaningful only when

$$-1 < (\lambda_{\parallel} - \lambda_{\perp})/\hat{\delta} < \frac{1}{2} \ . \tag{3.70}$$

If the above relation does not hold, the domain becomes a thin needle or a flat pancake for which the interface free energy is crucial.

The new characteristic length R_E divided by the interfacial width ξ strongly depends on materials and the distance from the critical point. In particular R_E is roughly a constant near the critical point. Hence there should be situations in which R_E is comparable to or even smaller than ξ. In such cases the elastic effects are strong and the usual critical behaviour should be drastically altered, although virtually nothing seems to be known in this regime.

(c) Interaction between Two Precipitates

Next we examine interactions between two ellipsoidal precipitates due to the elastic field. Let $\mathbf{r}_{AB}$ be the vector between the centers of the ellipsoids and V_A and V_B be the volumes. We can show that if $r_{AB}(= |\mathbf{r}_{AB}|)$ much exceeds the diameters, the interaction energies become independent of the shape of the ellipsoids. First f_d gives rise to

$$\begin{aligned} F_d^{AB} &\simeq -\left(\frac{3}{2\pi}\right)(\Delta\mu)\hat{\delta}V_A V_B \frac{1}{r_{AB}^5}(\mathbf{r}_{AB} \cdot \overleftrightarrow{S} \cdot \mathbf{r}_{AB}) \ , \\ &\simeq -\left(\frac{3}{\pi}\right)(\Delta\mu)\hat{\delta}(\lambda_{\parallel} - \lambda_{\perp})V_A V_B \frac{1}{r_{AB}^3}\left[\cos^2\theta_{AB} - \frac{1}{3}\right] \ . \end{aligned} \tag{3.71}$$

We have omitted the self-interaction terms given in (3.65). The second line is the expression for the uniaxial case, θ_{AB} being the angle between $\mathbf{r}_{AB}$ and the x axis.

(i) If $(\Delta\mu)\hat{\delta}(\lambda_{\parallel} - \lambda_{\perp}) < 0$, F_d^{AB} is repulsive for $|\theta_{AB}| < \cos^{-1}(1/\sqrt{3}) \simeq 55°$ and attractive for $|\theta_{AB}| > \cos^{-1}(1/\sqrt{3})$. Thus the external stress compresses each precipitate (as (3.65) implies) and gives rise to attraction among precipitates in the yz plane. As a result lamellar domain structures will be realized for not too small volume fractions. (ii) If $(\Delta\mu)\hat{\delta}(\lambda_{\parallel} - \lambda_{\perp}) > 0$, the reverse effect occurs. Precipitates elongate and tend to align along the x axis.

Second f_E gives rise to Eshalby's interaction;[42] for large r_{AB} it reads

$$F_E^{AB} \simeq (\frac{3}{8\pi^2})(\Delta\mu)\hat{\delta}^2 V_A V_B (V_A + V_B)\frac{1}{r_{AB}^6} \quad . \tag{3.72}$$

As is well-known this interaction is repulsive for $\Delta\mu > 0$ and attractive for $\Delta\mu < 0$. Recently Kawasaki and Enomoto examined the effect of this term on the coarsening process at small volume fractions.[46] Their theory is applicable to the isotropic case without external stress and they assumed spherical shapes for droplets even when the average radius much exceeds R_E, (3.68), by the factor of $1/\phi$! Strikingly the coarsening is predicted to slow down drastically when precipitates are softer than the matrix. We should reexamine and generalize their theory on the basis of our model.

3.6 Dynamic Model for Simulations

The dynamic equation describing the phase separation becomes very complicated because of (i) the cubic anisotropy, (ii) the anisotropic dipolar interaction due to external stress, and (iii) the Eshalby interaction. Let us express the dynamic equation for c, in which the elastic field has been eliminated, in the following form,

$$\frac{\partial}{\partial t}c = \lambda\nabla^2[f' - \nabla^2 c] + \lambda[I_a + I_E] \, . \tag{3.73}$$

Here I_a is linear in c and arises from the anisotropic $\hat{\mathbf{k}}$-dependence of $\tau_{el}(\hat{\mathbf{k}})$ (see (3.31)). Its real space representation is however not simple. Therefore, let us expand $\tau_{el}(\hat{\mathbf{k}})$ in powers of the elastic anisotropy $\xi_a = (C_{11} - C_{12} - 2C_{44})/C_{44}$ and the concentration-expansion coefficient C_{44}^1 for the shear modulus C_{44}. Then a very natural expansion form follows to first orders in ξ_a and C^1_{44},

$$\tau_{el}(\hat{\mathbf{k}}) = \text{const.} + \frac{1}{2}\tau_a \sum_{i\neq j} \hat{k}_i^2 \hat{k}_j^2 + g_{ex} \sum_{i,j} S_{ij} \hat{k}_i \hat{k}_j + \cdots , \tag{3.74}$$

where

$$\tau_a = 2\alpha^2 C_{44}\xi_a/(C_{11} + 2C_{44})^2 , \tag{3.75}$$

$$g_{ex} = -2\alpha C^1_{44}/(C_{11} + 2C_{44}) . \tag{3.76}$$

Then we can set up the following form,

$$I_a = \frac{1}{2}\tau_a \sum_{i\neq j} \nabla_i^2 \nabla_j^2 w + g_{ex} \sum_{i,j} S_{ij} \nabla_i \nabla_j c , \tag{3.77}$$

where ω is determined by

$$\nabla^2 \omega = c - \bar{c} , \tag{3.78}$$

$\bar{c}$ being the average. The above two terms in (3.77) will mimic the cubic anisotropy and the external stress.

The term I_E arises from the Eshalby interaction and is of the form,

$$I_E = g_E[\nabla^2\{Tr(\overleftrightarrow{W}^2)\} + 2\nabla\nabla : (c - \bar{c})\overleftrightarrow{W}] , \tag{3.79}$$

where

$$g_E = \mu_1(\alpha/K_{LO})^2 , \tag{3.80}$$

and W_{ij} is a traceless stensor defined by

$$W_{ij} = \frac{\partial}{\partial x_i} dj - \frac{1}{d}\delta_{ij}(c - \bar{c}) = \frac{\partial^2}{\partial x_i \partial x_j}\omega - \frac{1}{d}\delta_{ij}(c - \bar{c}) , \tag{3.81}$$

d being the skpatial dimensionality of the system (simulations are easy for $d = 2$). With (3.77) and (3.79) the equation (3.73) can be solved once ω is determined from (3.78). Numerical results will appear elsewhere.

3.7 Summary and Remarks

The results in this section are still those readily derivable from the model equations (3.1), (3.15) and (3.19). It then becomes a great challenge in the future to investigate the dynamical aspects of the elastic effects. Nevertheless we have obtained some new results and raised some new problems in addition to reproducing the previous results. (i) Our model takes into account of the effect of external stress in the simplest manner through the concentration-dependent stiffness constants (3.5). It has led to the general expressions for the effective temperature shift $\tau_{el}(\hat{\mathbf{k}})$, (3.30), and the structure factor (3.33). We can in principle calculate the spinodal curve and the habit planes for any type of external stress. (ii) The effective free energy for c only has been calculated in (3.60)$\sim$(3.63) in the isotropic case particularly when the concentration-dependence of the moduli is weak. Interestingly, the external stress gives rise to the dipolar interaction (3.62) and we have the cubic interaction (3.63) corresponding to Eshalby's interaction originally derived from the classical elastic theory. This cubic interaction destabilizes spherical shapes of droplets which are softer than the matrix as shown by (3.67). Its effect on the critical behavior is puzzling as stated below (3.70). (iii) Long-range interactions arise among precipitates due to the above-mentioned two interactions. They are crucial in the phase separation process at not too small volume fraction of precipitates.

Acknowledgements

I would like to thank Professor Toyoichi Tanaka for showing his fascinating experiments, which spurred me to study gels. My work on solid solutions was initiated as an effort to understand the pioneering work by Professor J.W. Cahn in my own way. I have been sharing common interests with and have learned much from Professor Kyozi Kawasaki. Valuable discussions were also made with Professor Shunsuke Hirotsu, Dr. Kenn Sekimoto, and Professor Takeji Hashimoto.

Note Added

After this work has been completed, I have noticed a number of intriguing problems of a gas-liquid transition of hydrogens absorbed in metals.[47] In metals, hydrogens give rise to a volume expansion of the lattice as a solute and, surprisingly, Wagner and Horner derived the free energy of such $H+$ metal systems in the same form as our free energy (3.1) for solid solutions. They furthermore found a long range interaction varying on the scale of sample dimensions, which arises from image forces caused by the free surface of the crystal. It should be noted that this paper neglects the corresponding long range interaction in the solid case. It should also be remarked that, due to this long range interaction, the spectrum of the hydrogen densisty fluctuations becomes discretized (as in the case of gels) and the spinodal curve depends on the sample shape (differently from gels).[48] Here the mean field critical exponents have been expected for the stress-free boundary case.

I would like to thank Professor H. Müller-Krumbhaar and Professor H. Zabel for introducing me to the problem of hydrogen-metal systems, who visited Kyoto just after this paper has been completed.

I have also noticed a paper by L. Golubović and Lubensky[49] on nonlinear elasticity of amorphous solids. Expecting the mean field behavior at $K = 0$ for $\mu \neq 0$, they have examined a very interesting case of vanishing μ. They have kindly informed me of an experiment of symmetry-nonbreaking transitions in KH_2PO_4 without critical microscopic fluctuations.[50] It constitutes the first observation of a Cowley type-0 transition.[51] It goes without saying that it has just been expected for gels at the swelling transition and solids at the valence transition in my papers.[23][21] I believe that Li and Tanaka's result,[35] on which we have discussed at the end of Section 2, should be reexamined in the light of these examples.

Appendix

The free energy density in ΔF, (2.3), is assumed to be of the form,

$$\begin{aligned} f =& a^{-3}[(1-\phi)\ln(1-\phi) + \chi\phi(1-\phi)] \\ &+ \nu_0(\frac{1}{2} + f_i)(\phi/\phi_0)\ln(\phi/\phi_0) \ , \end{aligned} \tag{A.1}$$

where a is the monomer size, χ is the interaction parameter representing the solvent quality with respect to the polymer, and f_i is the degree of ionization per effective chain. The χ may be assumed to depend on the temperature T by

$$u \equiv 1 - 2\chi = A(T - T_\theta) \ , \tag{A.2}$$

where A is a constant and T_θ is the θ temperature. We consider the isotropic, homogeneous case, where ΔF is given by (2.7). Let us define $g \equiv \Delta F/(V_0\phi_0 k_B T a^{-3})$, the free energy per monomer of the network divided by $k_B T$. In the semidilute case $\phi \ll 1$ we expand n$(1-\phi)$ in powers of ϕ in (A.1) to obtain

$$g = \frac{1}{2}u\phi + \frac{1}{6}\phi^2 + \gamma \ell n\phi + \frac{3}{2}\beta\phi^{-2/3} + \cdots \tag{A.3}$$

where

$$\gamma = \nu_0 a^3(f_i + \frac{1}{2})/\phi_0 \ , \beta = \nu_0 a^3 \phi_0^{-1/3} \ . \tag{A.4}$$

It is convenient to regard g as a function of $v = 1/\phi$. Then,

$$a^3\Pi/k_B T = -\frac{\partial g}{\partial v} = \frac{u}{2v^2} + \frac{1}{3v^3} + \frac{\gamma}{v} - \beta v^{-1/3} \ , \tag{A.5}$$

$$a^3 K/k_B T = v\frac{\partial^2 g}{\partial v^2} = \frac{u}{v^2} + \frac{1}{v^3} + \frac{\gamma}{v} - \frac{1}{3}\beta v^{-1/3} \ , \tag{A.6}$$

while $a^3\mu/k_B T = \beta v^{-1/3}$ from (2.13). Thus we find

$$u = -\frac{2}{3}\phi - 2\gamma v + 2\beta v^{5/3} \quad \text{for} \quad \Pi = 0 \ , \tag{A.7}$$

$$u = -\phi - \gamma v + \frac{1}{3}\beta v^{5/3} \quad \text{for} \quad K = 0 \ , \tag{A.8}$$

$$u = -\phi - \gamma v - \beta v^{5/3} \quad \text{for} \quad K + \frac{4}{3}\mu = 0 \ . \tag{A.9}$$

Note that the curve of $K = 0$ and that of $K + \frac{4}{3}\mu = 0$ are very different at large

v due to the different signs of the elastic contributions ($\propto \beta$). The critical point at $\Pi = 0$ is obtained from $\Pi = K = \partial K/\partial v = 0$, which is realized only when

$$p \equiv \beta\gamma^{-4/3} = \frac{3}{5}(\frac{3}{4})^{4/3} \cong 0.409 \ . \qquad (A.10)$$

For $p < 0.409$ the gel undergoes a discontinuous phase change along the curve $\Pi = 0$, while for $p > 0.409$ the volume change is gradual along the curve. In Fig.3 we have chosen $p = 0.3$ in (a), $p = 0.409$ in (b), and $p = 0.6$ in (c).

References

1) P.J. Flory, *Principles of Polymer Chemistry* (Cornell Univ., Ithaca, N.Y., 1953).

2) K. Dušek and D. Patterson, J. Polym. Sci. Part A-2, **6** (1968) 1209.

3) T. Tanaka, Physica **140A** (1986) 261; Nature **325**, (1987) 26; in *Molecular Conformation and Dynamics of Macromolecules in Condensed Systems*, edited by M. Nagasawa (Elsevier, 1988); E.S. Matsuo and T. Tanaka, J. Chem. Phys. **89** (1988) 1695.

4) T. Tanaka, Phys. Rev. Lett. **20** (1978) 820.

5) A Onuki, in *Dynamics of Ordering Processes in Condensed Matter*, edited by S. Komura and H. Furukawa (Plenum, N.Y., 1988), p.71; J. Phys. Soc. Jpn. **57** (1988) 699.

6) K. Sekimoto and K. Kawasaki, Physica **A154** (1989) 384.

7) A.G. Khachaturyan, *Theory of Structural Transformations in Solids* (John Wiley & Sons, New York, 1983).

8) J.W. Cahn, Acta Metall. **9** (1961) 795.

9) P.W. Voorhees, J. Stat. Phys. **38** (1985) 231 and references quoted therein.

10) A. Onuki, J. Phys. Soc. Jpn. **58** (1989) 3065, 3069.

11) A. Onuki, J. Phys. Soc. Jpn. **57** (1988) 703; Phys. Rev. **A39** (1989) 5932.

12) R. Kubo, J. Phys. Soc. Jpn. **3** (1948) 312.

13) R.S. Rivlin, in *Rheology*, edited by F. Eirich (Academic Press, 1956), vol.1.

14) P.G. de Gennes, *Scaling Concepts in Polymer Physics* (Cornell Univ. Press, Ithaca, N.Y., 1980).

15) L.D. Landau and E.M. Lifshitz, *Theory of Elasticity*, (Pergamon, N.Y., 1973).

16) T. Tanaka, L.O. Hocker and G.B. Benedek, J. Chem. Phys. **59** (1973) 5151.

17) A-M. Hecht and E. Geissler, J. Physique (Paris) **45** (1984) 309.

18) J.C. Bacri, J. Dumas and A. Levelut, J. Physique Lett. (Paris) **40** (1979) L-231.

19) T. Tanaka, S. Ishiwata and C. Ishimoto, Phys. Rev. Lett. **38** (1977) 771; T. Tanaka, Phys. Rev. A17 (1978) 763.

20) A. Onuki, in *Space-Time Organization in Macromolecular Fluids*, (the Proceedings of the 11th Taniguchi International Conference, Nov., 1988).

21) A. Onuki, Phys. Rev. **B39** (1989) 12308.

22) Y. Li and T. Tanaka, preprint.

23) A. Onuki, Phys. Rev. **A38** (1988) 2192.

24) T. Tanaka and D. Fillmore, J. Chem. Phys. **59** (1973) 5151; T. Tanaka et al. Phys. Rev. Lett. **55** (1985) 2455; A. Peters and S.J. Candau, Macromolecules **19** (1986) 1952.

25) S. Hirotsu, J. Chem. Phys. **88** (1988) 703; K. Sekimoto and K. Kawasaki, J. Phys. Soc. Jpn. **57** (1988) 2591, 2594.

26) K. Binder and H.L. Frisch, J. Chem. Phys. **81** (1984) 2126.

27) B.J. Bauer, R.M. Briber and C.C. Han, Macromolecules **22** (1989) 940.

28) S. Hirotsu and A. Onuki, J. Phys. Soc. Jpn. **58** (1989) 1508.

29) T. Takebe and T. Hashimoto, J. Chem. Phys. **91** (1989) 4360.

30) K. Sekimoto and K. Kawasaki, in Ref.20; Phys. Rev. **A39** (1989) 4912.

31) S. Hirotsu and A. Kaneki, in *Dynamics of Ordering Processes in Condensed Matter*, edited by S. Komura and H. Furukawa (Plenum. N.Y., 1988), p.481, and private communication.

32) K. Sekimoto and K. Kawasaki, J. Phys. Soc. Jpn. **56** (1987) 2977.

33) T. Hwa and M. Kardar, Phys. Rev. Lett. **61** (1988) 106.

34) K. Sekimoto and K. Kawasaki, J. Phys. Soc. Jpn. **57** (1988) 2594.

35) Y. Li and T. Tanaka, J. Chem. Phys. **90** (1989) 561.

36) J.M. Lawrence, M.C. Croft and R.D. Parks, Phys. Rev. Lett. **35** (1975) 289.

37) J.W. Cahn, Acta Metall. **10** (1962) 179.

38) J.E. Hillard, in *Phase Transformations*, American Society for Metals, Cleveland OH, 1970, 497.

39) H. Yamanouchi and D. de Fontaine, Acta Metall. **27** (1979) 763.

40) T. Miyazaki, H. Imamura and T. Kozakai, J. Mater. Sci. **54** (1982) 9; T. Miyazaki, M. Doi and T. Kozakai, Proc. Intern. Symp. on Nonlinear Phenomena in Materials Science, (Trans. Tech., 1988).

41) J. Gayda and D.J. Srolovitz, Acta Metall. **37** (1989) 641.

42) A.J. Adrell, R.B. Nicholson and J.D. Eshalby, Acta Metall. **14**(1966) 1295.

43) L.D. Landau, E.M. Lifshitz and L.P. Pitaevskii, *Electodynamics of Continuous Media* (Pergamon Press, 1984).

44) A. Pineau, Acta Metall. **24** (1976) 559.

45) W.C. Johnson and J.W. Cahn, Acta Metall. **32** (1984) 1925.

46) K. Kawasaki and Y. Enomoto, Physica **A150** (1988) 463; Y. Enomoto and K. Kawasaki, Acta Metall. **37** (1989) 1399.

47) H. Wagner and H. Horner, Adv, Phys. **23** (1974) 587.

48) H. Zabel and H. Peisl, Phys. Rev. Lett. **42** (1979) 511 and references quoted therein.

49) L. Golubović and T.C. Lubensky, Phys. Rev. Lett. **63** (1989) 1082.

50) E. Courtens, R.W. Gammon and S. Alexandar, Phys. Rev. Lett. **43** (1979) 1026; E. Courtens and R.W. Gammon, Phys. Rev. B**24** (1981) 3890.

51) R.A. Cowley, Phys. Rev. B**13** (1976) 4877.

www.ingramcontent.com/pod-product-compliance
Ingram Content Group UK Ltd.
Pitfield, Milton Keynes, MK11 3LW, UK
UKHW020400150926
8624IPUK00004B/267